Engineering Materials

This series provides topical information on innovative, structural and functional materials and composites with applications in optical, electrical, mechanical, civil, aeronautical, medical, bio- and nano-engineering. The individual volumes are complete, comprehensive monographs covering the structure, properties, manufacturing process and applications of these materials. This multidisciplinary series is devoted to professionals, students and all those interested in the latest developments in the Materials Science field, that look for a carefully selected collection of high quality review articles on their respective field of expertise.

Indexed at Compendex (2021) and Scopus (2022)

Subash C. B. Gopinath ·
Muhammad Mahyiddin Ramli
Editors

Hybrid-Nanomaterials

Fabrication, Characterization and Applications

Springer

Editors
Subash C. B. Gopinath
Faculty of Chemical Engineering and Technology
Universiti Malaysia Perlis
Arau, Perlis, Malaysia

Institute of Nano Electronic Engineering
Universiti Malaysia Perlis
Kangar, Perlis, Malaysia

Muhammad Mahyiddin Ramli
Institute of Nano Electronic Engineering
Universiti Malaysia Perlis
Kangar, Perlis, Malaysia

ISSN 1612-1317 ISSN 1868-1212 (electronic)
Engineering Materials
ISBN 978-981-97-9024-1 ISBN 978-981-97-9022-7 (eBook)
https://doi.org/10.1007/978-981-97-9022-7

This Springer imprint is published by the registered company Springer Nature Singapore Pte Ltd.
The registered company address is: 152 Beach Road, #21-01/04 Gateway East, Singapore 189721, Singapore

If disposing of this product, please recycle the paper.

Contents

Introduction to Hybrid Materials and Nanostructures

Hussaini Adam, Thangavel Lakshmipriya, Subash C. B. Gopinath, Tijjani Adam, Evan T. Salim, Makram A. Fakhri, and Thanganadar Ashokkumar

1 Introduction

The need for innovative materials has expanded dramatically due to technological improvements and the requirement for new functions [1]. Technological demands for a wide range of novel applications cannot be addressed by the materials already

---

H. Adam · S. C. B. Gopinath (✉) · T. Adam
Institute of Nano Electronic Engineering, Universiti Malaysia Perlis (UniMAP), 01000 Kangar, Perlis, Malaysia
e-mail: subash@unimap.edu.my

T. Lakshmipriya
Center for Global Health Research, Saveetha Medical College & Hospital, Saveetha Institute of Medical and Technical Sciences (SIMATS), Thandalam, Chennai 602 105, Tamil Nadu, India

S. C. B. Gopinath
Faculty of Chemical Engineering & Technology, Universiti Malaysia Perlis (UniMAP), 02600 Arau, Perlis, Malaysia

S. C. B. Gopinath · T. Adam
Center of Excellence for Micro System Technology (MiCTEC), Universiti Malaysia Perlis (UniMAP), Pauh Campus, 02600 Arau, Perlis, Malaysia

T. Adam
Faculty of Electronic Engineering & Technology, Universiti Malaysia Perlis (UniMAP), 02600 Arau, Perlis, Malaysia

E. T. Salim
Applied Science Department, University of Technology-Iraq, Baghdad 10066, Iraq

M. A. Fakhri
Laser and Optoelectronics Eng. Department, University of Technology-Iraq, Baghdad 10066, Iraq

T. Ashokkumar
Haramaya University, P.O. Box 138, Dire Dawa, Harar, Ethiopia

Royal Global University, NH-37, Betkuchi, Guwahati 781035, Assam, India

S. C. B. Gopinath and M. M. Ramli (eds.), *Hybrid-Nanomaterials*, Engineering Materials, https://doi.org/10.1007/978-981-97-9022-7_1

in use, such as metals, ceramics, and plastics [2]. Scientists and engineers have discovered that mixtures of materials have superior properties compared to their pure counterparts [3]. Composites are excellent examples of this, made by mixing a basic structural material with another component [4]. Inorganic fiber-reinforced polymers are lightweight materials with high mechanical qualities that are utilized to build vehicles and sports equipment. Their structural building components are mostly inorganic and range from micrometers to millimeters in size [5]. More homogenous materials may arise from a reduction of inorganic unit size, allowing for the fine-tuning of properties at the molecular and nanoscale levels [6]. Hybrid materials and nanostructures are attracting interest due to their unique features and prospective applications in numerous industries, which highlight similarities and differences in their properties [7]. The nanoscale mixing of diverse materials offers significant potential for the development of novel materials with improved properties [8]. One of the primary benefits of hybrid materials and nanostructures is the flexibility to adapt their properties for specific purposes [9]. These materials' composition and structure may be carefully designed to obtain desired features like as mechanical strength, electrical conductivity, and optical properties. The complex interplay of diverse materials at the nanoscale level creates a fascinating landscape for scientists and engineers to investigate, with the potential to transform several sectors. The fusing of many components at the nanoscale not only provides exceptional mechanical and electrical performance, but it also confers optical characteristics that challenge conventional materials. The capacity to precisely alter the composition and structure of these materials allows researchers to tune their properties to the exact requirements of a wide range of applications. The potential for customization in nanoscale materials is vast, enhancing mechanical strength, electrical conductivity, and optical properties for various applications. This synergistic combination of materials creates hybrid materials with exceptional mechanical properties, enhancing the capabilities of advanced electronics and photonic devices. Hybrid materials and nanostructures offer significant improvements in electrical conductivity and tensile strength, making them ideal for advanced electronics. By integrating conductive nanomaterials like carbon nanotubes or graphene with other components, these materials can create materials with enhanced electrical properties, paving the way for high-performance electronic devices, flexible conductive films, and other applications requiring exceptional conductivity and reliability. Additionally, hybrid materials and nanostructures excel in manipulating optical properties, such as light absorption, emission, and transmission characteristics, opening up new vistas for advanced photonic devices, optical sensors, and novel optical applications. The combined effects of diverse materials in hybrid structures can lead to new optical functionalities that surpass individual components, marking a significant advancement in optics and photonics. As researchers and engineers continue to explore this fascinating landscape, the potential for revolutionizing various industries through hybrid materials and nanostructures remains profound.

2 Natural Origin

Natural materials often consist of inorganic and organic building blocks, with the inorganic part providing mechanical strength and structure, and the organic part bonding with soft tissue [10]. Scientists are using a biomimetic approach to study materials like bone and nacre to understand their formation processes and transfer ideas to artificial materials, aiming to mimic the unique circumstances of biological hybrid inorganic–organic materials [11]. The study of biomineralization and shape control reveals questions about the morphological diversity and functions produced by a few building blocks [12]. Natural materials' unique structures, often characterized by their beauty and diverse functions, drive further research due to their complex hierarchical construction, ranging from nanometers to millimeters, each serving a specific purpose [13]. This complexity is achieved through soft chemical self-assembly mechanisms over large dimensions, a major challenge in modern materials chemistry.

3 Definition: Hybrid Materials and Nanocomposites

Hybrid materials and nanocomposites are nanoscale materials that combine distinct components, resulting in unique properties [14]. These materials are extremely adaptable and suited for a wide range of applications. Hybrid materials and nanocomposites include particle and fiber composites, sandwich structures, foams, lattice structures, segmented structures, zero expansion materials, and more [15]. These materials might come from polymers, metals, ceramics, carbon-based compounds like carbon nanotubes or graphene, or natural materials like cellulose. Hybrid materials are classified into classes I and II [16]. Van der Waals, hydrogen bonds, or weak electrostatic interactions are examples of weak interactions between the two phases in Class I hybrid materials, whereas significant chemical interactions between the constituents are seen in Class II hybrid materials [17]. The shift between weak and strong interactions is a result of the gradual alteration in the strength of chemical interactions [18]. An organic moiety with a functional group that enables attachment to an inorganic network can act as a network modifying compound, like phenyltrialkoxysilanes, which alters the silicon network without additional functional groups [19]. A system is referred to as a network functionalizer if it incorporates a reactive functional group [20]. When two or three anchor groups modify an organic segment, the inorganic group becomes an integral part of the hybrid network [21]. Blends form when inorganic and organic building blocks don't have strong chemical interactions, like in a combination of inorganic clusters or particles with organic polymers. Interpenetrating networks (IPNs) form when an inorganic and an organic network interpenetrate without strong chemical interaction [22]. Class I hybrids form when inorganic building blocks are covalently bonded to organic polymers, while Class II hybrids form when these inorganic building blocks are covalently connected [23].

4 The Development of Hybrid Material

Hybrid materials or nanotechnology have been used since ancient times to create vibrant paints by combining dyes, inorganic pigments, and other inorganic and organic components [24]. The sol–gel method, developed in the 1930s, was one of the first techniques to use ambient conditions for ceramic production [25]. The method gained industrial interest due to its ability to create multicomponent systems using a mild reaction mechanism, and its ease of preparation and molecular design, making it applicable in various materials science fields [26]. The synthesis of inorganic–organic materials involves four key areas: molecular engineering, nanometer and micrometer scale organization, transitioning from functional to multifunctional hybrids, and integration with bioactive components [27]. Gold colloids, a stable metal sol and colloidal, were formed centuries ago and consist of nanoparticles enclosed in a solvent matrix [28]. Metal colloids, used for optical purposes in nanocomposites and centuries ago for coloring glass, have been studied in scientific records since the 18th century. Gold colloids, produced by reducing gold salts, are often used in nanocomposites [29]. Advanced experimental methods are being used to review numerous known colloidal systems to understand the chemistry and physics of their unique origins [30]. In recent years, the focus has been on the creation of hybrid materials, which combine multiple materials to enhance their properties and functionalities (Fig. 1).

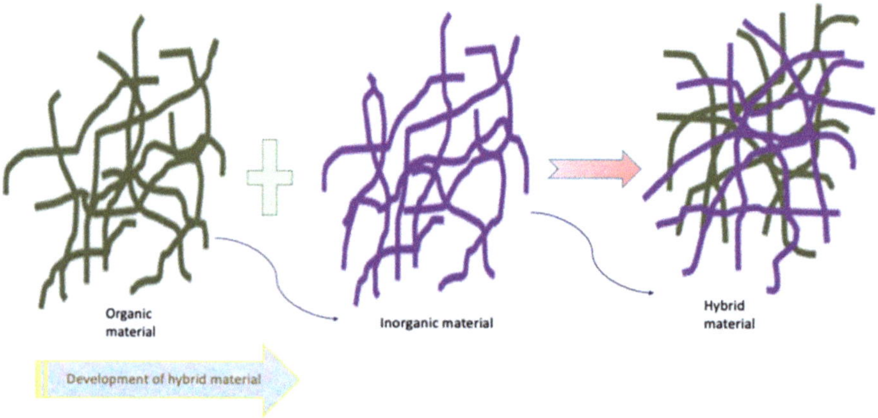

Fig. 1 Development of hybrid material: The earliest hybrid materials were paints formed from inorganic and organic components that were used thousands of years before. Rubber is an example of how inorganic materials are used as fillers in organic polymers

5 Basic Concepts

Hybrid nanomaterials, composed of various types of nanoparticles or structures, are gaining significant research due to their unique properties and potential applications in various fields. These materials, such as metallic and semiconductor nanoparticles, combine desired features, leading in materials with improved optical and electrical properties. Hybrid nanomaterials aim to develop novel materials with enhanced properties and potential applications. Hybrid nanomaterials are distinct chemical conjugates of organic and/or inorganic materials, resulting in synergistic materials with properties determined by the molecular or supramolecular interface [31]. These materials are utilized in various fields such as optics, electronics, sensors, ionics, energy conversion, mechanics, membranes, protective coatings, and catalysis. The incorporation of nano-sized materials has significantly enhanced the unique properties of hybrid materials, offering a wider range of multifunctional applications [32]. Over the past two decades, the development of multifunctional applications has been influenced by the chemical and physical properties of these materials, leading to significant value addition [33]. Hybrid materials are innovative, advanced materials used in various applications like optics, electronics, sensors, ionics, energy conversion, mechanics, membranes, protective coatings, and catalysis [34]. Nanomaterials and supramolecular materials offer diverse applications, ranging from molecules to solid-state materials [35]. The structure, organization, and properties of these materials are influenced by their chemical nature and their interactions with each other [36]. The design process involves adjusting the nature, extent, and accessibility of internal surfaces [37]. Nanochemistry and molecular engineering are closely related to predicting and controlling the chemical and physical properties of nano-sized materials [38]. Hybrid materials combine the properties of various monolithic or single materials, such as fibrous and particulate composites, foams, lattices, sandwiches, and natural materials (Fig. 2).

6 Methods of Hybrid Material Preparation

Hybrid materials are gaining attention due to their unique properties and combination of various material characteristics [39]. Sol–gel processes are a widely used method for preparing hybrid materials [40]. Sol–gel processes convert solutions or dispersions into solid materials through chemical reactions, offering benefits like precise refractive index regulation, good film-forming ability, low processing temperature, and uniform coating thickness [41]. Photochemical methods are another method used to prepare hybrid materials [42]. Photochemical methods utilize light-induced reactions to create hybrid materials with specific properties, while seeding growth is another technique used for this purpose [43]. Seeding growth involves introducing seed particles into a solution, acting as templates for hybrid material growth, while

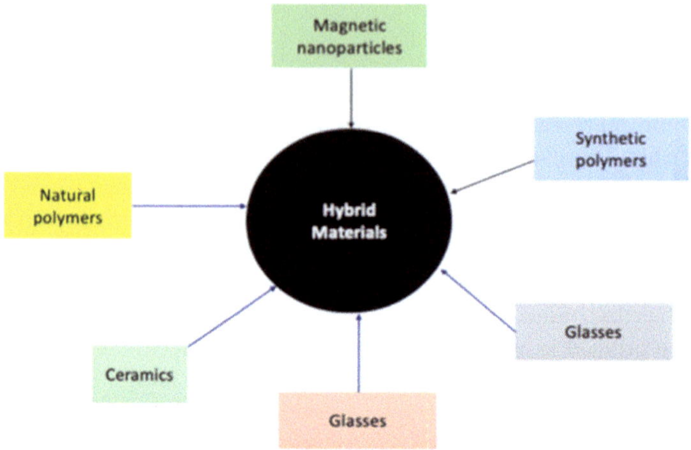

Fig. 2 **Hybrid material properties**: Hybrid materials combine the properties of monolithic or single materials, including fibrous and particulate composites, foams, lattices, sandwiches, and natural materials. Hybrid composites offer advantages over conventional materials such as balanced strength, stiffness, bending and membrane mechanical properties, thermal distortion stability, reduced weight and cost, improved fatigue resistance, reduced notch sensitivity, and improved fracture resistance

coprecipitation is a common method for preparing such materials [44]. Coprecipitation and sonochemical synthesis are methods used to create homogeneous hybrid materials by precipitating components from a solution simultaneously [45]. Sonochemical synthesis is a method that uses ultrasound waves to induce chemical reactions, resulting in the creation of hybrid materials with distinctive properties [46]. Green synthesis is another emerging method for preparing hybrid materials [46]. Green synthesis is the use of eco-friendly materials and methods, like plant extracts or biodegradable polymers, to create hybrid materials The choice of method for preparing hybrid materials depends on factors such as the desired properties, scalability, cost-effectiveness, and environmental considerations [47].

7 Characterization Techniques of Hybrid Material

Hybrid materials are gaining increasing importance in various industries such as aerospace, manufacturing, and composites [48]. These materials consist of advanced high-strength steels, non-ferrous alloys like aluminum, magnesium, and titanium alloys, and composites like carbon fiber, metal matrix composites, and nanocomposites. Microscopically analyzing hybrid materials involves examining their structure and morphology at a microscopic level to gain a deeper understanding of their properties [49]. This can be achieved through optical microscopy, scanning electron microscopy, or transmission electron microscopy. Researchers use methods like chemical analysis to understand the internal structure and surface characteristics of hybrid materials [50]. The analysis of a material's chemical composition using techniques like energy-dispersive X-ray spectroscopy, X-ray diffraction, or Fourier-transform infrared spectroscopy provides crucial insights into its elemental composition and molecular structure. Mechanical testing is a vital method for evaluating the mechanical properties of hybrid materials, such as strength, stiffness, and toughness. Tensile, flexural, impact, and hardness testing are used to assess these properties [50]. Researchers are conducting tests on the hybrid material to assess its mechanical behavior and performance under various conditions. Thermal analysis is a method used to analyze the thermal properties of hybrid materials, including their melting point, thermal conductivity, and coefficient of thermal expansion [51]. Differential scanning calorimetry, thermogravimetric analysis, and thermal conductivity measurement are utilized to study the properties of hybrid materials, enabling a better understanding of their response to temperature changes and their suitability for specific applications. Electrical analysis is a method used to assess the electrical properties of hybrid materials, including conductivity, resistivity, and dielectric constant. This purpose is achieved through techniques such as conductivity measurements, dielectric spectroscopy, and impedance analysis [52]. Understanding the electrical behavior of hybrid materials is crucial for applications involving electrical conductivity or insulation. Rheological characterization is a method utilized to analyze the flow behavior and viscosity of hybrid materials [53]. This can be achieved through various techniques, such as rheometry or viscometry. Researchers can understand the flow characteristics and suitability of a hybrid material by analyzing its rheological properties.

8 Applications

Hybrid materials, which are composed of both organic and inorganic components, have gained significant attention in various industries due to their unique properties and diverse applications [54]. One of the applications of hybrid materials is in the field of gas barriers [55]. Gas barriers play a crucial role in industries such as packaging, automotive, and electronics by preventing the permeation of gases like oxygen, moisture, and volatile organic compounds [56]. These barriers ensure the quality and shelf life of packaged products, protect electronic components from moisture damage, and maintain the integrity of automotive fuel systems. Another application of hybrid materials is in the development of sustainable packaging alternatives [57]. Polysaccharide-based packaging materials, which are biodegradable and derived from renewable sources, have emerged as promising candidates in the quest for sustainable and eco-friendly packaging solutions [58]. They have excellent gas barrier properties, making them efficient in preventing the transfer of gases such as oxygen and carbon dioxide. Moreover, hybrid materials find applications in the field of coatings [59]. Coatings are used in various industries for decorative purposes, protection against corrosion and wear, and to provide specific properties such as scratch resistance. Hybrid coatings, made from organic–inorganic materials, offer enhanced performance and durability compared to traditional coatings. Furthermore, hybrid materials are also used in the development of smart textiles [60]. Smart textiles are fabrics that can sense and respond to environmental stimuli, such as temperature, moisture, or light. Hybrid materials can be integrated into textiles to create fabrics that have functional properties, such as heat regulation or moisture-wicking capabilities. Acoustic and thermal insulators are used in construction to improve soundproofing and insulation properties. Hybrid materials are also being explored for their potential use in electrical insulators to enhance system performance and safety [61].

9 Nanostructures

Nanostructures are materials with structural components ranging from 1 to 100 nm, exhibiting unique physical, chemical, and biological properties compared to macro-scale counterparts [62]. These materials have a high surface area to volume ratio and quantum effects, allowing for precise control over their properties. They have potential applications in electronics, aerospace, medicine, and sensing. In electronics, nanostructures offer smaller, more powerful devices and new ways to manipulate electrons [34]. In aerospace, their lightweight and high-strength properties make them ideal for aircraft and spacecraft development [63]. In medicine, nanostructures hold promise for targeted drug delivery, imaging, and diagnostics [64]. In sensing, their high sensitivity to small changes makes them ideal for detecting and measuring substances and environmental conditions [65]. Nanostructures have revolutionized fields such as electronics, medicine, materials science, and engineering by showing unique properties and behavior due to quantum effects and surface-to-volume ratio [66]. By manipulating and controlling the size, shape, and composition of nanostructures, researchers can tailor their properties to meet specific requirements, leading to the development of high-performance devices and materials with enhanced properties [67]. In materials science and engineering, nanostructures have paved the way for the production of stronger, lighter, and more durable materials [68]. Researchers have created innovative materials with applications ranging from aerospace to construction, such as nanocomposites, which exhibit exceptional mechanical and thermal properties [69]. The exploration of nanostructures has extended to environmental and energy-related fields, such as energy storage, conversion, and harvesting [70]. Nanostructures hold promise in environmental remediation by enabling the development of advanced filtration systems and catalysts for pollution control [71]. Table 1 provides an overview of the different nanostructures and their assemblies.

Table 1 Nanostructures and their assemblies

Nanostructure	Size	Material	References
Clusters	Radius: 1–10 nm		
Nanocrystals		Poly-crystalline	[119]
Quantum dots	–	Semiconductors, Insulators, metals magnetic materials	[120]
Other nanoparticles	Radius: 1–100 nm	Ceramic oxides	[121]
Nanobiomaterials	Radius: 5–10 nm	Membrane protein	[122]
Nanowires	–	Metals, semiconductors, oxides, sulfides, nitride	[123]
Nanotubes	Diameter: 1–100 nm	Carbon, layered chalcogenide	[124]
Nanobiorods	Diameter: 5 nm	DNA	
2D arrays of Area: several nm^2–μm^2 Metals, semiconductors, nanoparticles	Thickness: 1–1000 nm	Metals, semiconductors, nanoparticles magnetic materials	[125]
Surfaces and thin films	–	Insulators, semiconductors, thin films metals, DNA	
3D superlattices of nanoparticles	Radius: several nm	Metals, semiconductors, nanoparticles magnetic materials	[126]

10 Types of Nanostructures

Nanostructures are materials that are engineered and controlled at the nanoscale, typically falling within the size range of 1–100 nm [72]. These nanostructures can be categorized into various groups based on their composition, shape, and characteristics. Examples of common nanostructures include nanoparticles, nanotubes, nanowires, nanorods, nanosheets, and nanocomposites [73]. These nanostructures can be further divided into classes such as carbon-based nanomaterials (e.g., carbon nanotubes, graphene), metallic nanoparticles (e.g., gold nanoparticles, silver nanoparticles), semiconductor nanoparticles (e.g., quantum dots), magnetic nanoparticles, and polymer-based nanostructures (e.g., polymer nanoparticles, hydrogels) [74]. Each type of nanostructure possesses distinct properties and features that render them suitable for specific applications [75]. Nanostructures with at least one dimension of 100 nm or smaller are highly valuable in numerous sectors, including electronics, aerospace, defense, pharmaceuticals, healthcare, and more [76]. For instance, nanoparticles, characterized by their nanoscale dimensions, can be utilized in drug delivery systems, solar cells, and catalyst supports [77]. Conversely, nanotubes, which are cylindrical structures with hollow interiors, exhibit exceptional mechanical and electrical properties, making them well-suited for applications in electronics and

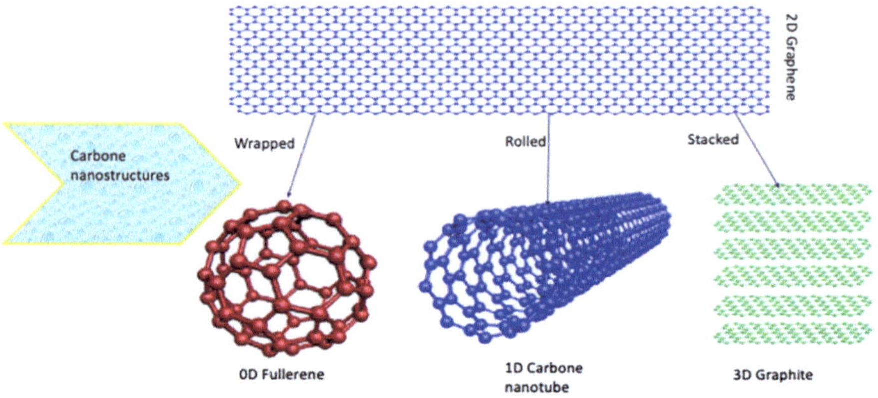

Fig. 3 **Carbon nanostructures**: Carbon nanostructures, including fullerene, carbon nanotubes, and graphite, are one-dimensional, three-dimensional, and two-dimensional structures with various applications. Carbon nanotubes, composed of single-walled and multiwalled parts, are widely utilized in electro-analytical approaches for electrode modification, due to their exceptional electrocatalytic activity, biocompatibility, and high sensitivity

optoelectronics [78]. Additionally, other nanostructures like nanowires, nanorods, and nanosheets find widespread use across various industries due to their unique characteristics [79]. Certain nanostructures, such as nanocomposites, combine different types of nanostructures, resulting in materials with enhanced versatility and tailored properties [76]. Figure 3 illustrates various carbon nanostructures. Fullerene is a one-dimensional carbon nanostructure, while carbon nanotubes are three-dimensional. Graphite and graphene are both two-dimensional carbon nanostructures commonly known for their applications.

11 Synthesis Techniques of Nanostructures

Various methods can be employed to synthesize nanostructures, including chemical vapor deposition, electrochemical deposition, sol–gel synthesis, and template-based synthesis [80]. Chemical vapor deposition involves introducing precursor gases into a chamber and inducing a chemical reaction to deposit the desired material onto a substrate [81]. On the other hand, electrochemical deposition utilizes electrical current to drive the deposition of nanostructures onto a substrate. Sol–gel synthesis, on the other hand, transforms a sol (a colloidal suspension of nanoparticles) into a gel through hydrolysis and condensation reactions [82]. This gel can then be further processed to form various nanostructures. Template-based synthesis, as the name suggests, utilizes a template or scaffold to guide the arrangement and growth of nanomaterials or nanoparticles [83]. These synthesis techniques enable the creation of nanostructures in different sizes and forms, such as nanoparticles, nanowires, nanotubes, and nanosheets. Due to their small size, these nanostructures possess unique properties and functionalities, including enhanced mechanical strength, improved electrical conductivity, and increased surface area-to-volume ratio [84]. Moreover, the synthesis of nanostructures can be tailored to achieve specific properties and applications. For instance, metallic nanoparticles can be synthesized through methods like chemical reduction or physical vapor deposition, while nanotubes can be produced using techniques such as chemical vapor deposition or template-based synthesis with carbon nanotubes as the template [85]. Nanostructures can be fabricated using Nanoimprint lithography. Nanoimprint lithography is a versatile and efficient technique for fabricating nanostructures with high resolution and precision. It involves pressing a mold onto a polymer or substrate material, causing it to deform and transfer the pattern of the mold onto the surface. The process of nanoimprint lithography begins with the selection and preparation of the mold, which is typically made of a rigid material such as silicon. The mold is then coated with a release agent to prevent sticking, and the desired pattern is etched or formed on the surface of the mold using techniques such as electron beam lithography or focused ion beam milling. Once the mold is prepared, it is aligned with the polymer or substrate material and pressed against it with controlled pressure and temperature. This causes the material to flow and conform to the shape of the mold, replicating the pattern in high fidelity (Fig. 4).

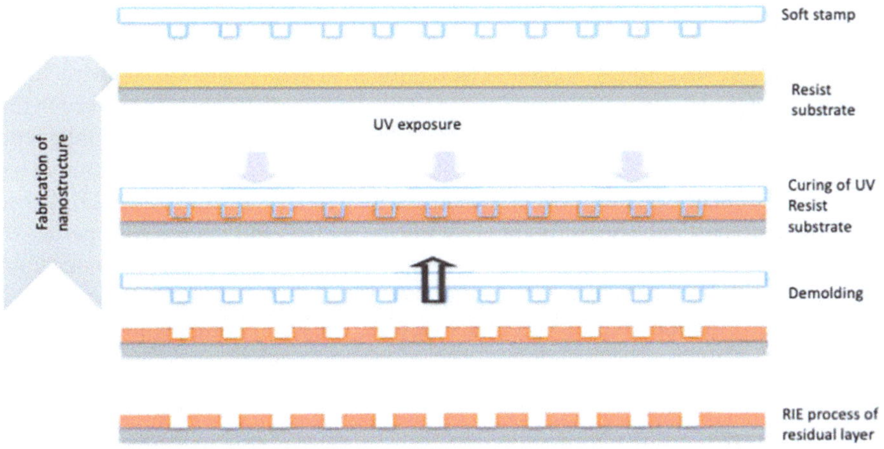

Fig. 4 Fabrication of nanostructure: Nanofabrication is the production of nanometer-sized materials for large-scale processing, aiming to create nanoscale structures that are part of a system, device, or component in large quantities and at a low cost. Nanoimprint lithography is used to fabricate nanostructures. It consists of three primary steps: origination, replication, and pattern transmission. The nanoimprint lithography (NIL) technique, which includes molding and thin film pattern transfer, is a well-established method for resist-based patterning of hard surfaces

12 Characterization Techniques of Nanostructures

Characterization techniques are used to analyze and understand the properties, composition, morphology, and behavior of these nanostructures [86]. There are several characterization techniques that are commonly used for nanostructures. These techniques include microscopy techniques such as scanning electron microscopy, transmission electron microscopy, atomic force microscopy, and scanning probe microscopy [87]. Spectroscopy techniques such as X-ray photoelectron spectroscopy, Raman spectroscopy, and Fourier transform infrared spectroscopy. Diffraction techniques such as X-ray diffraction and electron diffraction [88]. These characterization techniques enable researchers to gain insights into the structural and chemical properties of nanostructures, providing valuable information for further research and development in various fields such as materials science, nanotechnology, electronics, and biomedicine. Scanning electron microscopy is a powerful tool for imaging nanostructures, providing high-resolution images of the surface topography [89]. It uses a focused beam of electrons to scan the sample, and the resulting backscattered or secondary electrons are detected and used to create an image. Atomic force microscopy is another important characterization technique for nanostructures [90]. It uses a small probe to scan the surface of the sample, measuring the forces between the probe and the sample to create a high-resolution image. Furthermore, transmission electron microscopy is widely used for the characterization of nanostructures [91]. It uses a beam of electrons transmitted through a thin sample to obtain detailed information about the structure, composition, and morphology of nanostructures. Scanning probe microscopy is another technique used to characterize nanostructures [92]. It uses a probe, such as a sharp tip or a small cantilever, to scan the surface of the sample and measure properties such as surface roughness, conductivity, and magnetic fields at the nanoscale. X-ray photoelectron spectroscopy is a powerful technique for analyzing the chemical composition and electronic state of nanostructures [93]. These techniques are particularly useful for studying the surface and interface properties of nanostructures, as well as their chemical composition, bonding, and electronic structure. Roman spectroscopy is used to study the vibrational modes of nanostructures, providing information about molecular structure, chemical composition, and crystallinity [94]. Fourier transform infrared spectroscopy is another common technique used to analyze the chemical composition and bonding of nanostructures [95].

13 Characteristics and Properties of Nanostructures

At this scale, nanostructures exhibit unique properties and behaviors that differ from those observed in macroscopic materials [96]. These unique properties and behaviors can include enhanced mechanical strength, improved electrical conductivity, increased surface area-to-volume ratio, and enhanced catalytic activity, among others. Additionally, nanostructures often possess quantum confinement effects, where the electronic and optical properties of the material are influenced by the confinement of electrons within nanoscale dimensions [97]. This leads to phenomena such as size-dependent melting points and optical properties that can be tuned by controlling the size and shape of the nanostructure. Furthermore, the properties and characteristics of nanostructures are highly influenced by their composition [98]. Different types of nanostructures, such as carbon-based nanomaterials, hydrogels, magnetic nanoparticles, metallic nanoparticles, polymer nanoparticles, and nanocomposites, exhibit distinct properties and behaviors due to their unique chemical compositions [99]. Nanomaterials also have a high surface-to-volume ratio, which allows for greater interaction with the surrounding environment and makes them ideal for applications such as sensing, catalysis, drug delivery, and energy storage [100]. In addition to their unique properties, nanostructures also exhibit high reactivity and adsorption capabilities [101]. This makes them valuable in fields such as catalysis, medicine, sensing, and biology.

14 Applications of Nanostructures

Nanostructures, materials and devices with nanometer scale dimensions, have gained significant attention due to their unique properties and potential applications in various fields, including medicine, environmental sectors, electronics, and energy [102]. Nanostructures have shown great promise in targeted drug delivery and imaging, allowing for precise delivery of medications to specific targets within the body, minimizing side effects, and improving treatment efficacy [103]. They have also been utilized in the development of advanced imaging techniques, enabling more accurate diagnosis and monitoring of diseases at the molecular level. In the environmental sector, nanostructures have been used for pollution remediation and water purification, contributing to environmental sustainability. Their high reactivity and large surface area make them effective in removing pollutants and contaminants from air and water, contributing to environmental sustainability. In electronics, nanostructures have revolutionized the industry by enabling the development of smaller, faster, and more efficient devices, paving the way for flexible electronics, high-density data storage, and advanced sensors [104]. In the energy sector, nanostructures have been employed in the development of efficient solar cells, lightweight and high-capacity batteries, and catalytic converters for clean energy production and storage [105]. This has the potential to address global energy challenges and reduce reliance on traditional fossil fuels. Ongoing research and innovation in nanotechnology continue to expand the horizon of possibilities for nanostructures, making them a vital area of exploration and development for the future. Nanostructures operate by engaging with molecules and surfaces at the nanoscale, resulting in increased reactivity and selectivity. In the realm of sensing, nanostructures are capable of identifying and converting signals through interactions with analytes at the nanoscale. They can experience transition states, involving the creation and breaking of bonds between atoms or molecules, facilitating processes like catalysis. Moreover, nanostructures

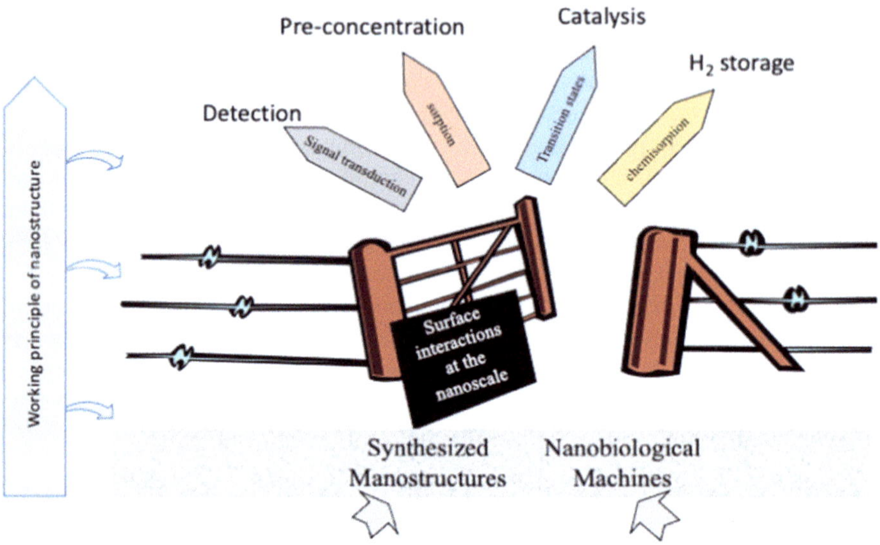

Fig. 5 Working principle of nanostructure: Nanostructures, based on quantum mechanics principles, exhibit unique physical and chemical properties at the nanoscale due to their high surface-to-volume ratio. These properties can be enhanced by functionalizing the surface, altering their reactivity and interaction with the environment, thus enhancing their working principle

possess notable sorption capacities, allowing them to effectively adsorb and retain molecules on their surfaces. This attribute proves valuable in scenarios such as environmental cleanup, where nanostructures can eliminate contaminants from water or air. Furthermore, nanostructures display chemisorption, forming robust chemical bonds with molecules, enabling precise and reversible interactions (Fig. 5).

15 Limitations of Hybrid Materials and Nanostructures

Hybrid materials, composed of various compounds and materials, have gained attention due to their unique properties and potential applications in fields like electronics, energy storage, and biomedicine [106]. However, there are several limitations associated with hybrid materials. Compatibility between constituent materials can be challenging, as the properties, processing conditions, and chemical reactivity may not be compatible, leading to difficulties in achieving a uniform and homogeneous material [107]. Structural integrity can also be affected by the combination of different materials, leading to stress concentrations and potential failure points. Degradation and stability are also potential limitations of hybrid performance trade-offs [108]. The optimum properties of individual materials may be compromised due to compatibility and integration [109]. Functional limitations may arise, as the properties and functions of constituent materials may not synergize effectively [27].

Chemical reactions may also occur, potentially degrading the overall performance of the hybrid material [110]. Processing challenges are also present, as the synthesis and fabrication of hybrid materials can be complex and require specialized techniques [111]. Achieving uniform distribution and proper bonding between materials can be challenging, leading to inconsistencies in the final material's properties [112]. Therefore, while hybrid materials offer exciting opportunities and unique properties, they must be carefully considered and addressed to ensure their successful application in these fields. Nanostructures have attracted significant attention in various fields due to their unique properties and potential applications [113]. However, they also come with certain limitations that must be considered. One limitation is their potential cytotoxicity, which can be caused by their structure, chemical content, or features [114]. Another limitation is the requirement for thorough assessment of their biocompatibility before using them in biomedical applications [115]. Additionally, the cost-effective synthesis of nanostructures poses a challenge that needs to be addressed [116]. Moreover, there is a lack of comprehensive understanding regarding the mechanism of action and safety evaluation of bioactive nanomaterials [117]. Furthermore, the distribution, metabolism, and clearance process of nanostructures in vivo are still not fully understood, necessitating further research in this area [118].

16 Future Perspective of Hybrid Materials and Nanostructures

Hybrid materials are essential for the development of advanced tools, manufacturing processes, sustainability, and applications across various fields. Research in hybrid materials and nanotechnology is expected to accelerate science and innovation breakthroughs, leading to new designs and applications guided by societal needs. Advancements in hybrid materials and nanotechnology may significantly lower costs of materials and technologies, such as solar energy storage, energy conversion costs, and water desalinization. In the upcoming decade, challenges must be addressed, including building systems from the nanoscale using nanoscale laws, information technology, biological principles, and system integration. Nanotechnology research has transformed areas to understand nanoscale phenomena and processes, using measurements and simulations, the classical/quantum physics transition in nanostructures and devices, and the multiscale self-assembly of materials. Advancements in nanostructures can enhance the performance and functionality of materials and devices, leading to smaller, faster, and more efficient technologies. Nanomaterials can be used for targeted drug delivery, diagnostics, and personalized treatments. Overall, nanotechnology holds great promise for shaping the future of technology.

17 Conclusion

Hybrid materials and nanostructures have gained attention due to their unique properties and potential applications. These materials, composed of a combination of different nanoscale materials, exhibit novel and enhanced properties compared to their individual components. Nanostructures, typically with dimensions ranging from 1 to 100 nm, can have different shapes and structures, such as nanoparticles, nanowires, or nanotubes. These nanostructures can be integrated into hybrid materials, which combine different types of materials, such as metals, polymers, ceramics, or composites, to create a new material with tailored properties. The development and study of hybrid materials and nanostructures are driven by the need for improved performance in various fields, such as electronics, energy storage, catalysis, sensing, and biomedical applications. These materials hold great promise for technological advancements and have the potential to revolutionize industries. Research groups worldwide are constantly developing new strategies and applications for hybrid materials, promoting innovation in various sectors. As researchers continue to explore and harness the potential of nanostructures, their impact on technology and society is expected to grow significantly in the coming years.

References

1. Wellmann, P.: The search for new materials and the role of novel processing routes. Discov. Mater. **1** (2021). https://doi.org/10.1007/s43939-021-00014-y
2. Tofail, S.A.M., Koumoulos, E.P., Bandyopadhyay, A., Bose, S., O'Donoghue, L., Charitidis, C.: Additive manufacturing: scientific and technological challenges, market uptake and opportunities. Mater. Today **21**(1), 22–37 (2018). https://doi.org/10.1016/j.mattod.2017.07.001
3. Zhu, Y., et al.: Heterostructured materials: superior properties from hetero-zone interaction. Mater. Res. Lett. **9**(1), 1–31 (2021). https://doi.org/10.1080/21663831.2020.1796836
4. Improvement, R.: Resilience Improvement (2023)
5. Fang, Z., Tang, S., Wang, Z., An, M., Yu, G.: General synthesis of large inorganic nanosheets via 2D confined assembly of nanoparticles. ACS Cent. Sci. **8**(5), 627–635 (2022). https://doi.org/10.1021/acscentsci.2c00252
6. Szczyglewska, P., Feliczak-guzik, A.: Nanotechnology—general aspects : a chemical reduction approach to the synthesis of nanoparticles (2023)
7. Mehta, S., Suresh, A., Nayak, Y., Narayan, R., Nayak, U.Y.: Hybrid nanostructures: versatile systems for biomedical applications. Coord. Chem. Rev. **460**, 214482 (2022). https://doi.org/10.1016/j.ccr.2022.214482
8. Anjum, A., Garg, R., Kashif, M., Eddy, N.O.: Nano-scale innovations in packaging: properties, types, and applications of nanomaterials for the future. Food Chem. Adv. **3**, 100560 (2023). https://doi.org/10.1016/j.focha.2023.100560
9. Das, P.P., Chaudhary, V., Singh, R., Singh, D., Bachchan, A.A.: Advancement in hybrid materials, its applications and future challenges: a review. Mater. Today: Proc. **47** (2021). https://doi.org/10.1016/j.matpr.2021.03.009
10. Sharma, A., et al.: Inorganic/organic combination: inorganic particles/polymer composites for tissue engineering applications. Bioactive Mater. **24**, 535–550 (2023). https://doi.org/10.1016/j.bioactmat.2023.01.003

11. Overview, A.: Polymer-based hydrogels applied in drug delivery, pp. 1–38 (2023)
12. Deng, Z., Jia, Z., Li, L.: Biomineralized materials as model systems for structural composites : intracrystalline structural features and their strengthening and toughening mechanisms. **2103524**, 1–32 (2022). https://doi.org/10.1002/advs.202103524
13. Zhong, J., Huang, W.: Multifunctionality in nature : structure—function relationships in biological materials (2023)
14. Gu, H., et al.: Introducing advanced composites and hybrid materials. Adv. Compos. Hybrid Mater. **1**(1), 1–5 (2018). https://doi.org/10.1007/s42114-017-0017-y
15. Reduwan Billah, S.M.: Composites and nanocomposites BT—functional polymers. In: Jafar Mazumder, M.A., Sheardown, H., Al-Ahmed, A. (eds.) Springer International Publishing, Cham, pp. 1–67 (2019)
16. Song, J., Vikulina, A., Parakhonskiy, B., Skirtach, A.: Hierarchy of hybrid materials. Part-II: the place of organics-on-inorganics in it, their composition and applications. Front. Chem. **11**, 1078840 (2023). https://doi.org/10.3389/fchem.2023.1078840
17. Beyou, E., Bourgeat-Lami, E.: Organic–inorganic hybrid functional materials by nitroxide-mediated polymerization. Prog. Polym. Sci. **121**, 101434 (2021). https://doi.org/10.1016/j.progpolymsci.2021.101434
18. Gomez-romero, P., Pokhriyal, A., Rueda-garc, D., Bengoa, L.N., Gonz, R.M.: Hybrid Materials : A Metareview (2024). https://doi.org/10.1021/acs.chemmater.3c01878
19. Ikake, H., Hara, S., Shimizu, S.: Skillful control of dispersion and 3D network structures : advances in functional organic—inorganic nano-hybrid materials prepared using the sol-gel method (2022)
20. Latz, M., Böhme, A., Ulrich, N.: Reactivity-based identification of oxygen containing functional groups of chemicals applied as potential classifier in non-target analysis. Sci. Rep. **13** (2023). https://doi.org/10.1038/s41598-023-50240-y
21. Zhang, F., Sherrell, P.C., Luo, W., Chen, J., Li, W., Yang, J.: Organic/inorganic hybrid fibers : controllable architectures for electrochemical energy applications. **2102859**, 1–30 (2021). https://doi.org/10.1002/advs.202102859
22. Fan, W., et al.: Flexible inorganic–organic hybrids with dual inorganic components. Mater. Today Chem. **22**, 100584 (2021). https://doi.org/10.1016/j.mtchem.2021.100584
23. Arya, M., Heera, S., Meenu, P., Deepa, K.G.: Organic-inorganic hybrid materials and architectures in optoelectronic devices: recent advancements. ChemPhysMater (2024). https://doi.org/10.1016/j.chphma.2024.03.004
24. Faustini, M., Nicole, L., Ruiz-Hitzky, E., Sanchez, C.: History of organic–inorganic hybrid materials: prehistory, art, science, and advanced applications. Adv. Funct. Mater. **28** (2018). https://doi.org/10.1002/adfm.201704158
25. Alabada, R., et al.: Investigation of effective parameters in the production of alumina gel through the sol-gel method. Case Stud. Chem. Environ. Eng. **8**, 100405 (2023). https://doi.org/10.1016/j.cscee.2023.100405
26. Gomez-Romero, P., Pokhriyal, A., Rueda-García, D., Bengoa, L.N., González-Gil, R.M.: Hybrid materials: a metareview. Chem. Mater. **36**(1), 8–27 (2024). https://doi.org/10.1021/acs.chemmater.3c01878
27. Song, J., Vikulina, A.S., Parakhonskiy, B.V.: Hierarchy of hybrid materials . Part-II : the place of organics-on-inorganics in it, their composition and applications, pp 1–23 (2023). https://doi.org/10.3389/fchem.2023.1078840
28. Sharma, G., et al.: Novel development of nanoparticles to bimetallic nanoparticles and their composites: a review. J. King Saud Univ. Sci. **31**(2), 257–269 (2019). https://doi.org/10.1016/j.jksus.2017.06.012
29. Dykman, L., Khlebtsov, N.: Methods for chemical synthesis of colloidal gold. Russian Chem. Rev. **88** (2019). https://doi.org/10.1070/RCR4843
30. Gołębiowski, A., Buszewski, B.: Characterization of colloidal particles of a biological and metallic nature. Microchem. J. **191**, 108864 (2023). https://doi.org/10.1016/j.microc.2023.108864

31. Sanchis-Gual, R., Coronado-Puchau, M., Mallah, T., Coronado, E.: Hybrid nanostructures based on gold nanoparticles and functional coordination polymers: chemistry, physics and applications in biomedicine, catalysis and magnetism. Coord. Chem. Rev. **480**, 215025 (2023). https://doi.org/10.1016/j.ccr.2023.215025
32. Fibers, O., et al.: Review on hybrid reinforced polymer matrix composites with (2023)
33. Jiao, L., Seow, J.Y.R., Skinner, W.S., Wang, Z.U., Jiang, H.-L.: Metal–organic frameworks: structures and functional applications. Mater. Today **27**, 43–68 (2019). https://doi.org/10.1016/j.mattod.2018.10.038
34. Hossain, N., et al.: Advances and significances of nanoparticles in semiconductor applications—a review. Results Eng. **19**, 101347 (2023). https://doi.org/10.1016/j.rineng.2023.101347
35. Fracaroli, A., de Rossi, R.: Supramolecular chemistry in solid state materials such as metal-organic frameworks. Israel J. Chem. **58** (2018). https://doi.org/10.1002/ijch.201800114
36. Materials, C.O.F., Ploetz, E., Engelke, H., Lächelt, U., Wuttke, S.: The chemistry of reticular framework nanoparticles (2020). https://doi.org/10.1002/adfm.201909062
37. Meti, P., et al.: Overview of organic–inorganic hybrid silica aerogels: progress and perspectives. Mater. Des. **222**, 111091 (2022). https://doi.org/10.1016/j.matdes.2022.111091
38. Khan, Y., et al.: Nanoparticles, and their applications in various fields of nanotechnology : a review (2022)
39. Dey, K., et al.: Graphene and its hybrid materials: properties and applications (2024)
40. Catauro, M., Vecchio Ciprioti, S.: Characterization of hybrid materials prepared by sol-gel method for biomedical implementations: a critical review. Materials **14**, 1788 (2021). https://doi.org/10.3390/ma14071788
41. Tao, C., Zou, X., Reddy, K.M., Zhang, L., Jiang, B.: A hydrophobic ultralow refractive-index silica coating towards double-layer broadband antireflective coating with exceptionally high vacuum stability and laser-induced damage threshold. Colloids Surf. A **563**, 340–349 (2019). https://doi.org/10.1016/j.colsurfa.2018.11.038
42. Moschetta, E.G., et al.: Photochemistry in pharmaceutical development: a survey of strategies and approaches to industry-wide implementation. Org. Process Res. Dev. **28**(4), 831–846 (2024). https://doi.org/10.1021/acs.oprd.3c00499
43. Nguyen, M.X.T., Nguyen, N.T., Dinh, H.N.H., Nguyen, N.N., Van Tran, T.T., Hoang, D.: Natural sourced and non-toxic hybrid materials for boosting the growth of lettuce in a hydroponic system. Plant Physiol. Biochem. **197**, 107652 (2023). https://doi.org/10.1016/j.plaphy.2023.107652
44. Li, Z., et al.: An overview of synthesis and structural regulation of magnetic nanomaterials prepared by chemical coprecipitation (2023)
45. Fragoso-Medina, A.J., López-Saucedo, F., Flores-Rojas, G.G., Bucio, E.: Sonochemical synthesis of inorganic nanomaterials, pp 263–279 (2021)
46. Li, Z., et al.: Nanoscale advances sonochemical catalysis as a unique strategy for the. Nanoscale Adv. **3**, 41–72 (2020). https://doi.org/10.1039/D0NA00753F
47. Bhiyan, M.M., Bhuiyan, A., Hammad, A.: A hybrid multi-criteria decision support system for selecting the most sustainable structural material for a multistory building construction. Sustainability **15** (2023). https://doi.org/10.3390/su15043128
48. Georgantzinos, S.K., Giannopoulos, G., Stamoulis, K., Markolefas, S.: Composites in aerospace and mechanical engineering. Materials **16**, 7230 (2023). https://doi.org/10.3390/ma16227230
49. Venkateshaiah, A., Padil, V., Nagalakshmaiah, M., Wacławek, S., Černík, M., Varma, R.: Microscopic techniques for the analysis of micro and nanostructures of biopolymers and their derivatives. Polymers **12**, 512 (2020). https://doi.org/10.3390/polym12030512
50. Ali, A., Chiang, Y.W., Santos, R.: X-ray diffraction techniques for mineral characterization: a review for engineers of the fundamentals, applications, and research directions. Minerals **12**, 205 (2022). https://doi.org/10.3390/min12020205
51. Saba, N., Jawaid, M.: A review on thermomechanical properties of polymers and fibers reinforced polymer composites. J. Ind. Eng. Chem. **67**, 1–11 (2018). https://doi.org/10.1016/j.jiec.2018.06.018

52. Ahmad, F., Maqsood, A.: Complex impedance, dielectric constant, electric modulus, and conductivity analysis of Cd doped ZnO nanostructures at high temperatures. Physica E **143**, 115353 (2022). https://doi.org/10.1016/j.physe.2022.115353
53. Ajeena, A.M., Farkas, I., Víg, P.: Characterization, rheological behaviour, and dynamic viscosity of ZrO2-SiC (50–50)/DW hybrid nanofluid under different temperatures and solid volume fractions: an experimental study and proposing a new correlation. Powder Technol. **431**, 119069 (2024). https://doi.org/10.1016/j.powtec.2023.119069
54. Saveleva, M.S., Eftekhari, K., Abalymov, A., Douglas, T.E.L.: Hierarchy of hybrid materials—the place of inorganics-in-organics in it, their composition and applications. **7**, pp. 1–21 (2019). https://doi.org/10.3389/fchem.2019.00179
55. Nandakumar, N., Nair, A.B.: Gas barrier properties and applications of nanocellulose-based materials BT—handbook of biopolymers. In: Thomas, A., AR, A., Jose Chirayil, C., Thomas, B. (eds.) Springer Nature Singapore, Singapore, pp. 1263–1279 (2023)
56. Perera, K.Y., Jaiswal, A.K., Jaiswal, S.: Biopolymer-Based Sustainable Food Packaging Materials, pp. 1–59 (2023)
57. Gamage, A., et al.: Polysaccharide-Based Bioplastics as Green and Sustainable Packaging Materials (2023)
58. Donkor, L., Kontoh, G., Yaya, A., Bediako, J.K., Apalangya, V.: Bio-based and sustainable food packaging systems: relevance, challenges, and prospects. Appl. Food Res. **3**(2), 100356 (2023). https://doi.org/10.1016/j.afres.2023.100356
59. I Alloys, and Its Alloys (2023)
60. Yu, Z., et al.: Organic-inorganic hybrid ZIF-8/MXene/cellulose-based textiles with improved antibacterial and electromagnetic interference shielding performance. Int. J. Biol. Macromol. **266**, 131080 (2024). https://doi.org/10.1016/j.ijbiomac.2024.131080
61. Saleem, M.Z., Akbar, M.: Review of the Performance of High-Voltage Composite Insulators (2022)
62. Mekuye, B., Abera, B.: Nanomaterials: an overview of synthesis, classification, characterization, and applications. Nano Select **4**, 486–501 (2023). https://doi.org/10.1002/nano.202300038
63. Li, S., et al.: Development and applications of aluminum alloys for aerospace industry. J. Market. Res. **27**, 944–983 (2023). https://doi.org/10.1016/j.jmrt.2023.09.274
64. Mishra, M., Prasad, K., Ramakrishn, S., Jha, A.: Nanomaterials in drug delivery—promises and limitations. Nano Med. Mater. **3**, 38 (2023). https://doi.org/10.59400/nmm.v3i1.38
65. Javaid, M., Haleem, A., Singh, R.P., Rab, S., Suman, R.: Exploring the potential of nanosensors: a brief overview. Sens. Int. **2**, 100130 (2021). https://doi.org/10.1016/j.sintl.2021.100130
66. Ahire, S., Bachhav, A., Pawar, T., Jagdale, B., Patil, A., Koli, P.: The augmentation of nanotechnology era: a concise review on fundamental concepts of nanotechnology and applications in material science and technology. Results Chem. 100633 (2022). https://doi.org/10.1016/j.rechem.2022.100633
67. Harish, V., et al.: Cutting-edge advances in tailoring size, shape, and functionality of nanoparticles and nanostructures: a review. J. Taiwan Inst. Chem. Eng. **149**, 105010 (2023). https://doi.org/10.1016/j.jtice.2023.105010
68. Malik, S., Muhammad, K.: Nanotechnology : A Revolution in Modern Industry (2023)
69. Bhat, A., et al.: Review on nanocomposites based on aerospace applications. Nanotechnol. Rev. **10**, 237–253 (2021). https://doi.org/10.1515/ntrev-2021-0018
70. Yu, H., et al.: Nature-resembled nanostructures for energy storage/conversion applications. J. Ind. Eng. Chem. **129**, 53–68 (2024). https://doi.org/10.1016/j.jiec.2023.08.041
71. Rather, M.A., Bhuyan, S., Chowdhury, R., Sarma, R., Roy, S., Neog, P.R.: Nanoremediation strategies to address environmental problems. Sci. Total. Environ. **886**, 163998 (2023). https://doi.org/10.1016/j.scitotenv.2023.163998
72. Pradhan, K.K., Chakraverty, S.: Chapter 1: Introduction to Nanostructures, pp. 1–10 (2021)
73. Noah, N.M.: Review article design and synthesis of nanostructured materials for sensor applications. **2020** (2020)

74. Applications, B., Fritea, L., Banica, F., Costea, T.O., Moldovan, L., Dobjanschi, L.: Metal nanoparticles and carbon-based nanomaterials for improved performances of electrochemical (Bio) sensors with (2021)
75. Rizvi, M., Gerengi, H., Gupta, P.: Functionalization of nanomaterials : synthesis and characterization. In: Functionalized Nanomaterials for Corrosion Mitigation: Synthesis, Characterization, and Applications, Part 1—Functionalization of Nanomaterials: Synthesis and Characterization, pp. 1–26 (2022). https://doi.org/10.1021/bk-2022-1418.ch001
76. Harish, V., Tewari, D., Gaur, M., Yadav, A.B., Swaroop, S.: Review on nanoparticles and nanostructured materials : bioimaging, biosensing, drug delivery, tissue engineering, antimicrobial, and agro-food applications (2022)
77. Altammar, K.A.: A review on nanoparticles : characteristics, synthesis, applications, and challenges, pp. 1–20 (2023). https://doi.org/10.3389/fmicb.2023.1155622
78. Slepi, N., Slepi, P.: Carbon Nanostructures, Nanolayers, and Their Composites (2021)
79. Ossai, C.I., Raghavan, N.: Nanostructure and nanomaterial characterization, growth mechanisms, and applications. **7**(2), 209–231 (2018)
80. Bokov, D., et al.: Nanomaterial by sol-gel method: synthesis and application. Adv. Mater. Sci. Eng. **2021** (2021). https://doi.org/10.1155/2021/5102014
81. Escorcia-d, D., Rend, L., Ram, M.: advancements in nanoparticle deposition techniques for diverse substrates : a review. 1–25 (2023)
82. Films, S., Superhydrophobicity, L., Zhou, B., Wu, Y.: Investigation of electrochemical assisted deposition of sol-gel (2023)
83. Sharma, R., Kumar, S., Gupta, A.: An insight of nanomaterials in tissue engineering from fabrication to applications. Tissue Eng. Regenerative Med. **19**(5), 927–960 (2022). https://doi.org/10.1007/s13770-022-00459-z
84. Joudeh, N., Linke, D.: Nanoparticle classification, physicochemical properties, characterization, and applications: a comprehensive review for biologists. J. Nanobiotechnol. **20**(1), 262 (2022). https://doi.org/10.1186/s12951-022-01477-8
85. Gacem, A., et al.: Review article recent advances in methods for synthesis of carbon nanotubes and carbon nanocomposite and their emerging applications : a descriptive review **2022** (2022)
86. Ramachandran, A., Joseph, R., Asok, A.: Characterization techniques for morphological and physicochemical evaluation of nanomaterials, pp. 21–50 (2021)
87. Kalinin, S.V., et al.: Deep learning for electron and scanning probe microscopy: from materials design to atomic fabrication. MRS Bull. **47**(9), 931–939 (2022). https://doi.org/10.1557/s43577-022-00413-3
88. Chiriac, A.E., et al.: Microscopy as Noninvasive Methods for Microstructural Alterations in Psoriatic Nails (2021)
89. Ali, A., Zhang, N., Santos, R.: Mineral characterization using Scanning Electron Microscopy (SEM): a review of the fundamentals, advancements, and research directions. Appl. Sci. **13**, 12600 (2023). https://doi.org/10.3390/app132312600
90. Bellotti, R., Picotto, G.B., Ribotta, L.: AFM measurements and tip characterization of nanoparticles with different shapes. Nanomanuf. Metrol. **5**(2), 127–138 (2022). https://doi.org/10.1007/s41871-022-00125-x
91. Lee, B., et al.: Statistical characterization of the morphologies of nanoparticles through machine learning based electron microscopy image analysis. ACS Nano **14**(12), 17125–17133 (2020). https://doi.org/10.1021/acsnano.0c06809
92. Feltin, N., Ducourtieux, S., Crouzier, L., Delvallée, A., Dirscherl, K., Zeng, G.: Chapter 2.1.3—Scanning probe microscopy (SPM). In: Hodoroaba, V.-D., Unger, W.E.S., A.G.B.T.-C. of Shard, N. (eds.) Micro and Nano Technologies, pp. 49–63. Elsevier (2020)
93. Jacob, G., Min, H.: A brief overview of X-ray photoelectron spectroscopy characterization of thin films. Res. J. Chem. Environ **28**, 142–160 (2024). https://doi.org/10.25303/281rjce1420160
94. Albuquerque Nogueira, B., Castiglioni, C.: Raman spectroscopy of crystalline materials and nanostructures. Crystals **14**, 251 (2024). https://doi.org/10.3390/cryst14030251

95. Khan, S., Khan, S., Khan, L., Farooq, A., Akhtar, K., Asiri, A.M.: Fourier transform infrared spectroscopy: fundamentals and application in functional groups and nanomaterials characterization. In: Handbook of Materials Characterization, pp. 317–344 (2018)
96. Rawat, N., et al.: Chapter Two—Interaction of cells with different types of TiO_2 nanostructured surfaces. In: Iglič, A., Rappolt, M., P. B. T.-A. in B., Losada Perez, L. S.-A. (eds.) Advances in Biomembranes and Lipid Self-Assembly, vol. 37. Academic, pp. 29–59 (2023)
97. Edvinsson, T.: Optical quantum confinement and photocatalytic properties in dimensional nanostructures (2018)
98. Shoukat, A., Rafique, M., Ayub, A., Razzaq, B., Tahir, M., Sagir, M.: An Insight into Properties and Characterization of Nanostructures, pp. 39–81 (2021)
99. Solangi, N.H., Karri, R.R., Mubarak, N.M., Mazari, S.A.: Mechanism of polymer composite-based nanomaterial for biomedical applications. Adv. Ind. Eng. Polymer Res. **7**(1), 1–19 (2024). https://doi.org/10.1016/j.aiepr.2023.09.002
100. Song, J.-H., Min, S.-H., Kim, S.-G., Cho, Y., Ahn, S.-H.: Multi-functionalization strategies using nanomaterials: a review and case study in sensing applications. Int. J. Precis. Eng. Manuf.-Green Technol. **9**(1), 323–347 (2022). https://doi.org/10.1007/s40684-021-00356-1
101. Ahmed, S., et al.: Nanomaterials as a sustainable choice for treating wastewater. Environ. Res. **214**, 113807 (2022). https://doi.org/10.1016/j.envres.2022.113807
102. Aghababai Beni, A., Jabbari, H.: Nanomaterials for environmental applications. Results Eng. **15**, 100467 (2022). https://doi.org/10.1016/j.rineng.2022.100467
103. Mitchell, M.J., Billingsley, M.M., Haley, R.M., Wechsler, M.E., Peppas, N.A., Langer, R.: Engineering precision nanoparticles for drug delivery. Nat. Rev. Drug Discov. **20**(2), 101–124 (2021). https://doi.org/10.1038/s41573-020-0090-8
104. Devices, E.S.: A Review of Manufacturing Methods for Flexible Devices and Energy Storage Devices (2023)
105. Banin, U., et al.: Nanotechnology for catalysis and solar energy conversion. Nanotechnology **32**, 42003 (2020). https://doi.org/10.1088/1361-6528/abbce8
106. Worku, A.K., Ayele, D.W.: Recent advances of graphene-based materials for emerging technologies. Results Chem. **5**, 100971 (2023). https://doi.org/10.1016/j.rechem.2023.100971
107. Asyraf, M.R.M., et al.: Mechanical properties of oil palm fibre-reinforced polymer composites: a review. J. Market. Res. **17**, 33–65 (2022). https://doi.org/10.1016/j.jmrt.2021.12.122
108. Ince, J.C., et al.: Overview of emerging hybrid and composite materials for space applications. Adv. Compos. Hybrid Mater. **6**(4), 130 (2023). https://doi.org/10.1007/s42114-023-00678-5
109. Mohammed, M., et al.: Comprehensive insights on mechanical attributes of natural-synthetic fibres in polymer composites. J. Market. Res. **25**, 4960–4988 (2023). https://doi.org/10.1016/j.jmrt.2023.06.148
110. Kimbi Yaah, V.B., Ojala, S., Khallok, H., Laitinen, T., Botelho de Oliveira, S.: Hybrid carbon materials: synthesis, characterization, and application in the removal of pharmaceuticals from water. J. Water Process Eng. **43**, 102279 (2021). https://doi.org/10.1016/j.jwpe.2021.102279
111. Agma, O., Basturk, S.: Synthetic, Hybrid and Natural Composite Fabrication Processes, pp. 115–137 (2023)
112. Malekinejad, H., Carbas, R.J.C., Akhavan-safar, A., Marques, E.A.S., Sousa, F.C., Silva, L.F.M.: Enhancing Fatigue Life and Strength of Adhesively Bonded Composite Joints : A Comprehensive Review (2023)
113. Taha, T.B., Barzinjy, A.A., Hussain, F.H.S., Nurtayeva, T.: Nanotechnology and computer science: trends and advances. Memories Mater. Dev. Circuits Syst. **2**, 100011 (2022). https://doi.org/10.1016/j.memori.2022.100011
114. Egbuna, C., et al.: Toxicity of nanoparticles in biomedical application : nanotoxicology (2021)
115. Kyriakides, T.R., et al.: Biocompatibility of nanomaterials and their immunological properties. **16**(4), 1–49 (2022). https://doi.org/10.1088/1748-605X/abe5fa.Biocompatibility
116. Javed, M., et al.: Potential and challenges in green synthesis of nanoparticles: a review. Xi'an Shiyou Daxue Xuebao (Ziran Kexue Ban)/J. Xi'an Shiyou Univ. **19**, 1155–1165 (2023)
117. Singh, R., Kumar, S.: Regulatory and Safety Concerns Regarding the Use of Active Nanomaterials in Food Industry, pp. 269–306 (2023)

118. Zhang, A., et al.: Absorption, distribution, metabolism, and excretion of nanocarriers in vivo and their influences. Adv. Coll. Interface. Sci. **284**, 102261 (2020). https://doi.org/10.1016/j.cis.2020.102261
119. Zaccone, M., et al.: Influence of Chitin Nanocrystals on the Crystallinity and Mechanical Properties of Poly (hydroxybutyrate) Biopolymer (2022)
120. Osman, W., et al.: Electronic and magnetic properties of graphene quantum dots doped with alkali metals. J. Mater. Res. Technol. **11**. https://doi.org/10.1016/j.jmrt.2021.01.119
121. Items, R., Rose, W., Rose, W., If, T., Rose, W.: Multiscale modelling of ceramic nanoparticle interactions and their influence on the thermal conductivity of nanofluids (2020)
122. Ye, X., Lendel, C., Langton, M., Olsson, R.T., Hedenqvist, M.S.: Chapter 2—Protein nanofibrils: Preparation, properties, and possible applications in industrial nanomaterials. Elsevier Inc. (2019)
123. Comini, E.: Metal oxides nanowires chemical/gas sensors: recent advances. Mater. Today Adv. **7**, 100099 (2020). https://doi.org/10.1016/j.mtadv.2020.100099
124. Zhang, X., Chu, H., Li, Y., Zhao, S., Li, D.: Diameter-selected single-walled carbon nanotubes for the passive Q-switching operation at 2 μm. Opt. Mater. **100**, 109627 (2020). https://doi.org/10.1016/j.optmat.2019.109627
125. Koyappayil, A., Yagati, A.K.: Recent Trends in Metal Nanoparticles Decorated 2D Materials for Electrochemical Biomarker Detection (2023)
126. Yun, H., Paik, T.: Colloidal Self-assembly of Inorganic Nanocrystals into Superlattice Thin-Films and Multiscale Nanostructures (2019)

Analytical Tools for Characterizing Nanomaterials

Shazlina Johari, Muhammad Mahyiddin Ramli, Mohd Fairus Ahmad, Nurul Huda Osman, and Ali H. Reshak

1 Introduction

Nanomaterials are substance that existed or synthesized in the size of between ~1 and 100 nm (nm) physically. A nanometer is one millionth of a millimetre—approximately 100,000 times smaller than the diameter of a human hair. Nanomaterials exhibit unique physical, chemical, and biological properties thus make it very attractive and popular among researchers. These nanomaterials are widely used in various fields due to their high surface area, remarkable biocompatibility, optical and electrical properties. Some examples of nanomaterials include nanoparticles, nanofibers, nanotubes, carbon, and quantum dots. In general, it can be broadly categorized into four different type of nanomaterials which are; Nanoparticles: particles with at least one dimension measuring less than 100 nm (gold nanoparticles, silver nanoparticles, zinc oxide nanoparticles and etc.); Nanofibers: long, thin fibres with diameters in the nanoscale range (polymers, ceramics, composites, or natural materials such as proteins and DNA); Nanotubes: tubes with walls made of a single layer of atoms with diameters in the nanoscale range (carbon nanotubes, titanium nanotubes) and;

S. Johari · M. M. Ramli (✉) · M. F. Ahmad
Faculty of Electronic Engineering & Technology, Universiti Malaysia Perlis (UniMAP), 02600 Arau, Perlis, Malaysia
e-mail: mmahyiddin@unimap.edu.my

Institute of Nano Electronic Engineering, Universiti Malaysia Perlis (UniMAP), 01000 Kangar, Perlis, Malaysia

N. H. Osman
Applied Electromagnetic Laboratory, Department of Physics, Faculty of Science, Universiti Putra Malaysia, 43400 Serdang, Selangor, Malaysia

A. H. Reshak
Physics Department, College of Science, University of Basrah, Basrah 61004, Iraq

Department of Instrumentation and Control Engineering, Faculty of Mechanical Engineering, Czech Technical University in Prague, Technicka 46 166 07, Prague, Czech Republic

S. C. B. Gopinath and M. M. Ramli (eds.), *Hybrid-Nanomaterials*, Engineering Materials, https://doi.org/10.1007/978-981-97-9022-7_2

Nanolaminates: thin layers of different materials stacked on top of each other to form a composite material (Molybdenum disulfide, Hafnium(IV) oxide, aluminium oxide). Over the years, researchers have discovered numerous methods to synthesize these nanomaterials. It is often sophisticated and complicated. Some of the methods are already commercialized and produced in bulk, while some of the methods are still in the lab scale production. However, the second most important aspect of nanomaterials research is how to measure or characterize these materials. It can be physically and chemically characterized to determine their size, shape, composition, and structure. These characterization techniques or methods are extremely important for the researchers to thoroughly understand the characteristics, properties and behaviour of these nanomaterials. Subsequently, it will greatly help researchers to fully utilize these nanomaterials.

2 Scanning Electron Microscopy (SEM)

In optical microscopy, images are magnified using light and lenses. This enables the observation of small objects like cells, but there are limits to the achievable magnification and analyzable materials. Electron microscopy is different. Instead of light, it uses a beam of electrons. SEM produces detailed, magnified images of an object by scanning its surface with a focused beam of electrons to create a high-resolution image. Electron microscopes can surpass the limits of optical ones because they use a shorter wavelength, giving a clearer image. SEM operates by using electron beams to scan a sample. An electron gun releases these beams, and the beam acceleration occur through a high voltage system (up to 20 kV). The electron beam gets narrowed and move down passing through the apertures and electromagnetic lenses of the scanning electron microscope. This occurs in a vacuum to prevent any molecules or atoms in the microscope column from interfering with the electron beam, ensuring high-quality imaging. The vacuum also protects the electron source from vibrations and noise. The electron beams scan the sample in a pattern, moving across the surface in lines from side to side and top to bottom. As the electrons interact with atoms on the sample's surface, they generate signals in the form of secondary electrons, backscattered electrons, and characteristic rays.

Backscattered electrons are reflected when the primary electron beam interacts with the sample object. These are elastic interactions. Elastic interactions happen when the primary electron doesn't lose energy. In this case, the electrons can alter their direction but maintain the same wavelength. Inelastic interactions occur when an interaction causes the primary electron to lose energy. Backscattered electrons and secondary electrons carry distinct information as backscattered electrons come from deeper parts of the sample, while secondary electrons come from the surface. Typically, images produced by backscattered electrons show high sensitivity to variations in atomic numbers, which appear as brighter or darker areas. In order to capture SEM images, the sample needs to be conductive on its surface and grounded to prevent charge buildup. Samples are attached to the specimen holder using a conductive adhesive. Metal samples usually require minimal preparation—just securing them to the stage with conductivity in mind. However, nonconductive materials often undergo a process where they are thinly coated with conductive substances like gold, platinum, chromium, or graphite. This coating is applied using methods such as sputter-coating or evaporation under high vacuum conditions.

SEM is a commonly used tool to study nanomaterials and nanostructures as it provides information on the surface morphology of a certain material structure. Surface morphology is important because it refers to the structure and features of a material's surface. Understanding surface morphology provides valuable information about the physical characteristics, texture, and topography of a material at the micro or nanoscale. In addition, SEM is often used for particle size analysis due to its resolution of 10 nm, that is, 100 Å.

Figure 1 shows examples of SEM images ZnO thin film doped with tin (Sn). The image clearly shows the hexagonal wurtzite structure of the ZnO with tin dopants wrapped around the ZnO particle. Another example of SEM image is shown in Fig. 2, where micropillars made of polydimethylsiloxane (PDMS) was captured. SEM imaging involves scanning the sample surface using an electron beam, and the sample surface will accumulate charge if there is no conductive path for the electrons to escape. Consequently, the image formed by SEM will be very poor if no conductive path exists. Since PDMS is an insulating material, the polymer device was initially coated with thin metal layer (10 nm Gold Palladium) using a sputtering machine to prevent surface charging (Fig. 3).

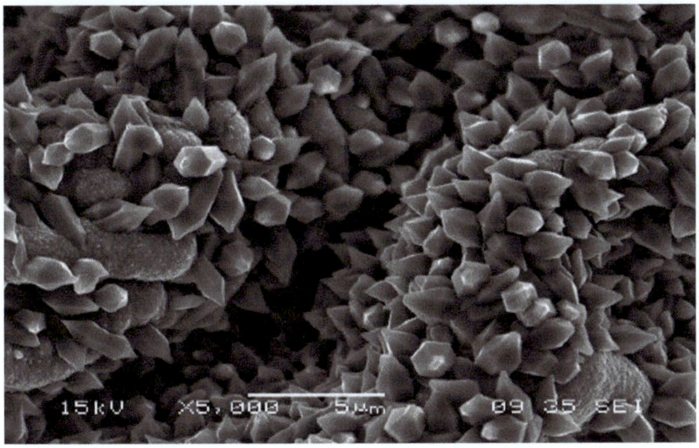

Fig. 1 SEM image of Sn doped ZnO thin film [1]

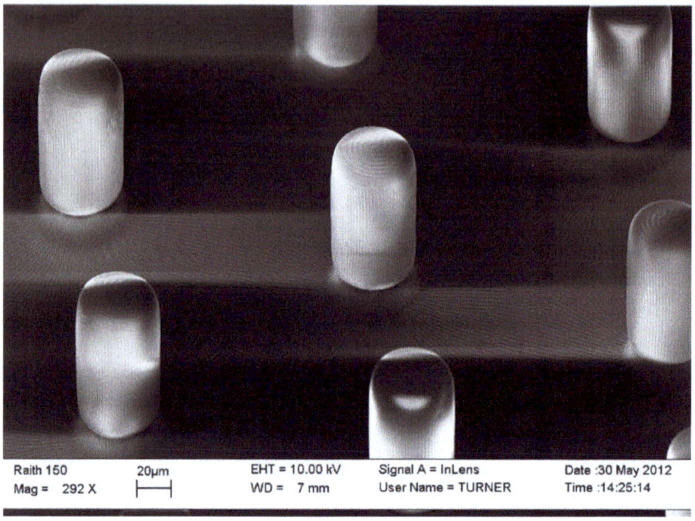

Fig. 2 SEM image of PDMS micropillar [2]

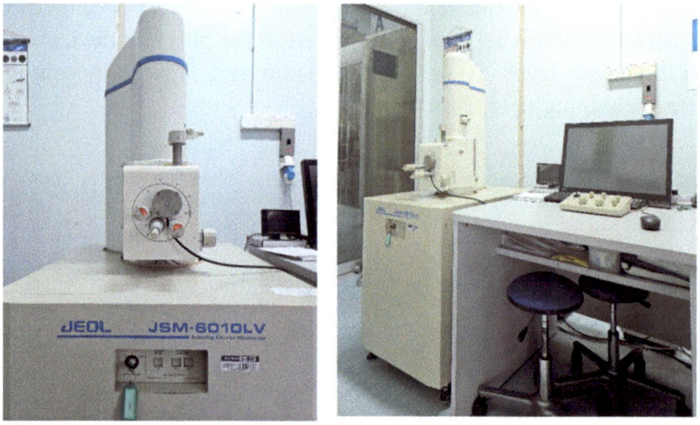

Fig. 3 SEM machine and the associated accessories

3 X-ray Diffraction Analysis (XRD)

In the field of nanomaterials and nanostructure research, X-ray diffraction analysis (XRD) is a technique used to characterize and determine the crystallographic structure of a material. Analyzing the crystallographic configuration of a substance is imperative as it yields invaluable explanations on its physical and chemical characteristics. X-ray beams are used in XRD due to their wavelength is equal to the spacing between atoms in the sample. This means that the diffraction angle will be affected by the spacing of the atoms in the molecule, as opposed to utilizing larger wavelengths, which would not be affected by the spacing between atoms. When the sample is penetrated by the x-rays, it will then interact with the atoms in the structure, causing the beam to "bounce" and change direction at a distinct angle, referred to as theta, θ deviating from the original trajectory. This phenomenon is known as diffraction angle. While some of the diffracted beams may counteract each other, constructive interference occurs when beams with similar wavelengths align.

Constructive interference happens when x-ray beams with whole numbers of the same length join together. This creates a new beam that is stronger. The stronger beam gives a bigger signal for that specific diffraction angle. This angle of diffraction helps figure out the difference between atomic planes using Bragg's law. Bragg's Law states that the angles at which X-rays are diffracted by a crystal are related to the spacing between atomic planes in the crystal lattice, which can be represented by the following equation;

$$n\lambda = 2d\sin\theta$$

where λ is the wavelength of the x-ray, d is the spacing of the crystal layers (path difference), θ is the incident angle (the angle between incident ray and the scatter plane), and n is an integer. Constructive interference occurs when the path difference given by 2dsin θ is a multiple of the used X-ray beam's wavelength. Since the wavelength is known, and the angles at which constructive interference occurs could be measured, using Bragg's equation, one can determine the distance between the lattice planes of the material. The measurement results obtained from XRD is represented by the intensity of the signal for various angles of diffraction at their respective two theta positions. These positions indicate the spacing between crystals or atoms in the samples, determined by the angle of diffraction from the incident x-ray beam. Peak intensity correlates with the quantity of molecules in the respective phase or possessing that spacing. Higher peak intensity signifies a larger quantity of crystals or molecules characterized by that distinct spacing.

The width of the peaks is inversely proportional to the crystal size. A narrower peak indicates a larger crystal, while a broader peak suggests a smaller crystal, a defect in the crystalline structure, or the sample's amorphous nature—characterized by imperfect crystallinity. When dealing with smaller samples, XRD analysis can be employed to deduce the composition of a sample. A comprehensive database contains diffraction patterns for various elements, compounds, and minerals. Comparing the diffraction pattern of an unknown compound to literature and experimentally derived values allows for the verification of the element's identity, considering the location, width, and relative heights of the diffraction patterns.

Figure 4 show an example of XRD machine used to characterize the structure of commercial graphite [3]. The diffraction pattern was observed in the range of 2θ from 10 to 90°, covering a wide angular range. The XRD analysis using the XPert HighScore Plus software provided valuable insights into the crystallographic properties of synthetic graphite samples. From the commercial graphite, the peak of the graphitic phase of graphite is shown at 26.5° in 2θ and 55° in 2θ, which represent the formation of the crystalline phase of graphite at (0, 0, 2) as stated in literature [4] (Fig. 5).

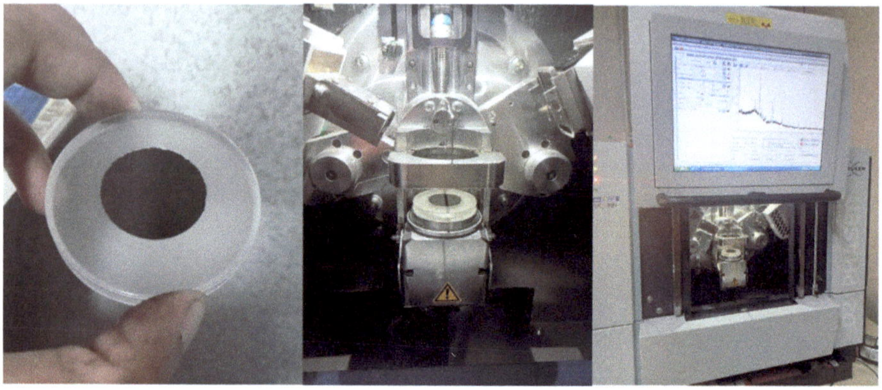

Fig. 4 **a** Example of prepared graphite sample **b** XRD machine **c** XRD machine with characterization results

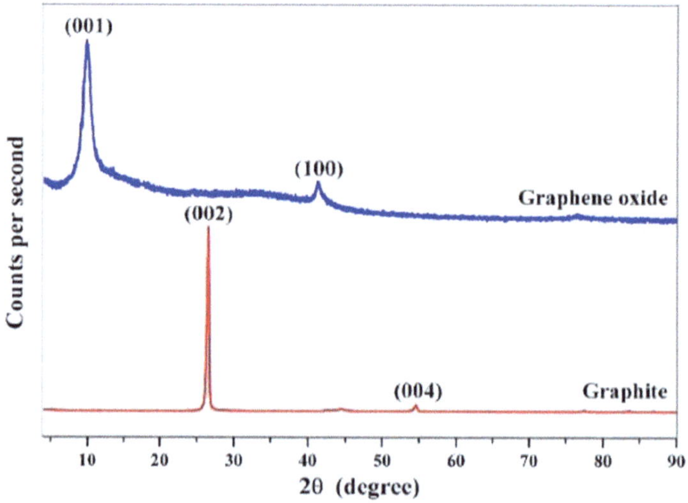

Fig. 5 XRD characterization results of commercial graphite and graphene oxide [5]

4 Energy-Dispersive X-ray Spectroscopy (EDX)

Energy-dispersive X-ray spectroscopy (EDS or EDX) is a useful tool for figuring out what elements are in a sample. Elemental analysis involves examining a sample of a substance (such as soil, waste, drinking water, bodily fluids, minerals, or chemical compounds) to understand its elemental and sometimes isotopic composition. This process can be both qualitative, determining which elements are present, and quantitative, figuring out the amount of each element present. The terms EDS and EDX are frequently used, sometimes interchangeably, to refer to either the same method or equipment. According to some perspectives, there is a subtle distinction: EDX specifically denotes the method, EDX analysis, whereas EDS pertains to the equipment, Energy Dispersive Spectrometer.

In general, the sample for EDS analysis is prepared by either cutting it into a thin section or coating it with a conductive layer, depending on the nature of the material and the analysis requirements. The sample is then bombarded with a focused electron beam, typically from a scanning electron microscope (SEM). This high-energy electron beam interacts with the atoms in the sample, causing them to become excited and release inner-shell electrons. As inner-shell electrons are ejected, electrons from higher energy levels fall to fill the vacancies. This process releases energy in the form of characteristic X-rays specific to each element. The energy of these emitted X-rays is unique to the element from which they originated. A detector placed near the sample collects the emitted X-rays. In EDX, a crystal or a solid-state detector is commonly used to separate and identify X-rays based on their energies.

The detected X-ray signals are then processed to create a spectrum that represents the elemental composition of the sample. Peaks in the spectrum correspond to the energy levels of X-rays emitted by specific elements present in the sample. The intensity of the X-ray peaks is proportional to the concentration of the sample's elements. By comparing the peak intensities to a calibrated standard, the quantitative elemental composition of the sample can be determined. In addition to elemental analysis, EDX can be used to create elemental maps and images of the sample. This provides spatial information about the distribution of different elements within the sample. By bombarding a sample with X-rays, we can see what elements are present and in what amounts, as the energy released during this process is specific to each element on the periodic table. So, EDS helps us analyze the elemental composition of a sample by using X-rays to identify the different elements present and their proportions.

The analysis obtained from EDX are usually represented by a table listing the elements present in a sample/material, or a spectrum plot detailing the elements present in the sample.

For instance, Table 1 shows EDX analysis of carbon and oxygen in graphite, graphite oxide, GO and rGO. From the table, it can be seen that the ratio of C/O for graphite is 9.81, and such a high C/O ratio indicates that pure graphite mainly contains carbon. In a different example, Fig. 6 shows the EDX spectrum of graphene oxide [6]. The elemental atom ratios for carbon and oxygen are 75.71 and 24.29%,

Table 1 EDX measurement of graphite, graphite oxide, GO and rGO

No.	Material	C (at %)	O (at %)	C/O ratio
1	Graphite	90.75	9.25	9.81
2	Graphite oxide	50.75	25.02	2.03
3	GO	44.9	23.59	1.90
4	rGO	60.39	22.34	2.70

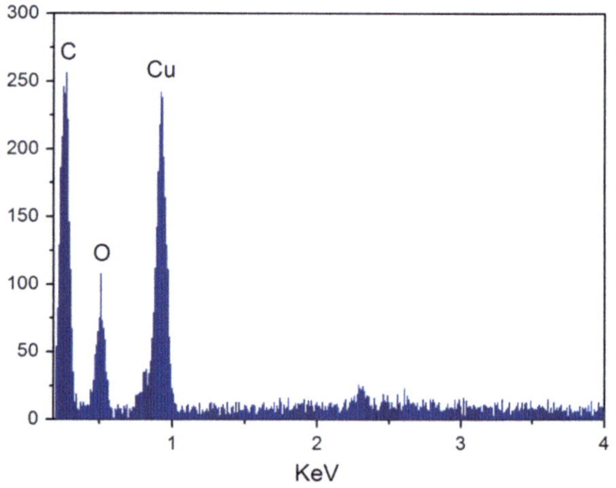

Fig. 6 EDX spectrum of graphene oxide [6]

respectively. It's important to mention that during the testing process, the GO sample was positioned on a copper sheet, leading to the detection of signals from copper as well, as shown in the spectrum.

5 Ultraviolet–Visible Sepctroscopy (UV–Vis)

Ultraviolet–visible (UV–vis) spectroscopy is a technique that are commonly used in characterizing the optical properties of nanomaterials. This technique is conducted by quantifying the absorption or transmission of specific wavelengths of UV or visible light in a sample, comparing it to a reference or blank sample. The sample's composition influences this property, offering insights into the sample's content and concentration. Given the reliance of this spectroscopy technique on light, it is important to understand the properties of light. Light contains certain energy and can be described by its wavelength. The energy of light is inversely proportional to the light's wavelength. Hence, light with shorter wavelengths has more energy than light with longer wavelengths. The visible light spectrum which approximately from the wavelength of 380 nm (violet) to the wavelength of 780 nm (red). UV light, with wavelengths shorter than visible light, extends to around 100 nm. Describing light by its wavelength proves valuable in UV–Vis spectroscopy, facilitating the analysis or identification of substances by identifying specific wavelengths linked to maximum absorbance.

The fundamental concept of UV–Visible Spectroscopy revolves around the absorption of ultraviolet or visible light by chemical compounds, leading to the generation of characteristic spectra. Spectroscopy, at its core, relies on the interplay between light and matter. As matter absorbs light, it experiences excitation and subsequent de-excitation, culminating in the creation of a spectrum. Upon absorbing ultraviolet radiation, the electrons within matter undergo excitation, transitioning from a ground state (characterized by a relatively low energy level) to an excited state (characterized by a comparatively higher energy level). It's crucial to emphasize that the energy difference between the ground state and the excited state of the electron always equals the amount of ultraviolet or visible radiation that was absorbed by the samples.

The principle behind absorbance in spectroscopy is related to the Beer-Lambert Law, where the law states that the amount of light absorbed by a sample is directly related to the volume of sample the light passes through and the concentration of the sample. This can be represented by equation $A = \varepsilon bc$, where A is the absorbance, ε is the molar absorptivity of the compound or molecule in solution, b is the path length of the cuvette or sample holder, and c is the concentration of the solution.

The main components of UV–vis are light source, wavelength selector, sample, detector and a computer for signal processing and output data collection, as illustrated in Fig. 7. Being a light-dependent method, it is crucial to have a consistent light source capable of emitting light across a broad spectrum of wavelengths. There is various light source that are used in uv–vis such as xenon lamp, deuterium lamp, hydrogen lamp, and mercury arc lamp. In devices utilizing dual lamps, a tungsten or halogen lamp is frequently utilized for generating visible light, while a deuterium lamp serves as the prevalent source for UV light.

As the name suggests, the wavelength selector consists of monochromator or filters, which functions to disperse polychromatic (multi-wavelength) light into its individual components, allowing only a selected wavelength or a narrow band of wavelengths to pass through. This process is essential for obtaining a pure and specific wavelength of light for analysis. The light subsequently traverses through

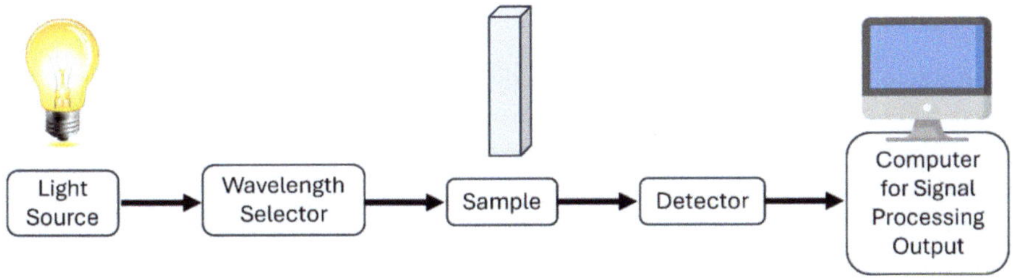

Fig. 7 A basic diagram illustrating the primary elements within the UV–VIS spectrophotometer

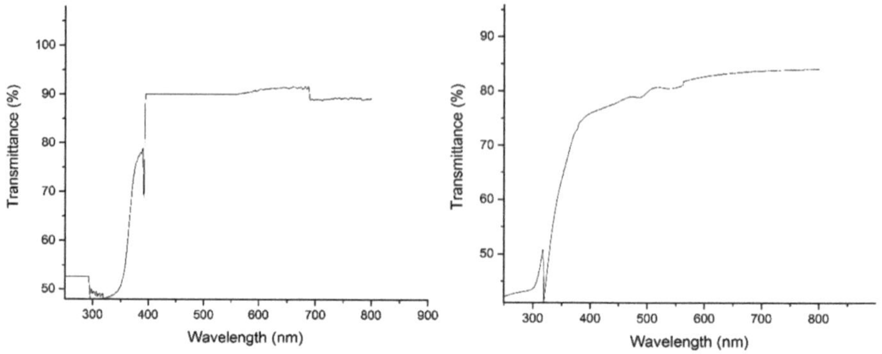

Fig. 8 Uv–vis spectrum of ZnO [7]

a sample. In all analytical procedures, it is essential to measure a reference sample, commonly known as the "blank sample." This may involve using a cuvette filled with a similar solvent as that used in preparing the sample. In the case of measurements conducted with an aqueous buffered solution containing the sample, the reference is the same solution without the substance of interest. After the light has passed through the sample, a detector is used to convert the light into a readable electronic signal. Generally, detectors are made from photoelectric coatings or semiconductors. Data from UV–Vis spectroscopy can be depicted in the form of a graph illustrating absorbance, optical density, or transmittance against wavelength. Nevertheless, it is more common to represent the data with absorbance or transmittance on the vertical y-axis, and wavelength on the horizontal x-axis. This graphical representation is commonly known as an absorption spectrum, as illustrated in Fig. 8 [7].

UV–Vis spectroscopy was employed to record optical transmittance measurements of ZnO thin film using a single-beam spectrophotometer across the wavelength range of 250–800 nm. The researchers concluded that, ZnO deposited on quartz substrate achieved the highest transmittance percentage, which is 90% as opposed to the transmittance of 80% for ZnO deposited on glass substrate.

6 Transmission Electron Microscopy (TEM)

In principle, chemical substances or materials of which a single unit is sized between 1 and 100 nm in general regarded as nanomaterials. One of the tools to investigate the thickness of the substance is transmission electron microscopy (TEM). It is an analytical technique used to visualize or zooming of smallest structure utilising the movement or bombardment of the electrons. A high energy beam of electrons is shone through a very thin sample, and the interactions between the electrons and the atoms can be used to observe features such as the crystal structure and features in the structure like dislocations and grain boundaries. The TEM operates on the same

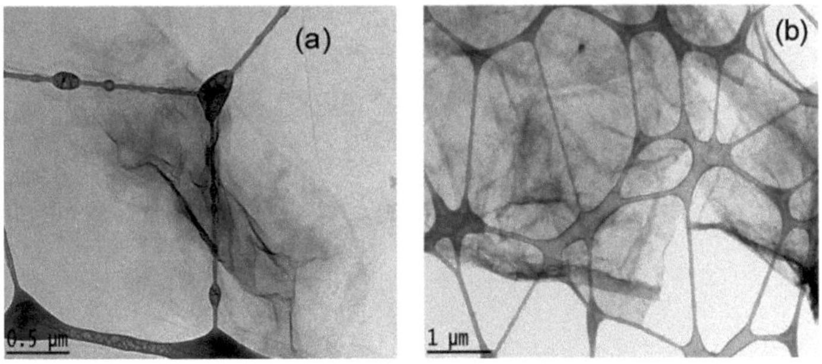

Fig. 9 Shows the images of graphene sheet using TEM [8]

basic principles as the light microscope but uses electrons instead of light. Because the wavelength of electrons is much smaller than that of light, the optimal resolution attainable for TEM images is many orders of magnitude better than that from a light microscope. Thus, TEM is the ideal tool to investigate the nanomaterials down to the level of atomic sizes.

This is how the image is acquired. First, the beam of electrons from the electron gun is focused into a small, thin, coherent beam using the condenser lens. This beam is restricted by the condenser aperture, which excludes high angle electrons. The beam then strikes the specimen and parts of it are transmitted depending upon the thickness and electron transparency of the specimen. This transmitted portion is focused by the objective lens into an image on phosphor screen or charge coupled device (CCD) camera. The image then passed down the column through the intermediate and projector lenses, is enlarged all the way. The image strikes the phosphor screen and light is generated, allowing the user to see the image. The darker areas of the image represent those areas of the sample that fewer electrons are transmitted through while the lighter areas of the image represent those areas of the sample that more electrons were transmitted through. Figure 9 shows the TEM images of nanomaterials namely graphene sheets [8]. The morphology of of the graphene is folded, rough wrinkled and aggregated surface which attributed by the strong π–π stacking interaction. It also shows some of the graphene sheets are overlapped on each other with the average thickness of around 1–5 nm and the size of the sheet is around 2–5 μm.

7 Atomic Force Microscopy (AFM)

One of the important technique to observe the morphology of nanomaterials is atomic force microscopy. It can investigate the lateral size, thicknesses and surface roughness through 3D topographical map across nanomaterials. Figure 10 shows the basic principle of how this high-resolution scanning probe microscopy works [9]. It is

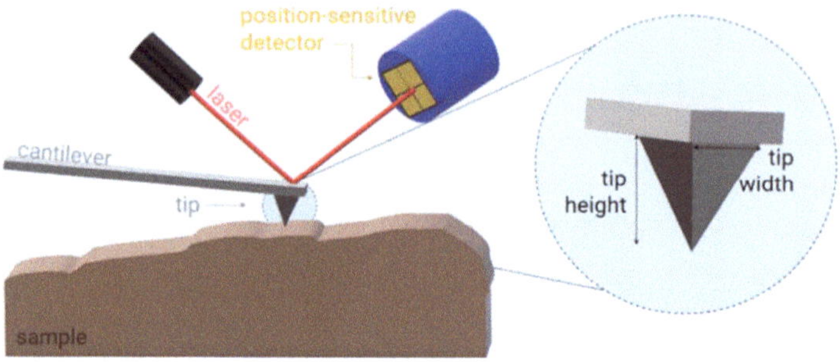

Fig. 10 Illustration of the basic principle for AFM [9]

divided into 3 main mechanism which are surface detection; detection technique and imaging technique [10].

Surface detection: An AFM uses a cantilever with a very sharp tip to scan over a sample surface. As the tip approaches the surface, the close-range, attractive force between the surface and the tip cause the cantilever to deflect towards the surface. However, as the cantilever is brought even closer to the surface, such that the tip makes contact with it, increasingly repulsive force takes over and causes the cantilever to deflect away from the surface.

Detection Technique: A laser beam is used to detect cantilever deflections towards or away from the surface. By reflecting an incident beam off the flat top of the cantilever, any cantilever deflection will cause slight changes in the direction of the reflected beam. A position-sensitive photo diode (PSPD) can be used to track these changes. Thus, if an AFM tip passes over a raised surface feature, the resulting cantilever deflection (and the subsequent change in direction of reflected beam) is recorded by the PSPD.

Imaging Technique: An AFM images the topography of a sample surface by scanning the cantilever over a region of interest. The raised and lowered features on the sample surface influence the deflection of the cantilever, which is monitored by the PSPD. By using a feedback loop to control the height of the tip above the surface—thus maintaining constant laser position—the AFM can generate an accurate topographic map of the surface features (Fig. 11).

8 Raman Spectroscopy

Another important tool for nanomaterials investigation is Raman spectroscopy. It is an analytical technique where scattered light is used to measure the vibrational energy modes of a sample. It provides detailed information about chemical structure, phase and polymorphy, crystallinity and molecular interactions. It was named after the Indian physicist C. V. Raman, together with his research partner K. S. Krishnan, who was the first to observe a scattering phenomenon called Raman scattering in

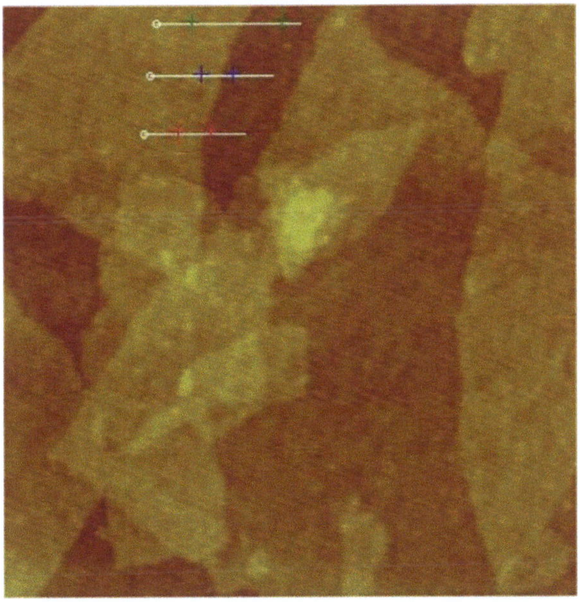

Fig. 11 Typical AFM images of graphene oxide sheets [8]

1928. When light is scattered off a sample there are two possible outcomes; (1) Elastic scattering, also known as Rayleigh scattering, that will occur when the scattered light has the same energy as the light that initially struck the sample. This means that the elastically scattered light will be the same frequency, wavelength, and color as the original beam of light; (2) Inelastic scattering, or Raman scattering, occurs when the scattered light has a different energy than the light that initially struck the sample. This means the inelastically scattered light will have a different frequency, wavelength, and color than the original beam of light. The constant motion of atoms in a sample molecule will makes the bonds in atoms constantly vibrating in different directions. These molecular vibrations happen at specific frequencies which are unique to the molecule and type of bond. This vibrational frequency of molecules is referred as wavenumbers, which have units of reciprocal centimetres [cm 1]. Every time incoming light has the same frequency as the vibration of the molecule, light can be absorbed, "exciting" the vibration further, and amplifying its amplitude. Since the different bonds in molecules vibrate at different frequencies, different molecules will absorb unique frequencies of light to excite vibrations.

Figure 12 shows the typical spectrum of Raman scattering for graphene oxide nanomaterials [11]. Based on that, Raman spectroscopy shows the presence of characteristic peaks of graphitic at 1367 cm^{-1}, the characteristic of D peak was shown, resulting from the disordered bands in sp^2 hybridized carbon materials and led to the disruption of the symmetrical hexagonal graphitic lattice as a result of edge defects, internal structural defects and dangling bonds while the G peak centered at 1599 cm^{-1}, refer to the first order of the Raman band of all sp^2 hybridized carbon materials related to the C–C vibrational mode. The shift in the G band from 1582 cm^{-1} of

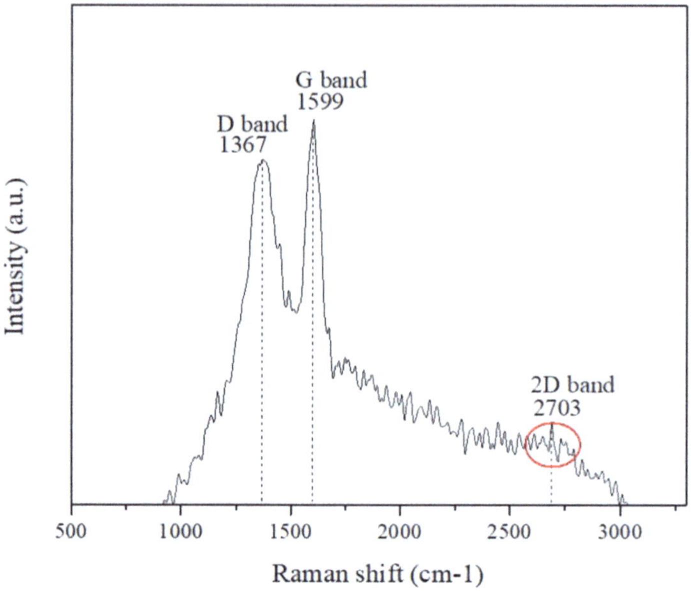

Fig. 12 Raman spectrum of GO [11]

graphite to 1599 cm^{-1} of graphene oxide was due to the presence of isolated double bonds on the carbon network of graphene oxide. The other bands were observed at 2703 cm^{-1} which is known as a 2D band. The broadened of the 2D band peak is an indicator of a few layers of graphene oxide.

9 Fourier Transform Infra-Red Spectroscopy (FTIR)

Infrared absorption spectroscopy is one of the techniques used to determine the structures of the molecules by absorption of infrared radiation. When exposed to infrared radiation, sample molecules selectively absorb radiation of specific wavelengths which causes the change of dipole moment of sample molecules. Consequently, the vibrational energy levels of sample molecules transfer from ground state to excited state. The frequency of the absorption peak is determined by the vibrational energy gap. The number of absorption peaks is related to the number of vibrational freedoms of the molecule. The intensity of absorption peaks is related to the change of dipole moment and the possibility of the transition of energy levels. Therefore, by analyzing the infrared spectrum, the structure information of a molecule can be obtained. Most molecules are infrared active except for several homonuclear diatomic molecules such as O_2, N_2 and Cl_2 due to the zero dipole change in the vibration and rotation of these molecules. The commonly used region for infrared absorption spectroscopy is 4000–400 cm^{-1} because the absorption radiation of most organic compounds is within this region.

Figure 13 shows the common characteristic of FT-IR spectrum for reduced-graphene oxide mixed with single-walled carbon nanotubes, called hybrid nano-materials [8]. According to this figure, it was clearly seen a broad peak in the high frequency area at 3365 cm^{-1} corresponds to the stretching of O–H groups which indicates that the oxygen containing groups. The absorption peak at 1640 and 1108 cm^{-1} represents $C = O$ and C–O, which result for stretching vibration of the carbonyl and carboxylic groups. The peak at 721 cm^{-1} can be assigned to $C = C$ stretching of aromatic rings.

The presence of functional groups reveals that the hybrid material was well oxidized. The functional groups are favourable to the formation of a stable colloidal suspension hybrid material due to highly negative surface charge of the as-prepared

Fig. 13 FT-IR spectra of hybrid material **a** before reduction **b** after reduction [8]

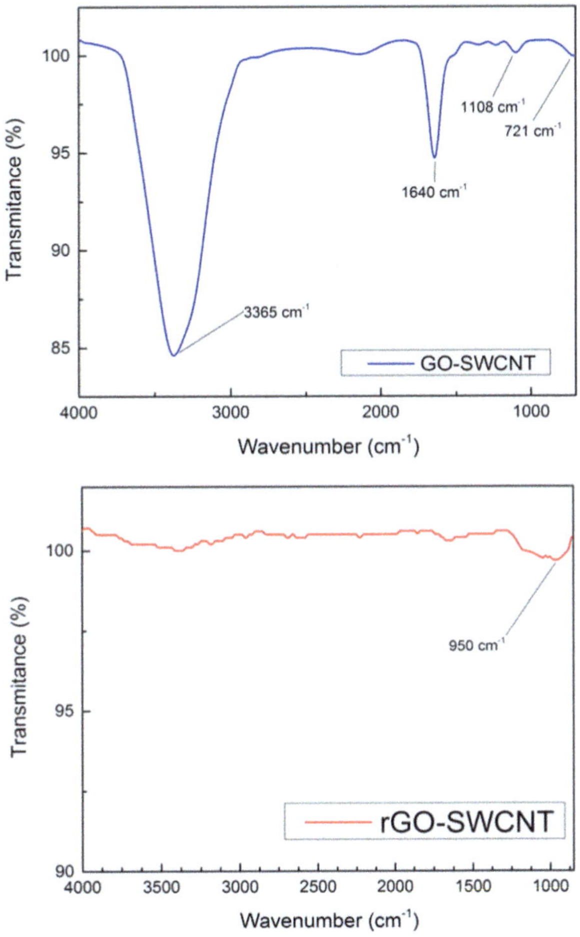

hybrid material. In Fig. 13b, those peaks for oxygen functional groups disappeared after thermal reduction process which because of abrupt elimination of intercalated water and oxygen group.

References

1. Ishak, S., Johari, S., Ramli, M.M.: Formaldehyde detection using Sn doped ZnO thin film. J. Sol-Gel Sci. Technol. **95**(2), 265–275 (2020). https://doi.org/10.1007/s10971-020-05318-8
2. Johari, S.: Microsystems for C. Elegans Mechanics and Locomotion Study (2013)
3. Atchutha Nair, K.: Production of Synthetic Graphite from Oil Palm Empty Fruit Bunch (2023)
4. Ain, Q.T., Haq, S.H., Alshammari, A., Al-Mutlaq, M.A., Anjum, M.N.: The systemic effect of PEG-nGO-induced oxidative stress in vivo in a rodent model. Beilstein J. Nanotechnol. **10**, 901–911 (2019). https://doi.org/10.3762/bjnano.10.91
5. Rosman, A.S., Navalan, R., Ramli, M.M., Abd Karim, N., Ahmad, M.F., Johari, S., Ali@Hashim, N., Osman, N.H.: Production of low temperature synthetic graphite. Int. J. Nanoelectron. Mater. **16**(2), 469–447 (2023)
6. Zhao, Q., Chen, D., Li, Y., Zhang, G., Zhang, F., Fan, X.: Rhodium complex immobilized on graphene oxide as an efficient and recyclable catalyst for hydrogenation of cyclohexene. Nanoscale **5**(3), 882–885 (2013). https://doi.org/10.1039/C2NR33290F
7. Shazlina, J., Yazmin, M.N., Rosydi, Z.M.: Study of zinc oxide thin film characteristics. EPJ Web Conf. **162**, 1057 (2017). https://doi.org/10.1051/epjconf/201716201057
8. Muda, M.R., Ramli, M.M., Isa, S.S.M., Jamlos, M.F., Murad, S.A.Z., Norhanisah, Z., Mohamad Isa, M., Kasjoo, S.R., Ahmad, N., Nor, N.I.M., Khalid, N.: Fundamental study of reduction graphene oxide by sodium borohydride for gas sensor application. AIP Conf. Proc. **1808**(1), 020034 (2017). https://doi.org/10.1063/1.4975267
9. Adapted from Bruker. https://www.bruker.com/en/products-and-solutions/microscopes/materials-afm/faq.html
10. Adapted from Park System. https://www.parksystems.com/en/learning-center/how-afm-works
11. Ramli, M.M., Rosman, A.S., Mazlan, N.S., et al.: Cell viability and electrical response of breast cancer cells treated in aqueous graphene oxide solution deposition on interdigitated electrode. Sci. Rep. **11**, 20702 (2021). https://doi.org/10.1038/s41598-021-00171-3

Methods for Fabrication of Metal Nanoparticles

Periasamy Anbu, Vinitha Packirisamy, and Senthil Kumar Bavan

1 Introduction

Particles with at least one dimension less than 100 nm are referred to as nanoparticles. They have quite different physical and chemical properties from their counterparts in bulk materials because of their small size and high surface area-to-volume ratio. Furthermore, their high surface area to volume ratio improves their reactivity and ability to interact with other substances [1]. Materials can exhibit quantum mechanical phenomena at the nanoscale, which can drastically change their optical, electrical, and magnetic properties. For a variety of uses, nanoparticles can be created to have functional groups on their surface to enhance compatibility, targeting, and reactivity. Particularly the metal nanoparticles have significant surface plasmon resonance (SPR), resulting in unusual optical phenomena such as increased light absorption and scattering [2]. It also has superior electrical conductivity, making it valuable in electronic applications. In addition, certain metal nanoparticles have strong antimicrobial, anti-cancer, and antioxidant properties, making them valuable medical applications. Consequently, the design of metal nanoparticles has been utilized to deliver medications directly to target cells, increasing therapy efficacy and decreasing side effects [3].

The nanoparticles have also been applied to cancer treatment as diagnostic, therapeutic, and imaging tools. These precious metal nanoparticles were produced

P. Anbu (✉) · V. Packirisamy (✉)
Center for Global Health Research, Saveetha Medical College & Hospital, Saveetha Institute of Medical and Technical Sciences (SIMATS), Thandalam, Chennai 602 105, Tamil Nadu, India
e-mail: anbu25@yahoo.com

V. Packirisamy
e-mail: vinithaprs18@gmail.com

S. K. Bavan
Department of Pharmacology and Environmental Toxicology, Dr. ALM PG IBMS, Tharamani Campus, University of Madras, Tharamani, Chennai 600113, India

S. C. B. Gopinath and M. M. Ramli (eds.), *Hybrid-Nanomaterials*, Engineering Materials, https://doi.org/10.1007/978-981-97-9022-7_3

by chemical, physical, and biological approaches. Both chemical and physical approaches have many advantages including simple, cost effective for large scale production highly purity, uniform particles and easy to control the particle size [4]. However, these two methods involved costly equipment, high energy consumption, high heat generation, toxic solvents, reducing and stabilizing agents, posing risk to human health and environment. Numerous benefits come with both chemical and physical methods, such as being straightforward, affordable for large-scale production, producing homogeneous, and being simple to control particle size. Despite this, these two methods were toxic to the environment and public health, requiring costly machinery, a lot of energy, heat generation, hazardous solvents, and reducing and stabilizing agents. Biological approach for metal nanoparticles has a number of clear benefits including environmental sustainability, cost-effectiveness, biocompatibility, reduced energy consumption, and scalability over conventional chemical and physical techniques [5].

2 Nanoparticles Synthesis Methods

(a) **Physical Methods**

Bulk materials are the basis of this top-down approach. (Shown in Fig. 1) This strategy reduces particle size to nanometers by employing mechanical and physical techniques. One may readily perform physical approaches. Still, this approach has certain drawbacks. This approach is not appropriate to attain lower particle sizes and defined forms. This approach does not readily produce an alteration of surface chemistry and physicochemical characteristics.

(i) *Physical/Chemical Vapor Deposition (PVD/CVD)*

This method is extensively applied for thin film deposition on carbon nanotubes and metal nanoparticles. Thin lanthanum, strontium, and cobalt films are produced using these methods regularly. This approach is used to prepare a thin film of tungsten selenides, Pt-Ru nanoparticles, and Yttria-stabilized zirconia nanoparticles [6]. Fabricated thin films of copper indium gallium selenide produced by pulsed laser deposition proved to be most successful. Physical deposition is the framework of material on a surface either as thin films or nanoparticles [7]. Before depositing it on a substrate, vacuum processes including thermal evaporation and sputtering deposition evaporate the substance [8]. Pulsed laser deposition describes laser ablation on a solid target that generates a plasma of ablated species subsequently deposited on a substrate to create a film. This approach has certain drawbacks: the amount of samples will be small and it is very expensive.

The CVD method comprises the deposition of a thin film of the material of interest on a surface through a chemical reaction that involves atoms and volatile molecules [9]. This method can be used to produce zirconia-alumina nanopowder and SnO_2 nanoparticles. The primary application of this method is the production of

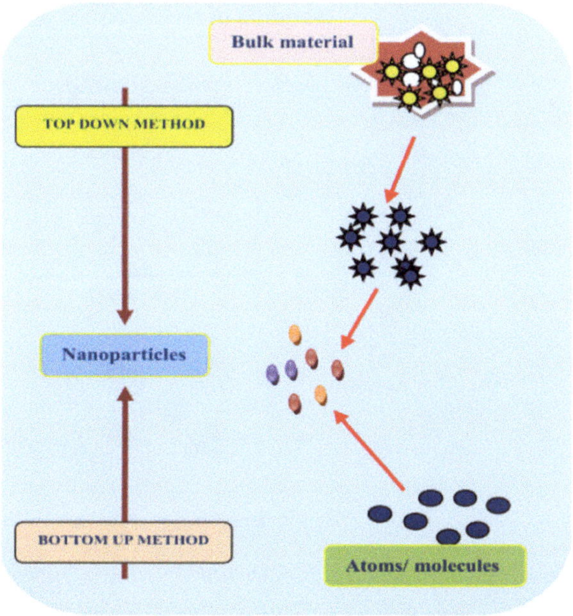

Fig. 1 Pictorial representation of top-down and bottom-up approach to synthesizing nanoparticles

graphene. A precursor is formed when the material that is being targeted is released as a volatile molecule. A thin coating is formed as a result of chemical interactions between the precursor fragment, precursor, and host surface. Chemical reactions on the surface generate thin coatings through atomic layer deposition (ALD). The chemical vapour deposition method consisted of thermally active, plasma-enhanced, and photo-initiated procedures for the deposition of materials. This method facilitates the examination of nanoparticles' surface morphology. The discharge of toxic, explosive gases is the primary disadvantage of this method [10].

(ii) *Laser Ablation*

When a high-intensity laser beam is directed at a metal target in a liquid or gas medium, the metal disintegrates and condenses into nanoparticles. Laser irradiation is employed in laser ablation to reduce the dimensions of particles to the nanoscale. A thin layer is applied to the hard material of concern, which is then subjected to pulsed laser irradiation. This method primarily employs copper vapor lasers, Nd:YAG lasers, and Ti: Sapphire lasers [11]. Solid material is fragmented into nanoparticles by laser irradiation, which then remain in the liquid protecting the target, resulting in a colloidal solution. The quantity of ablated atoms and particles is influenced by the intensity and duration of laser pulses [12]. This method is capable of synthesizing silicon and alumina nanoparticles. This method generates a greater quantity of nanoparticles, and its properties can be easily modified. The general methods employed for the synthesize of nanoparticles was shown in Fig. 2.

(iii) *Mechanical Milling*

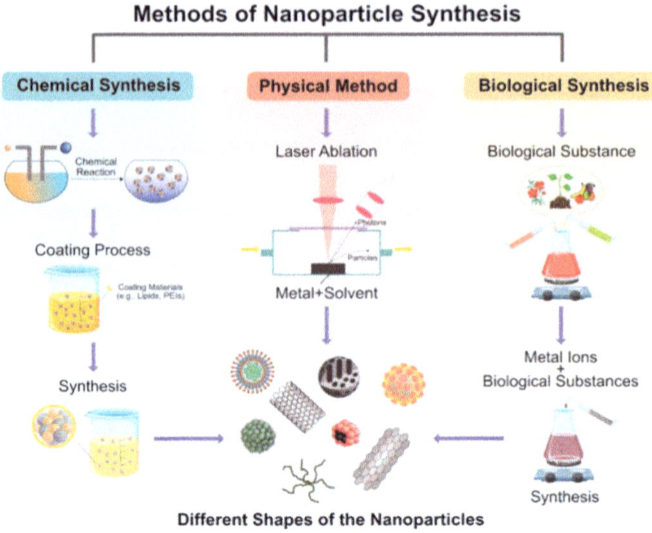

Fig. 2 Pictorial representation of method of synthesize nanoparticles by different approaches

Mechanical milling involves reducing particle size using high-powered ball milling. Mechanical milling performance depends on process variables and powder qualities. Milling can be either high or low in energy, depending on the amount of mechanical force applied to the powder composition. Nanosized particles are often generated via high-energy ball milling. This approach is commonly used for synthesizing intermetallic nanoparticles. This approach involves adding large amounts of powder to a container that holds along with heavy balls. Great-speed rotating balls apply great mechanical energy to bulk powder materials. Various high-energy mills, including attrition, planetary, vibrating, low-energy tumble, and high-energy ball mills, can be used to reduce particle size. Bulky, free-moving, high-powered balls can either roll downward on the chamber's surface in parallel layers or contact the powder directly [13]. Through this method, ceramic nanoparticles, alloy nanoparticles, silicon and alumina nanoparticles can be prepared. This method offers a large-scale production of nanoparticles with high purity, which possess superior physical properties. The drawbacks of this method are, that it needs higher energy and prolonged time to synthesize nanoparticles.

(b) **Chemical Methods**

(i) *Sol–Gel Process*

In 1988, Livage et al. conducted the first assessment of sol–gel chemistry for oxides of transition metals. This method is easy to perform to prepare nanoparticles. The physicochemical properties and morphologies of nanoparticles can be easily tailored through this method (as shown in Fig. 3). To synthesize nanoparticles using the sol–gel approach, either premade colloids metal (oxide) are mixed with a sol incorporating matrix-forming species, resulting in gel formation, or metal and metal oxide or nanoparticles are directly mixed in a hydrolyzed silica sol. Metals are complexed with silone and reduced before hydrolysis. Sol–gel formation consists of four steps: hydrolysis, condensation, growth of particles, and particle aggregation. This method involves preparing a homogenous solution of one or more specified alkoxides and adding a catalyst to commence the chemical process at a predetermined pH. Metallic oxide particles emerge from silica sol via low-temperature heat treatment. This process is commonly used for preparing thin films [14]. Transition metal oxides such as ZnO, TiO2, oxides of Ni, Co, Al, and others can be successfully synthesized by this method [15].

(ii) *Chemical Reduction*

The chemical reduction method involves reducing ionic salt in a suitable medium with surfactants and various reducing agents. Metal nanoparticles are prepared in an aqueous solution using a reducing agent, such as sodium borohydride, trisodium citrate, etc. The stabilizing agent is employed with a reducing agent. Stabilizing agents such as thiolates, carbonyls, and phosphines can be used to obtain smaller nanoparticles in 1–3 nm size regime with enhanced surface properties. The long-term persistence of metal nanoparticles in the dispersion was assessed using absorbance measuremen. Silver nanoparticles can be synthesized using reducing agents such as sodium borohydrate ($NaBH_4$), glucose, ethylene glycol, ethanol, sodium citrate, and hydrazine hydrate [16]. Nanoclusters such as Au, Ag, Cu, Ni, Co, etc. can be synthesized by this method.

(iii) *Thermal Decomposition*

The hydrothermal process involves reacting water-soluble vapours with solid material at high pressure and temperature, resulting in the deposition of tiny particles.

This approach precipitates cations as polymeric hydroxides, which are then dehydrated to promote the production of crystal structures of nanoparticles. When a base is introduced to a salt or metal solution, the production of a second metal cation prevents the formation of complex hydroxide, which helps regulate particle formation [17]. Highly crystalline metal composites were synthesized through this method. The specific shape and size of nanoparticles with high purity can be attained in this method. However, the reproducibility of the product was quite difficult.

(iv) *Electrochemical Methods*

Nanomaterials can be produced via electrodeposition or template synthesis methods. Electrochemical deposition methods, such as anode dissolving techniques, are preferred for producing porous semiconductor structures [18]. Metal nanoparticles are produced using electrochemical deposition (ED) in baths with metal salts. The baths use a three-terminal potentiostat and can be acidic or basic. This approach requires a cathode electrode to deposit metal nanoparticles, while an anode electrode with AgCl or calomel serves as the reference electrode. In an electrolytic bath containing metal salt, a low voltage is delivered for a set period of time. By this method, Au, Ni, Co and Pt nanowires can be prepared. This method is inexpensive and the size and morphology of nanoparticles can be regulated by this method.

(c) **Biological Methods**

Plant extracts, microbes, or enzymes are used in biological processes to create nanoparticles, which provide environmentally favourable alternatives (Fig. 2). Microbial synthesis makes use of the metabolic pathways found in bacteria, fungi, or algae, whereas plant-mediated synthesis uses the naturally occurring reducing agents found in plant extracts to transform metal ions into nanoparticles [19].

(i) *Microbial Synthesis*

Based on the intended type of nanoparticle, the microorganism (algae, fungi, or bacteria) was chosen for its ability to reduce metals. Selective culture media were used to cultivate the microorganisms in order to obtain microbial cells. After that, the microbial culture was combined with the metal ion aqueous solution and incubated at the proper temperature and duration.

Microbial synthesis is the process of creating metal nanoparticles from microorganisms including bacteria, fungi, and algae. These microorganisms can secrete enzymes or use their metabolic processes can to convert metal ions into nanoparticles [20]. The nanoparticles can be synthesized by microorganisms through intracellular and extracellularly. For extracellular synthesis, the microbial cells can produce the enzyme and metabolic substances and release into the culture media. Then the metal ions reduced into the nanoparticles (Fig. 4).

(ii) *Plant mediated synthesis*

From the resource region, the plant's leaves, stem, root, or seeds were harvested and brought to the laboratory. After carefully cleaning the gathered plant materials

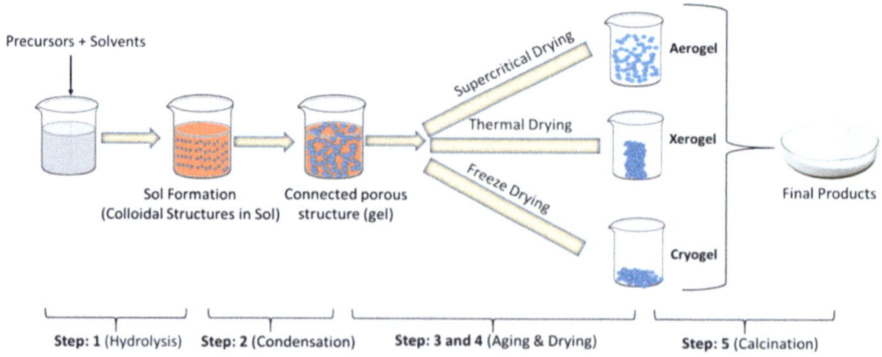

Fig. 3 Synthesize of nanoparticles through sol–gel processes

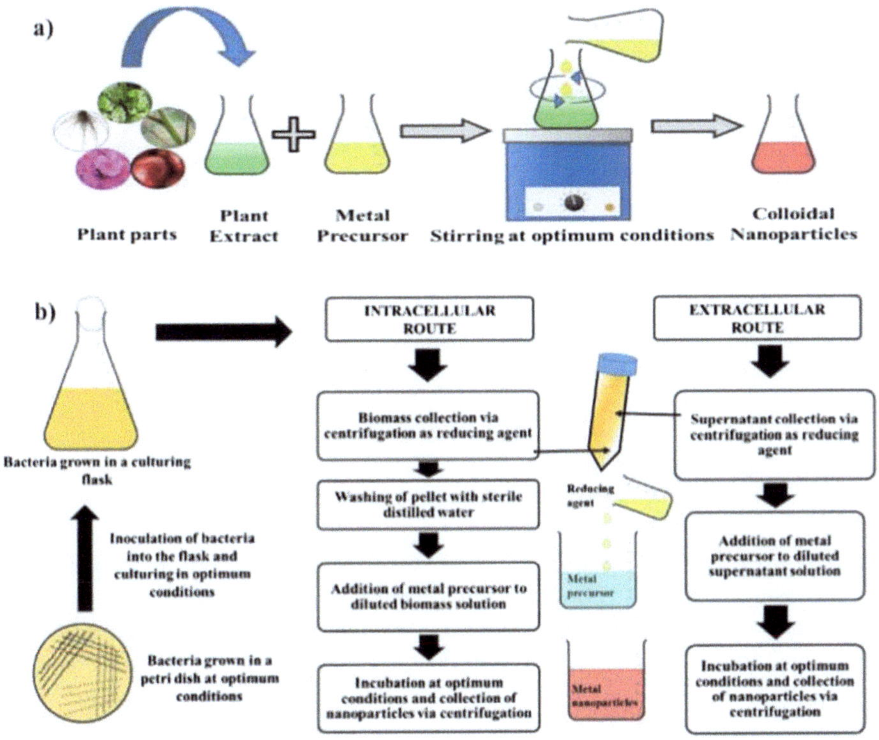

Fig. 4 Synthesize nanoparticles by using biogenic materials as a reducing and stabilizing agents

with sterile distilled water, they were allowed to dry for a few days. The dried plant materials, weighing about 10 g, were combined with 100 mL of distilled water and brought to a boil using a water bath for 20 min. Subsequently, the boiling plant extract was allowed to cool down and filtered using Whatman No. 1 filter paper, which was followed by filtering through a 0.45 μm filter membrane. Next, the filtered extract was preserved and employed in the production of metal nanoparticles [21]. Metal nanoparticles are made using plant extracts that naturally include reducing agents, such polyphenols (Fig. 4a).

(iii) *Enzyme-Mediated Synthesis*

Enzyme-mediated synthesis of nanoparticles converts metal ions into nanoparticles by taking advantage of the catalytic abilities of enzymes. This approach is thought to be environmentally benign and biocompatible, and it provides exact control over the size, shape, and functionalization of nanoparticles. Enzymes are biological catalysts that help reduce metal ions so they can create nanoparticles. The process typically involves reduction reaction, capping and stabilization. The creation of nanoparticles occurs when metal ions (such as Ag^{+}, Au^{3+}, and Cu^{2+}) are reduced to their elemental form by enzymes [22]. Biomolecules that are linked to enzymes can function as capping agents, avoiding agglomeration and stabilizing the nanoparticles.

Various enzymes required to synthesis different kinds of metal nanoparticles. In particular, the silver ions were reduced to silver nanoparticles (AgNPs) by the enzyme nitrate reductase. Furthermore, laccase catalyses the reduction of gold to gold nanoparticles [23]. Platinum was converted to platinum nanoparticles using hydrogenase. Due to its capacity to reduce the required metal ions, the particular enzyme was chosen. After that, the enzyme was isolated and purified from bacteria or enzyme that was sold commercially. Then the enzyme solution was mixed with the metal ion solution and incubate under suitable conditions including temperature, pH, incubation time. Allow the reduction of metal ions to occur, leading to the formation of nanoparticles.

3 Hybrid Methods

In order to maximize the production of nanoparticles, hybrid approaches such as combustion synthesis, plasma-based synthesis, and electrospinning—combine components from top-down and bottom-up techniques. The chosen methodology will depend on the desired properties of the nanoparticles and the intended uses.

(i) *Plasma-Based Synthesis*

Metal nanoparticles with exceptional purity, regulated size, and distinctive characteristics can be created by the sophisticated process of plasma-based synthesis. In order to start the chemical reactions that result in the production of nanoparticles, this approach uses plasma, a highly charged gas that contains electrons, ions, and neutral particles. Plasma is generated by applying a high voltage or electromagnetic field to a gas, creating a highly reactive environment of ionized particles. Through interactions with reactive species and energetic electrons in the plasma, metal ions are reduced to their metallic state in the plasma environment. This process leads to the nucleation and growth of metal nanoparticles. By modifying the plasma characteristics (e.g., power, pressure, gas composition), the resulting nanoparticles' size, shape, and dispersion can be adjusted. The introduction of stabilizing chemicals or surfactants into the plasma can help to stabilize the nanoparticles and prevent agglomeration.

(ii) *Electrospinning*

One efficient and adaptable way to create metal nanoparticles within polymer nanofibers is through electrospinning. This method has many benefits, such as scalable morphological control, large surface area, and scalability, which makes it appropriate for a variety of uses in environmental research, medical, catalysis, and sensing. By applying a high-voltage electric field to a melt or polymer solution, continuous nanofibers can be created with this technique [24]. During the electrospinning process, metal ions in the polymer matrix are reduced, forming metal nanoparticles immediately within the nanofibers. In addition, by giving the electrospun nanofibers additional treatments like heat reduction, chemical reduction, or photoreduction, metal nanoparticles can develop on the surface or within the nanofibers.

(iii) *Atomic Layer Deposition*

ALD, or atomic layer deposition, is a precise and regulated technique for producing metal nanoparticles. This technique involves the sequential use of gas-phase chemical processes to deposit thin layers of material with atomic-level precision. ALD is particularly effective for producing uniform and well-dispersed metal nanoparticles on various substrates. To build nanostructures with exact control over thickness and composition, layers of material are deposited one atomic layer at a time. It is extremely appropriate for many applications in catalysis, electronics, energy, medicine, and sensing due to its capacity to provide atomic-level control.

(iv) *Combustion Synthesis*

Combustion synthesis is a method that uses exothermic chemical reactions to create metal nanoparticles. The creation of nanoparticles is facilitated by the high temperatures and rapid reaction speeds of this approach. The process of combustion synthesis is economical, efficient, and amenable to industrial scale-up. Combustion synthesis is a quick, scalable, and economical way to make metal nanoparticles with precise size and shape. Its efficiency and adaptability make it appropriate for a range of uses in environmental science, energy, medicine, and catalysis.

4 Summary and Future Perspective

The global application of metal nanoparticles is experiencing a significant increase in various disciplines, including energy storage, medical fields, and environmental remediation. The synthesis of nanostructures is a highly dynamic and intricate process. In order to synthesize nanoparticles with the appropriate morphology, a variety of methods have been employed, including physical vapor deposition, chemical vapor deposition, sol–gel methods, chemical reduction, hydrothermal and solvothermal methods, green synthesis, and electrodeposition process. It is important to note that each method has its own unique advantages and limitations. The green synthesis of nanoparticles is a sustainable, environmentally benign, and cost-effective method that is generally free of chemical contaminants. Despite the successful emergence of numerous methods for nanoparticle synthesis, the need for more precise techniques is necessary to address the limitations of existing methods and ensure their effective implementation at the commercial level.

References

1. Selmani, A., Kovacevic, D., Bohinc, K.: Nanoparticles: From synthesis to applications and beyond. Adv. Coll. Interface. Sci. **303**, 102640 (2022)
2. Philip, A., Ruban Kumar, A.: The performance enhancement of surface plasmon resonance optical sensors using nanomaterials: a review. Coord. Chem. Rev.. Chem. Rev. **458**, 214424 (2022)
3. Elumalai, K., Srinivasan, S., Shanmugam, A.: Review of the efficacy of nanoparticle-based drug delivery systems for cancer treatment. Biomed. Technol. **5**, 109–122 (2024)
4. Kumari, S., Raturi, S., Kulshrestha, S., Chauhan, K., Dhingra, S., Andras, K., Thu, K., Khargotra, R., Singh, T.: A comprehensive review on various techniques used for synthesizing nanoparticles. J. Market. Res. **27**, 1739–1763 (2023)
5. Saravanan, A., Kumar, P.S., Karishma, S., Vo, D.V., Jeevanantham, S., Yaashikaa, P.R., George, C.S.: A review on biosynthesis of metal nanoparticles and its environmental applications. Chemosphere 264:128580
6. Gondoni, P., Ghidelli, M., Di Fonzo, F., Li Bassi, A., Casari, C.S.: Fabrication of nano-engineered transparent conducting oxides by pulsed laser deposition. J. Visualized Exp. : JoVE (72) (2013). https://doi.org/10.3791/50297
7. Park, J.S., Chung, W.H., Kim, H.S., Kim, Y.B.: Rapid fabrication of chemical-solution-deposited La0.6Sr0.4CoO3$-\delta$thin films via flashlight sintering. J. Alloy. Compd. **696**, 102–108 (2017). https://doi.org/10.1016/j.jallcom.2016.11.074
8. Pandey, P.A., Bell, G.R., Rourke, J.P., Sanchez, A.M., Elkin, M.D., Hickey, B.J., Wilson, N.R.: Physical vapor deposition of metal nanoparticles on chemically modified graphene: observations on metal-graphene interactions. Small **7**(22), 3202–3210 (2011). https://doi.org/10.1002/smll.201101430
9. Pedersen, H., Elliott, S.D.: Studying chemical vapor deposition processes with theoretical chemistry. Theoret. Chem. Acc.. Chem. Acc. **133**(5), 1–10 (2014). https://doi.org/10.1007/s00214-014-1476-7
10. Miikkulainen, V., Leskelä, M., Ritala, M., Puurunen, R.L.: Crystallinity of inorganic films grown by atomic layer deposition: overview and general trends. J. Appl. Phys. **10**(1063/1), 4757907 (2013)
11. Dolgaev, S.I., Simakin, A.V., Voronov, V.V., Shafeev, G.A., Bozon-Verduraz, F.: Nanoparticles produced by laser ablation of solids in liquid environment. Appl. Surf. Sci. **186**, 546–551 (2002). https://doi.org/10.1016/S0169-4332(01)00634-1
12. Kruis, F.E., Fissan, H., Peled, A.: Synthesis of nanoparticles in the gas phase for electronic, optical and magnetic applications—a review. J. Aerosol Sci. (1998). https://doi.org/10.1016/S0021-8502(97)10032-5
13. Koul, B., Poonia, A.K., Yadav, D., Jin, J.O.: Microbe-mediated biosynthesis of nanoparticles: applications and future prospects. Biomolecules 11(6) (2021). https://doi.org/10.3390/biom11060886
14. Cushing, B.L., Kolesnichenko, V.L., O'Connor, C.J.: Recent advances in the liquid-phase syntheses of inorganic nanoparticles. Chem. Rev. **104**(9), 3893–3946 (2004). https://doi.org/10.1021/cr030027b
15. Zhang, Q., Yin, Y.: Nanomaterials engineering and applications in catalysis. Pure Appl. Chem. **86**(1), 53–69 (2014). https://doi.org/10.1515/pac-2014-5000
16. Landage, S., Wasif, A., Dhuppe, P.: Synthesis of nanosilver using chemical reduction methods. Int. J. Adv. Res. Eng. Appl. Sci. **3**(5), 14–22 (2014)
17. Tavakoli, A., Sohrabi, M., Kargari, A.: A review of methods for synthesis of nanostructured metals with emphasis on iron compounds. Chem. Pap. (2007). https://doi.org/10.2478/s11696-007-0014-7
18. Dikusar, A.I., Globa, P.G., Belevskii, S.S., Sidel'nikova, S.P.: On limiting rate of dimensional electrodeposition at meso- and nanomaterial manufacturing by template synthesis. Surf. Eng. Appl. Electrochem. **45**(3), 171–179 (2009). https://doi.org/10.3103/S1068375509030016

19. Qamar, S.U.R., Ahmad, J.N.: Nanoparticles: mechanism of biosynthesis using plant extracts, bacteria, fungi, and their applications. J. Mol. Liq. **334**, 116040 (2021)
20. Tiquia-Arashiro, S., Rodrigues, D.: Nanoparticles synthesized by microorganisms. In: Extremophiles: Applications in Nanotechnology. Springer, Cham (2016)
21. Anbu, P., Gopinath, S.C.B., Jayanthi, S.: Synthesis of gold nanoparticles using *Platycodon grandiflorum* extract and its antipathogenic activity under optimal conditions. Nanomater. Nanotechnol. **10**, 1–9 (2020)
22. Mikhailova, E.O.: Silver nanoparticles: mechanism of action and probable bio-application. J. Funct. Biomater. **26**, 84 (2020)
23. Hammami, I., Alabdallah, N.M., Al Jomaa, A., Kamoun, M.: Gold nanoparticles: Synthesis properties and applications. J. King Saud Univ. Sci. 33:101560
24. Chattopadhyay, D.P., Patel, B.H.: Nano metal particles : Synthesi, , characterization and application to textiles. Manuf. Nanostruct. 184–215 (2014)

Fabrication and Characterization of Hybrids with Graphene Derivatives/ Analogues for Analytical Applications

Theivasanthi Thirugnanasambandan, Balaguru Venkatesh Sankar, Manimehalai Muthumurugan, and Subash C. B. Gopinath

1 Introduction

Hybridization of nanomaterials is advantageous in terms of providing synergistic properties to the resultant hybrid material. The process of hybridization of graphene with other nanomaterials is performed to prevent the stacking of graphene layers that arises due to the van der Waals interaction between the layers. The performance of transition metal oxide nanoparticles can be enhanced through the addition of graphene [1]. Graphene consists of a single or a few layers of sp2 carbon atoms with large surface area (2600 m^2/g). Among the graphene derivatives graphene oxide and reduced Graphene oxide shows excellent sensing performance. The reason is their easy processing and amphiphilicity. Their surface can be functionalized with more functional groups and in addition these materials show a strong fluorescence quenching. The properties of metal and metal oxide nanoparticles such as interaction with light and electrochemical performance can be synergistically improved when hybridized with graphene [2]. Detection of flammable and explosive gases is important issue for industrial processes. Metal oxide nanoparticles are well- known for making chemical sensors. To overcome the challenges such as their low selectivity

T. Thirugnanasambandan (✉) · M. Muthumurugan
Department of Physics, Kalasalingam Academy of Research and Education (Deemed University), Krishnankoil, Virudhunagar, India
e-mail: t.theivasanthi@klu.ac.in

B. V. Sankar
Department of Power Engineering and Management, Anna University, Guindy, Chennai 600025, Tamil Nadu, India

S. C. B. Gopinath
Center for Global Health Research, Saveetha Medical College & Hospital, Saveetha Institute of Medical and Technical Sciences (SIMATS), Thandalam, Chennai 602 105, Tamil Nadu, India

Faculty of Chemical Engineering & Technology and Institute of Nano Electronic Engineering, Universiti Malaysia Perlis (UniMAP), 02600 Arau, Persil, Malaysia

S. C. B. Gopinath and M. M. Ramli (eds.), *Hybrid-Nanomaterials*, Engineering Materials, https://doi.org/10.1007/978-981-97-9022-7_4

and slow response, they are hybridized with other 2D nanomaterials. Graphene with large surface area high mobility of charge carriers when hybridized with metal oxide nanoparticles are suitable for making electrochemical sensors.

2 Graphene/Metal Nanoparticles Hybrids

Gold nanoparticle-reduced graphene oxide hybrid is synthesized by electrostatic self-assembly. This hybrid system possesses a high stability by controlling the distribution of gold nanoparticles on the surface of reduced graphene oxide. Gold nanoparticles are prepared by adding the precursor $HAuCl_4$ (in water) with tetra octyl ammonium bromide (in toluene) and sodium borohydride. After the reaction, the toluene phase is washed away using sulphuric acid, sodium hydroxide and water. The sample is dried over anhydrous Na_2SO_4 and is dispersed in aqueous 4-dimethylaminopyridine. Reduced graphene oxide synthesized through modified Hummer's method is added to this solution and stirred for three hours to obtain the hybrid nanoparticles [3]. Vapors of ethanol, 2-propanol, and toluene are detected by optical sensing using graphene/gold nanoparticles hybrid materials. The shift in the surface plasmon resonance peak of the gold nanoparticles is visible for various concentration of the analytes. Gold nanoparticles are deposited over silica substrate by vacuum evaporation technique. Graphene synthesized by chemical vapor deposition on a copper foil is transferred to the silica substrate by thermal release tape and an etchant ($CuCl_2$ solution and HCl). The substrate is annealed to achieve a great attachment of graphene over gold nanoparticles [4]. NO_2 gas sensors are developed using gold nanoparticle/graphene hybrid as sensing element and the electrical transduction concepts. Gold nanoparticle/graphene hybrid is drop casted over glass substrate. When the electrical characteristics are analyzed using two probe instruments, the device respond quickly with a high recovery time with a limit of detection for 50 ppm NO_2 gas. Even though graphene oxide shows an insulating behavior, when hybridized with gold nanoparticles, the functional groups on the surface of graphene oxide are largely reduced and thereby an enhancement in the electrical properties is achieved [5]. Hybrid graphene/gold nanoparticle is immobilized with ssDNA as a bio-recognition element for sensitive detection of hepatitis B. The constructed biosensor shows resistance changes in a 2-point resistance measuring device for different concentrations of hepatitis B antigen and the limit of detection of the analyte is in the range of picogram per ml [6].

A non-enzymatic sensing platform is developed using gold nanoparticles/graphene hybrids coated on Surface Enhanced Raman scattering substrate for the detection of bacteria or tumor cells such as *Staphylococcus aureus* and adenine. Graphene is functionalized with poly (diallyl dimethylammonium chloride) to achieve enhancement in the intensity of Raman signal. Gold nanoparticles prepared by citrate thermal reduction method is adsorbed on the surface of graphene sheets due to electrostatic interaction which is shown in the Fig. 1. This hybrid system shows a strong SERS signal for the biomolecules. The characterizations such as transmission electron microscopic images and X-ray diffraction patterns of gold nanoparticles/graphene hybrids are shown in the Figs. 2 and 3. The identified Raman signal enhancement for *S. aureus* at is shown in the spectrum at 733 cm^{-1}. Similarly for adenine, the enhancement of Raman signal is shown in the Fig. 4 [7].

For Surface Enhanced Raman Scattering substrates, addition of graphene to the metal nanoparticles offers more advantages. The two-dimensional nature of graphene

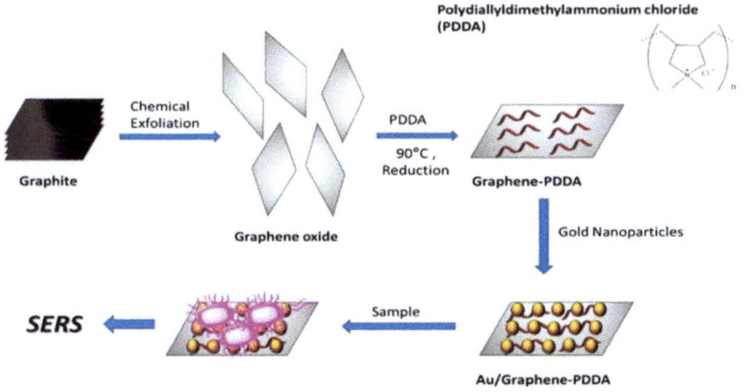

Fig. 1 Synthesis process of gold nanoparticles/graphene hybrids

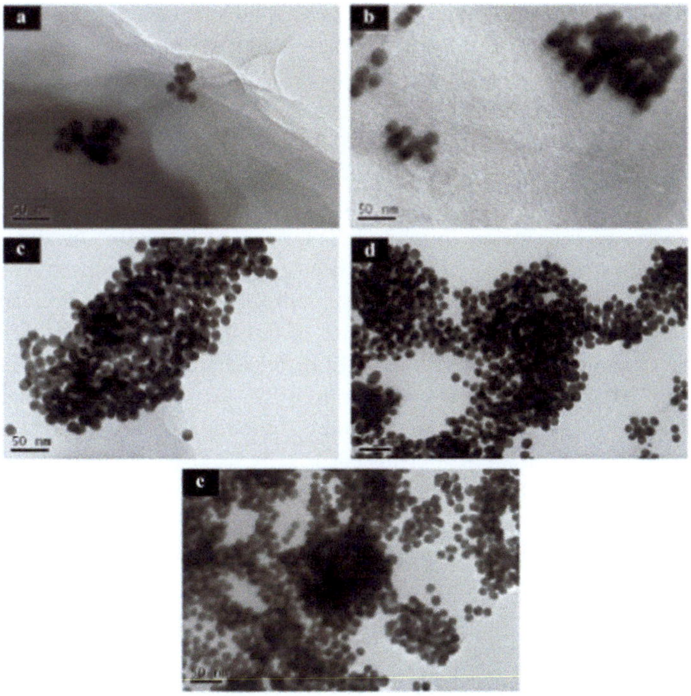

Fig. 2 TEM images of gold nanoparticles/graphene hybrids

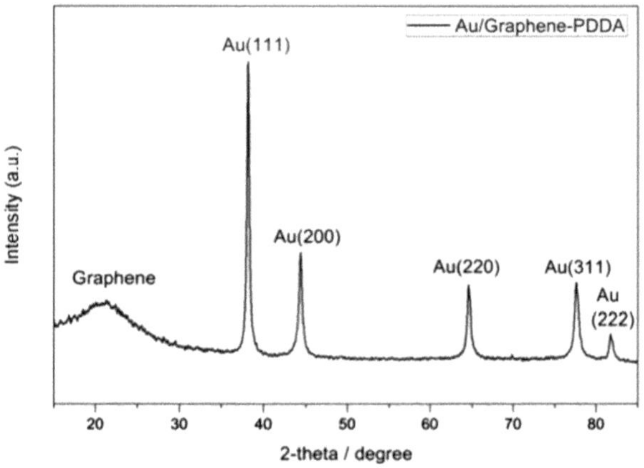

Fig. 3 X-ray diffraction patterns of gold nanoparticles/graphene hybrids

has been a barrier to the formation of graphene hybrid materials. Moving towards three-dimensional graphene, then the hybrid can detect the analytes with more sensitivity. Graphene is coated on a copper foil by chemical vapor deposition method. Gold nanoparticle solution is poured over it and dewetting is done through thermal evaporation. The size of the gold nanoparticles can be optimized by varying the dewetting temperature. This hybrid material is then transferred to polystyrene substrate which is heated to 100 °C. During this heating process, the formation of hybrid occurs. The enhancement in the intensity of Raman signal is high as compared to the bare metal SERS substrates which is suitable for sensor applications [8]. Graphene/Au nanoparticle hybrid film is used as sensing element in an electronic pH sensor. The variations in pH are measured from the capacitance–voltage curve. A high sensitivity of the sensor is attributed to the variation in the Fermi level of the graphene [9]. Reduced graphene oxide is functionalized with Luminol and gold nanoparticles and is used as sensing elements for the detection of cholesterol. The formation hybrid is through π–π conjugation and electrostatic interaction. It exhibits the property of chemiluminescence which is employed as sensing method. The presence of cholesterol in human serum samples is detected to a limit of 0.6 μM concentration [10]. Stable solutions of graphene oxide-gold nanohybrid is prepared for using as a chemical label. Figure 5 shows the stability of the solutions by UV–Visible spectroscopy and the transmission electron microscopic images of the hybrid material is shown [11].

Surface-enhanced Raman scattering (SERS) is used for analysis of chemicals where silver films are used conventionally as sensing platforms. Silver-nanoparticles/graphene hybrids are able to produce two to three times enhancement in the Raman signal during sensing. This technique can be applied to detect the presence of Polycyclic aromatic hydrocarbons in the environment even if they are present in feeble amount. Graphene interacts with these chemicals through π–π stacking and can detect the analytes without recognition elements. A very low concentration (up to 1 ppb) of pyrene, anthracene and phenanthrene can be detected with high sensitivity. These kinds of sensors find applications in the field of food safety [12]. The aggregation of silver nanoparticles reduces its active surface area which in turn decreases

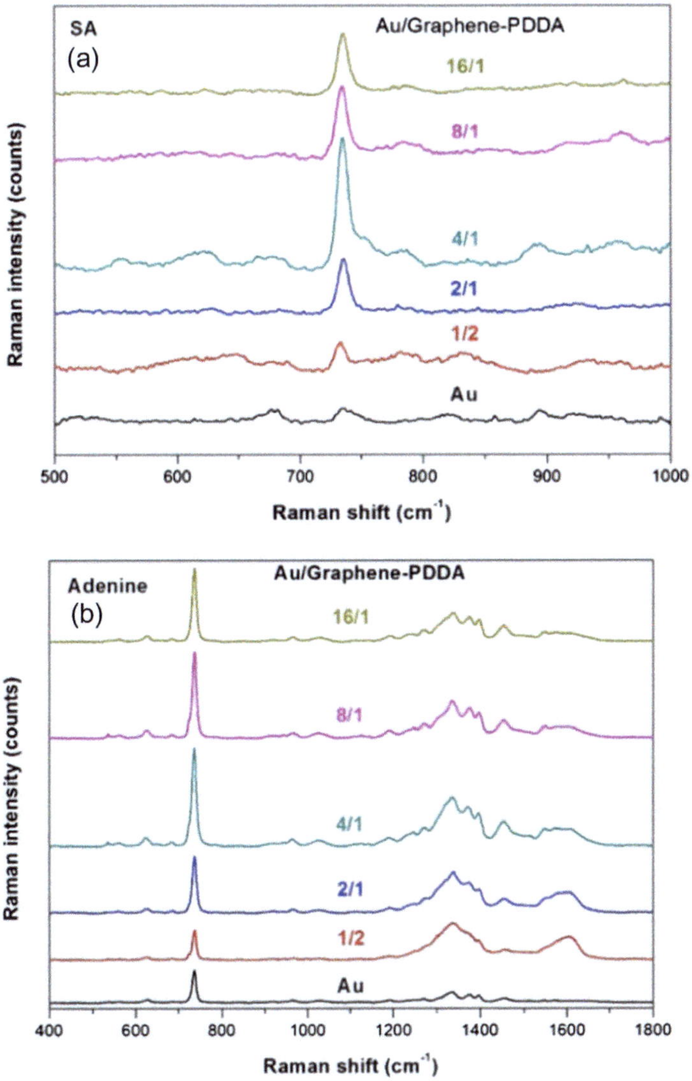

Fig. 4 Raman analysis of gold nanoparticles/graphene hybrids for different ratios of Au and graphene for Staphylococcus aureus and adenine

the antibacterial activity. Graphene oxide is a suitable candidate to disperse silver nanoparticles over its surface. This is due to the excellent physico-chemical properties of graphene oxide and its antibacterial activity. This hybrid system of graphene oxide and silver nanoparticles benefit from both the synergistic antibacterial effect of the constituents of the hybrid and the prevention of aggregation of graphene oxide nanosheets with the addition of silver nanoparticles. Silver nanoparticles are prepared using aqueous silver nitrate solution and Polyvinylpyrrolidone in ethanol.

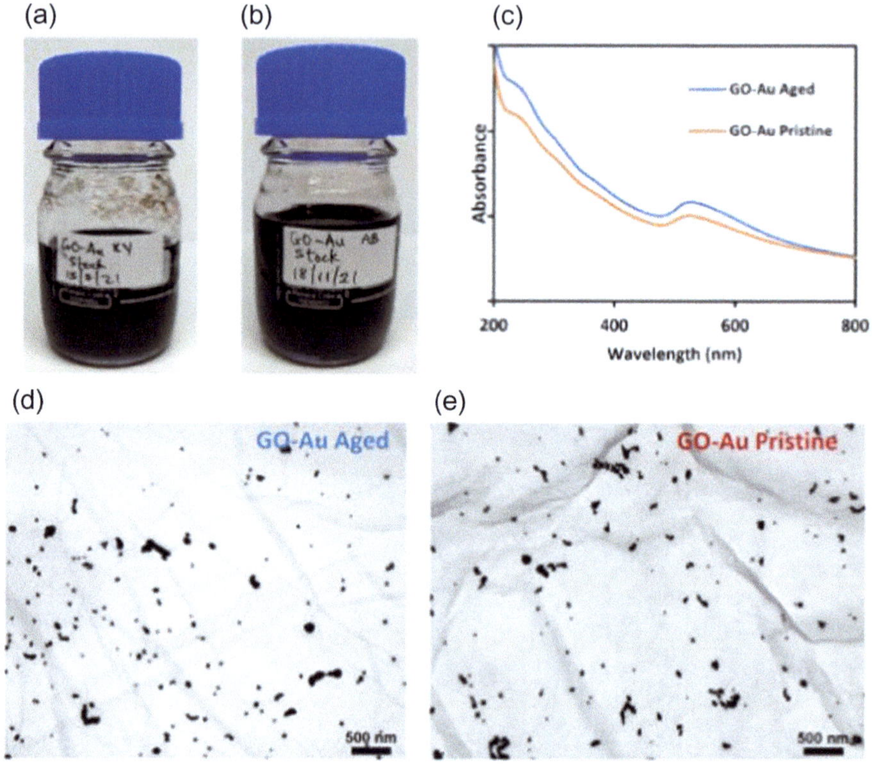

Fig. 5 UV–Visible spectroscopy and the transmission electron microscopic images of the graphene oxide-gold nanohybrid

Polyvinylpyrrolidone is used as a reducing and capping agent. The obtained silver nanoparticles are added dropwise to the graphene oxide dispersion and sonicated for one hour to form the hybrid material [13]. Graphene oxide/silver nanoparticle hybrid is prepared by adding graphene oxide and the silver precursor Ag (PPh3) Cl (triphenyl phosphine). Tetrakis hydroxy methyl phosphonium chloride is used as the reducing agent. Graphene oxide is prepared by the conventional Modified Hummers method. Reduced graphene oxide/zinc oxide nanoparticle hybrid is prepared by adding reduced graphene oxide and the metal–organic precursor of acetylacetonate complexes of Zn^{2+}. Tetrakis hydroxy methyl phosphonium chloride and KOH are added as the reducing agent and to change the pH value of the process [2].

The mechanical movements can be measured by detecting the changes in the resistance using strain sensors. The strain sensors can be developed using sensing elements made of electrically conductive materials embedded in elastomers. Graphene/silver

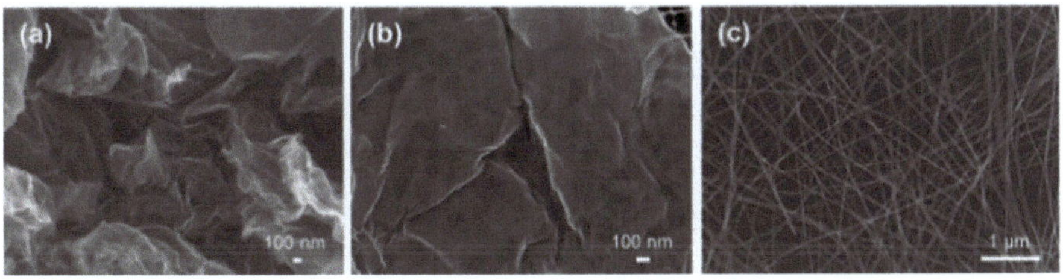

Fig. 6 Scanning electron microscopy SEM images of **a** graphene. **b** graphene after sonication. **c** silver nanowires

nanowire hybrid incorporated within spandex (a synthetic fiber having large elasticity) has been demonstrated as strain sensors. The device shows a very good performance with gauge factor of 150 and strain of 120% and can be used to monitor heartbeats and body movements. Figure 6 shows the SEM images of graphene sheets and silver nanowires [14].

Flexible and wearable strain sensors have become more important in health monitoring systems. Hybrid graphene/silver nanoparticle conductive inks are printed over flexible substrate such as polyimide to develop strain sensors. The device exhibits a high sensitivity and more linearity [15]. Crystal violet and rhodamine 6G are detected using surface enhanced Raman scattering. The substrate shows a very low limit of detection when graphene deposited on the bilayer silver/Cu sandwich structure is used as sensing element [16]. A highly selective detection of ammonia becomes possible at room temperature using silver nanoparticle/reduced graphene oxide hybrids as sensing elements. The sensing material is fabricated using interdigitated gold electrodes on a silicon substrate and a SiO_2 thin top layer using e-beam lithography. Reduced graphene oxide dispersion in dimethyl formamide is drop casted over this substrate and annealed to a temperature of 250 °C. Silver nanoparticles are coated on this substrate by physical vapor deposition using arc plasma reactor. The deposition of silver nanoparticles enhances the sensing performance of the substrate in a field-effect transistor device for the detection of ammonia gas. The electrical characteristics show a linear behavior indicating a high sensing ability [17]. Graphene oxide and silver nanoparticle hybrids are demonstrated for surface-enhanced Raman scattering (SERS) sensing. The hybrid shows enhancement in Raman signal for various concentration of the analyte. This enhancement is even higher than that from the individual graphene oxide or silver nanoparticles when used as sensing elements. In surface-enhanced Raman scattering (SERS) sensing, graphene oxide shows a large chemical enhancement. Ag nanoparticles with various morphology such as spheres, cubes and octahedra are tested for their interaction in making hybrid nanomaterials with graphene oxide nano sheets. The hybrid is able to detect aromatic molecules in nano molar level [18]. Silver nanoparticle-reduced graphene oxide hybrid is employed for the detection of Eriochrome Black T, an azo dye in water through non-enzymatic manner. The hybrid is prepared by green synthesis method using plant extracts followed by reflux method. The precursors used are silver nitrate and graphene oxide. Graphene oxide is transformed into graphene quantum dots due to the defects created by silver nitrate. The hybrid exhibits a luminescent "turn-off" mechanism for various concentrations of Eriochrome Black T [19].

3 Graphene/Metal Oxide Nanoparticles Hybrids

The exceptional physical and chemical properties of metal oxide nanoparticles make them suitable candidates for gas sensing. Reduced Graphene Oxide/Bimodal TiO_2 nanoparticles with 2.5% of graphene shows high sensitivity and selectivity for n-butanol at room temperature [20]. Reduced graphene oxide/TiO_2 nanoparticles hybrid is applied for the detection of hydrogen and oxygen gases as chemoresistive-type sensor [21]. Ternary hybrid, Ag–TiO_2–reduced graphene oxide is used for the electrochemical detection of 8-hydroxy-2′-deoxyguanosine a DNA damage biomarker. The sensor shows distinctive current response for the concentration of the analyte in nano molar level [22]. TiO_2 nanoparticles-reduced graphene oxide hybrids participate in electrochemical reduction N_2 to NH_3 that exhibit a high selectivity [16]. Graphene and titanium dioxide nanoparticles hybridization causes a built-in electric field that results in improved electrical conductivity through interfacial charge transfer. Titanium dioxide TiO_2 nanoparticles are prepared by the hydrolysis of titanium iso-propoxide and the addition of isopropyl alcohol. Graphene nanoplatelets are mixed with this solution and heated to various temperatures to obtain the final product [23].

TiO_2 nanoparticles hybrids with nitrogen-doped carbon quantum dots and reduced graphene oxide are developed that can be used for sensor applications. Carbon quantum dots have a tunable size-dependent bandgap that results in the emission of light upon excitation [24]. During the synthesis of graphene–TiO_2 hybrids, the morphology of titanium dioxide nanoparticles can be controlled through the functionalization of graphene. The functionalization of graphene using carboxyl groups results in anatase truncated bipyramids which is bonded to graphene through {101} whereas the functionalization of graphene using NH_2 groups leads to the formation of belted truncated bipyramids that is bonded to graphene through {100} facets. Graphene–TiO_2 hybrids are prepared by adding Titanium isopropoxide, triethanolamine and graphene nanoplatelets and the mixture is heated to 400 K for three days [25]. Polyaniline, a conducting polymer is added with TiO_2/graphene oxide hybrid for fabricating resistance-based ammonia sensors. The amide bonds in polyaniline are used to chemically conjugate the two constituents of the hybrid. The sensor shows a detection limit of 5 ppm with fast response and recovery times to ammonia at room temperature. It is also highly selective thus preventing the interference of other gases. The high stability of the hybrid material is attributed to the synthesis of titanium dioxide from bulk titanium. The functional groups on graphene oxide sheets favor for effective adsorption and diffusion of ammonia [26]. Graphene oxide/niobium-doped titanium dioxide nanotubes hybrid is used to detect hydrogen. Niobium doping and hybridization of titanium dioxide nanotubes improves the selectivity and response to hydrogen gas [27].

Photoelectrochemical sensors are developed using Graphene Oxide/TiO_2 nanoparticles hybrid for quantifying glutathione. Zinc oxide nanorods well-known for its band gap and light absorption grown on the indium tin oxide substrates are used for making the photoanode in the sensor which is more critical for the performance of the sensor. During photochemical reactions, separation of charge carriers such as electrons and holes and their transport to the electrodes from the active material is varied by an interfacial layer. Graphene Oxide/TiO_2 nanoparticles hybrid is introduced as an interfacial layer that supports for the growth of zinc oxide nanorods and for enhancing the sensing performance [28].

The chemical sensors particularly designed for gas sensing find applications in electronic nose technology. Graphene-zinc oxide hybrid is used for the detection of NO_2 gas. Zinc films are coated on alumina substrates by radio frequency magnetron sputtering technique. The deposited zinc is converted into zinc oxide by electrolysis process with the substrate as anode, platinum as cathode and oxalic acid with ethanol as electrolyte. Zinc oxalate is formed initially which is then annealed at 400 °C to form zinc oxide films. Graphene oxide is poured over the substrate by drop casting which is then heated at 300 °C to form reduced graphene oxide [29]. Three-dimensional graphene foam is formed by chemical vapor deposition over nickel foam as the substrate and ethanol as the carbon source. Zinc oxide is coated on this substrate by placing the nickel foam in an autoclave containing zinc chloride alkaline solution and hydro thermal operation is performed. After two hours, zinc oxide nanorods are formed and they cluster on graphene sheets. The obtained graphene/ZnO hybrids are tested for the detection of hexacyanoferrate (III) and dopamine using electrochemical impedance spectroscopy. Highly sensitive electrochemical sensors can be developed with these hybrid materials with a limit of detection of 1 μM and 10 nM for hexacyanoferrate (III) and dopamine respectively [30]. Zinc Oxide nanoclusters $Zn_{12}O_{12}$/graphene hybrid has been reported for its excellent gas sensing properties. The formation of hybrid nanomaterials is confirmed by scanning electron microscopic images and X-ray diffraction analysis. The detection limit for ethanol gas is reported up to 2,000 ppm for 5 wt% of graphene in the hybrid [31]. Hybrid materials of reduced graphene oxide with materials such as silver nanoparticles, zinc oxide nanoparticles, zirconium oxide nanoparticles and palladium nanoparticles have been demonstrated for hydrogen sensing. The hybridization process considerably improves the sensing performance and the hybrids exhibit p-type behavior during sensing [32].

Reduced graphene oxide/mesoporous ZnO hybrid nanostructures are synthesized for fabricating wearable E-textiles in which gas sensing application can be introduced. The hybrid materials are embedded in cotton/elastic threads to form the gas sensor. The sensor exhibits a high sensitivity, stability for more than 80 days, limit of detection 40 ppb for NO_2 gas, high mechanical deformation tolerance and durability with gas detection at room-temperature [33]. Electrochemical sensors are fabricated using Reduced graphene oxide/zinc oxide nanoparticles hybrid as sensing elements for measuring the level of acetylcholine, a neurotransmitter. An imbalance in this biochemical will result in Alzheimer's and Parkinson's diseases. Reduced graphene

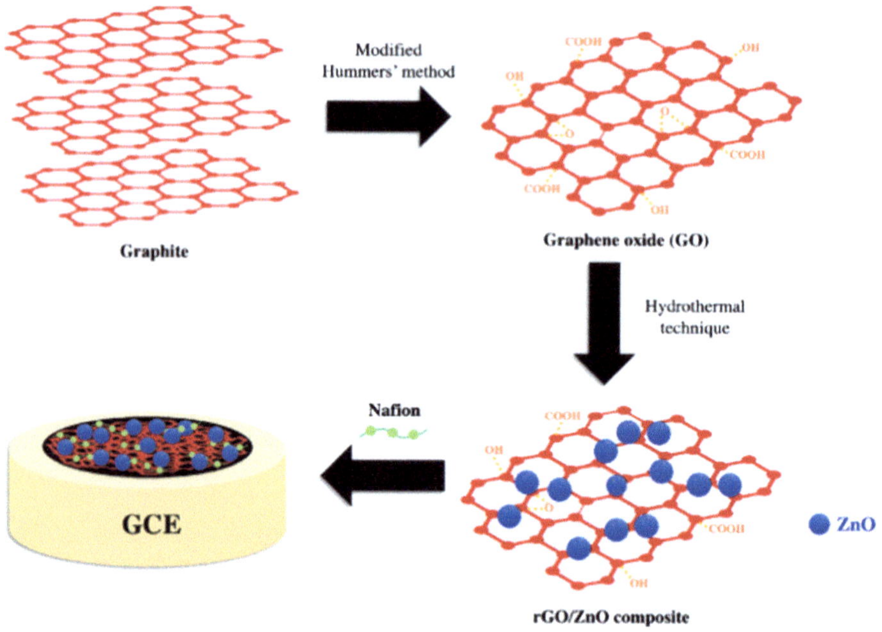

Fig. 7 Reduced graphene oxide/ZnO composite on a glassy carbon electrode

oxide/zinc oxide nanoparticles hybrid is fabricated through hydrothermal method and the hybridization occurs due to deoxygenation of zinc oxide nanoparticles on the graphene sheets. Apart from the interfering biochemicals such as glutamate and gamma-aminobutyric acid, acetylcholine is analyzed with high selectivity through electrochemical impedance spectroscopy and cyclic voltammetry. Figure 7 shows the fabrication of electrode for electrochemical sensors using Reduced graphene oxide/ zinc oxide nanoparticles hybrid [34].

Figures 8 and 9 show the characterization of nanomaterials using XRD and Raman analysis, respectively.

Naphthol is a pollutant that can cause damage to kidney and cornea and carcinogenic. Reduced graphene oxide is green synthesized using green tea extract and is decorated with Fe_3O_4 nanoparticles. This hybrid nanomaterial is applied for the electrochemical sensing of naphthol by coating over carbon paste electrode. The sensor shows a limit of detection for 70 nM of naphthol in real samples with high reproducibility [35]. Ferrites of nickel/zinc ($Ni0.5Zn0.5Fe_2O_4$) is prepared through gel combustion of the nitrate salts of iron, nickel and zinc. Citric acid and ammonia are added as a fuel and to change the pH value. Graphene oxide is added with this and after stirring for one-hour, dark brown viscous xerogel is obtained. The obtained product is heated and further calcined to get the final product in powder form. The sample is characterized using X-ray diffraction analysis and Fourier transform infrared spectroscopy. Figure 10a, b shows the X-ray diffraction analysis and

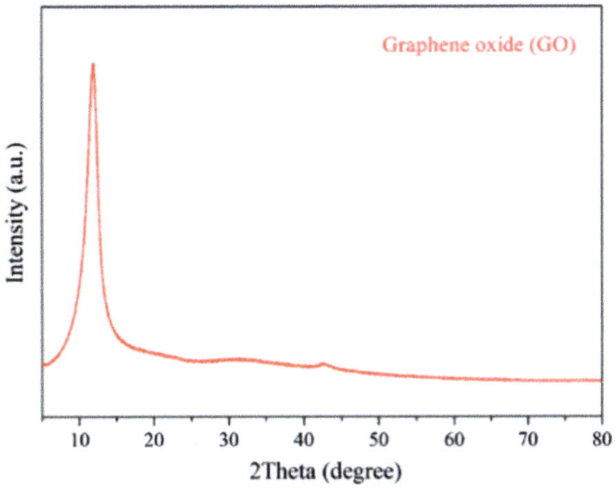

Fig. 8 XRD spectrum of graphene oxide

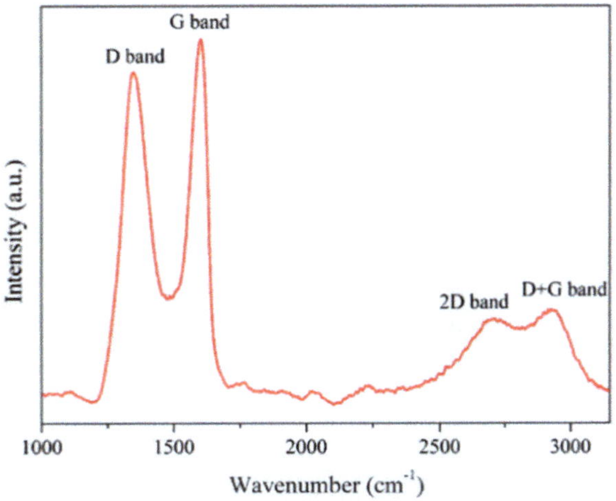

Fig. 9 Raman spectrum of graphene oxide

Fourier transform infrared spectrum of Graphene oxide-ferrite hybrid [36]. Further, graphene oxide-ferrite hybrid was confirmed by SEM analysis and displayed Fig. 11.

The addition of graphene oxide causes broadening of the peaks which confirms the distribution of ferrite nanoparticles over the surface of graphene oxide sheets. This also reduces the agglomeration of the ferrite nanoparticles. Nickel zinc ferrite nanoparticles possess cubic structure and for the hybrid samples, the peak for graphite is shown in the spectrum. The addition of graphene oxide supplies more surface area that leads to the formation of small nickel zinc ferrite particles. In the FTIR (Fourier transform infrared) spectrum, the peaks for nickel zinc ferrite particles shows a redshift for the increased content of graphene oxide. All the peaks are shifted by graphene oxide addition that confirms the interaction between the constituents of the hybrid.

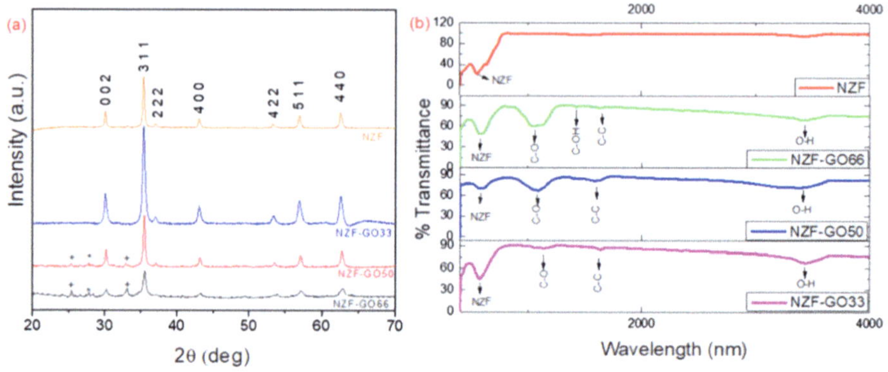

Fig. 10 **a** X-ray diffraction analysis and **b** Fourier transform infrared spectrum of Graphene oxide-ferrite hybrid

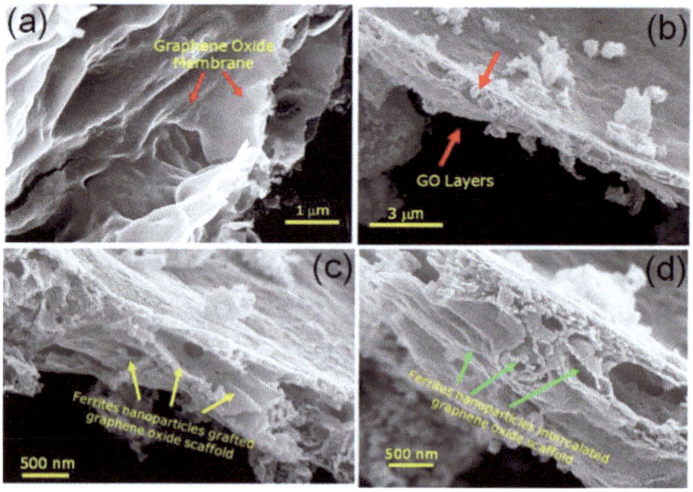

Fig. 11 SEM images of graphene oxide and Graphene oxide-ferrite hybrid under different magnifications

Nano manganite $La_{0.8}Sr_{0.2}MnO_3$/graphene hybrid films are shown as magnetic field sensors. The device is fabricated in a Wheatstone half-bridge in such a way that to induce a Lorentz force which is due to the positive magnetoresistance of graphene and colossal negative magnetoresistance of manganite [37]. SnO_2 Nanoparticles/Reduced Graphene Oxide hybrids are used to develop room-temperature NO_2 sensors. This hybrid system can be synthesized by solvothermal method and the sensing ability is high because of a strong gas interaction with the hybrid, quick charge transport, and electrostatic attraction due to Sn–C covalent bond [38]. SnO_2 Nanoparticle/Reduced Graphene Oxide Hybrids are used for the development of room-temperature NO_2 sensors. The sensing material is synthesized by solvothermal method and made into different layers using the Differential Centrifugation technique. Chemiresistive devices are then fabricated using these components in the form of 2×2 sensor array. The limit of detection of the device is up to 80 ppm for NO_2 gas and the humidity can be measured to the level of 30% with a quick response that can also be recovered in a fast manner. The device with SnO_2 Nanoparticle/Reduced Graphene Oxide Hybrids shows very good response for a concentration of 80 ppm of NO_2 gas at room-temperature within 6 s and can recover within 14 s. The Sn–C covalent bond in heterostructure allows for effective gas adsorption and high mobility of charge carriers [38]. Exposure to acetone gas causes damage to vision and central nervous system. Sensors for acetone gas find applications in air quality control and health monitoring. Graphene oxide-cobalt oxide hybrid materials are used to develop sensors for acetone gas. These hybrid materials prepared by hydrothermal method are porous in nature. The operating temperature for gas sensing of these materials is greatly reduced. The device can detect acetone gas up to 20 ppm [39].

4 Graphene/Semiconductor Nanoparticles Hybrids

CdS microparticles–graphene hybrid is synthesized by hydrothermal process using graphene oxide and cadmium chloride as precursors. The formation of hybrid does not affect the hexagonal phase of CdS and CdS is distributed uniformly over graphene sheets [40]. CdS nanoparticles/graphene nanoribbons hybrid is prepared by solvothermal method. Graphene Nanoribbons prevent the agglomeration of CdS nanoparticles [41]. Chemical vapor deposited graphene is transferred to the polyethylene terephthalate substrate. Hybrid structures of graphene/CdS quantum dots are fabricated using chemical bath deposition and are demonstrated as photodetectors. Since CdS quantum dots is a semiconducting material with a bandgap of 2.4 eV, the high electrical conductivity of graphene supports for the hybrid for its photoconductive applications in the visible region. Even though graphene possesses a high mobility for charge carriers, its optical absorption is less and so its hybridization with quantum dots will be more suitable for the fabrication of photodetectors [42]. The formation of graphene hybrid materials can be confirmed using Raman spectroscopy which is important for 2D materials. Graphene–CdS hybrid exhibit an absorption at 550 nm similar to CdS nanoparticles. Figure 12 shows the synthesis process of Graphene–CdS hybrid by chemical bath deposition method. XPS, XRD and Raman spectrum of Cd S nanoparticles and graphene/Cd S hybrid structures are shown in the Figs. 13, 14 and 15, respectively [43].

CdS quantum dots of size 6 nm is used to prepare Graphene-CdS quantum dots hybrid by hydrothermal method. This hybrid material with 0.5% of graphene has been demonstrated for electrochemical sensing of glucose with high selectivity and durability even for very low concentration of glucose [44]. CdS quantum dots/chitosan/

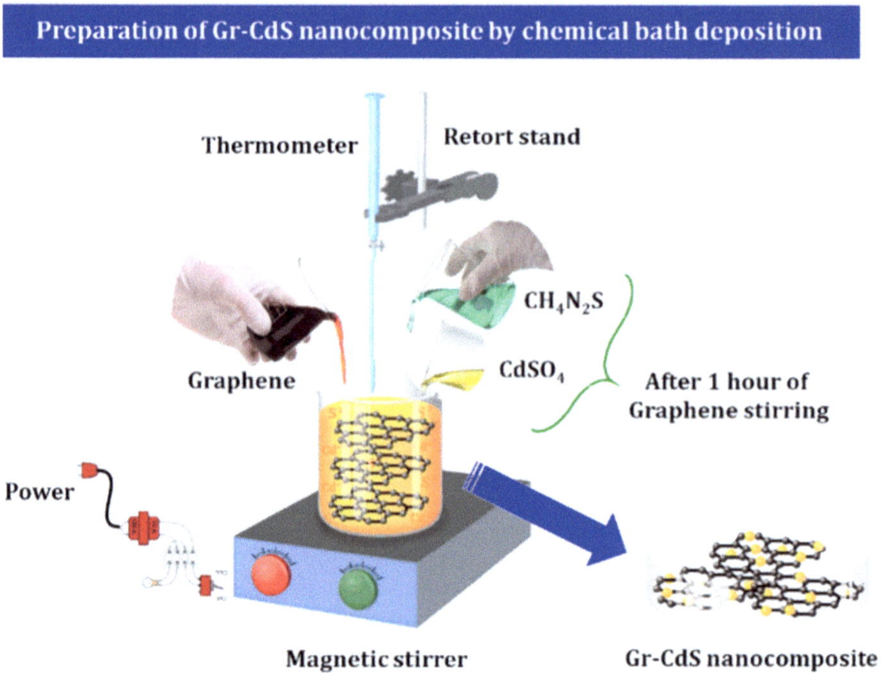

Fig. 12 Synthesis of Graphene–CdS hybrid by chemical bath deposition method

Fig. 13 XPS analysis of **a** Cd S nanoparticles and **b** Graphene/Cd S hybrid on soda-lime glass

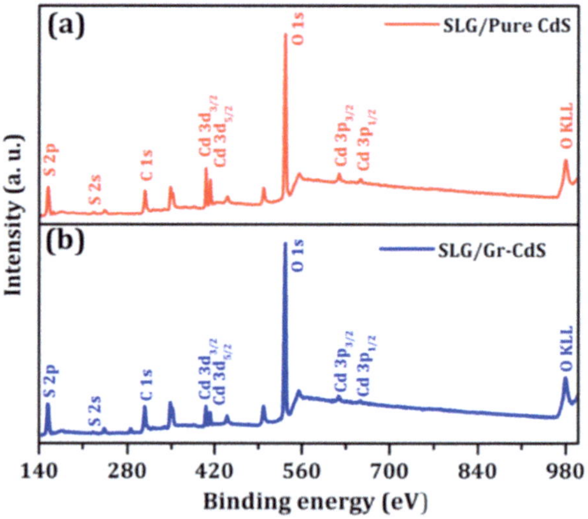

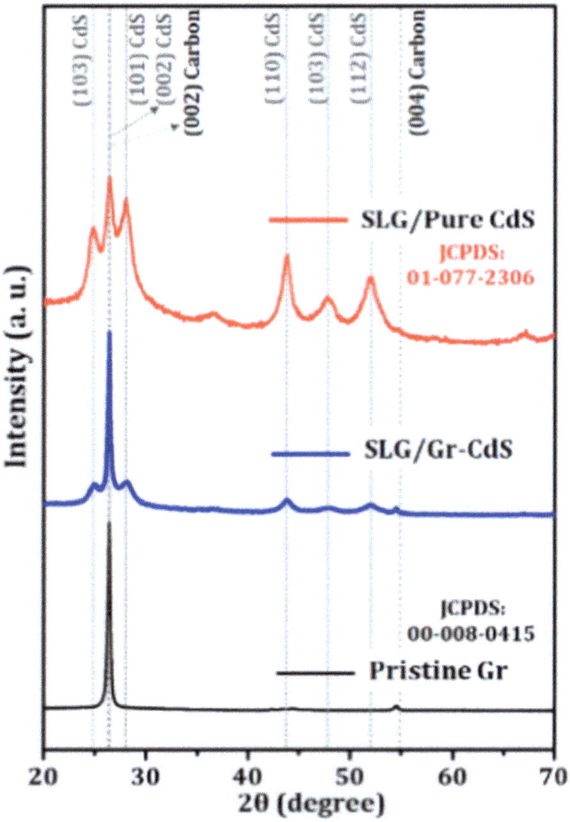

Fig. 14 XRD spectrum of graphene, Graphene/Cd S hybrid and Cd S nanoparticles on soda-lime glass

reduced graphene oxide hybrid coated on glassy carbon electrode is used for electrochemical sensing of Ceftazidime, an anti-bacterial infection drug. This is due to the production of high current caused by the electrooxidation toward Ceftazidime [45]. Nano-bismuth decorated CdS–graphene hybrid is used for fabricating highly sensitive photoelectrochemical sensors. The precursors bismuth nitrate and cadmium (II)acetylacetonate in polyether sulfone are drop casted over the surface of indium–tin oxide glass which is irradiated with laser. Graphene is formed due to the graphitization of polyether sulfone. This method is one step formation of the Bi_0/CdS/graphene hybrid. S2, a toxic pollutant in water is detected by measuring the photocurrent response in a photoelectrochemical cell [46]. Two dimensional materials such as graphene and molybdenum sulphide MoS_2 are exfoliated physically and is treated solvo-thermally to prepare MoS_2 quantum dots. Graphene–MoS_2 quantum dots hybrids show enormous fluorescence that can be used for sensing gases such as NO_2 and NH_3 [47].

Reduced graphene oxide/carbon dot hybrids show enhanced performance for NO_2 gas. A concentration of 10 ppb of NO_2 gas is measured and for 25 ppm of gas is detected with 120% of sensitivity with excellent selectivity and repeatability for more

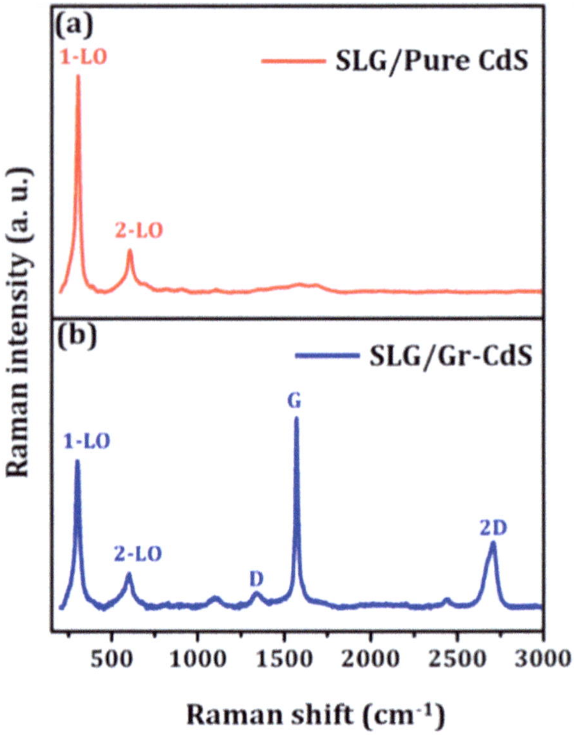

Fig. 15 Raman spectrum of graphene, Graphene/Cd S hybrid and Cd S nanoparticles on soda-lime glass

than six times. This is due to an improved charge transfer from a greater number of holes on the surface of reduced graphene oxide [48]. Reduced graphene oxide film is made by spin-coating the solution on a SiO_2/Si substrate. Graphene quantum dots obtained through the carbonization of citric acid is sprayed on the substrate. When graphene quantum dots/reduced graphene oxide hybrid film is irradiated with ultraviolet UV light, the electrons produced from quantum dots are transported by the highly conductive reduced graphene oxide. This hybrid material is applied to fabricate an UV sensor with photo responsibility of 8.7×10^2 A/W and detectivity of 7.7×10^{13} Jones at a low operating voltage [49]. An electrochemical sensor for the determination of Cd2+ and Pb2+ ions is fabricated with N-doped carbon quantum dots/graphene oxide hybrid coated glassy carbon electrode. The hybrid material is able to adsorb the heavy metal ions by electrostatic interaction. The sensor possesses a detection limit of 7 μg/L for Cd2+ ions and 1 μg/L for Pb2+ ions and is highly useful for water analysis [49]. Reduced graphene oxide quantum dots/zinc oxide hybrid nanofibers are produced to prepare the electrode for electrochemical sensing of intracellular H_2O_2 released from cancer cells. This biosensor differentiate can cancer and noncancer cells [50]. Graphene quantum dot-NaYF4: Yb, Er nanohybrid exhibit improved emission of fluorescence and finds application for the detection of carcinoembryonic antigen. The linker DNA is used as bio-recognition element and the concept of fluorescence resonance energy transfer is applied for the detection of biomolecules [51].

5 Graphene/Carbon Nanomaterials Hybrids

Three-dimensional graphene/carbon nanotube hybrids are formed through chemical vapor deposition method. The hybrid is a monolithic foam covered with carbon nanotubes that finds applications in electrochemical sensing for the detection of biomolecules such as dopamine. The working electrode is fabricated with this 3D graphene–carbon nanotube hybrid immobilized with horseradish peroxidase enzyme and nafion is used as a binder. The electrode possesses a high sensitivity with a detection limit of 20 n M of dopamine. It also detects H_2O_2 up to 1–10 mM [52]. Strain sensors are developed using carbon nanotubes and graphene nanoplatelets hybrid materials as sensing elements. These sensors possess a high sensitivity with gauge factor 200 at 10% strain and can be stretched up to 50% of its original length. The sensors are showing a tenability of that the material can be recovered to its original condition even after 50% strain. They show distinctive changes in the resistance values upon stretching that too highly reproducible even after 1000 loading cycles. Graphene, its high electrical conductivity and its hybridization with carbon nanotubes results in durable strain sensors. The applications include monitoring of finger bending and throat swallowing in human particularly for wearable electronic devices. The Fig. 16 shows the fabrication process of strain sensors. The spray coating method is employed for coating carbon nanotubes and graphene nanoplatelets hybrid on the substrate. The strain can be measured by observing the changes in the values of resistance (Fig. 17) [53].

Welded Carbon Nanotube–Graphene Hybrids are fabricated by thermal annealing that are used to make strain sensors. These sensors are used for sensing the human motion by collecting the epidermal bio-signals from the faint skin deformation and wrist pulse [54].

Graphene oxide/graphene hybrid films have been demonstrated for ammonia detection. The high sensitivity is due to the graphene oxide as molecular capture layer and a high electrical conductivity of graphene. Graphene is coated on a substrate by chemical vapor deposition over which graphene oxide layer is drop-casted. A thickness of 300 nm graphene oxide layer exhibits a high ammonia sensing with a detection limit of 1.5 ppb (parts per billion) [55]. Expanded graphite and cellulose are exfoliated in a ball-mill to form a hybrid material. A dispersion of this hybrid material with a concentration of 15 mg/ml possesses an in-plane conductivity of 6000 S/m. The hybridization of graphene and nanocellulose is due to the hydrophobic and hydrogen bond interaction [56].

Cellulose nanofiber/graphene oxide hybrid filaments are explored as humidity sensors. The filament is made by wet-spinning of the constituents of the hybrid and possesses a high mechanical strength. The filament can be embedded in smart

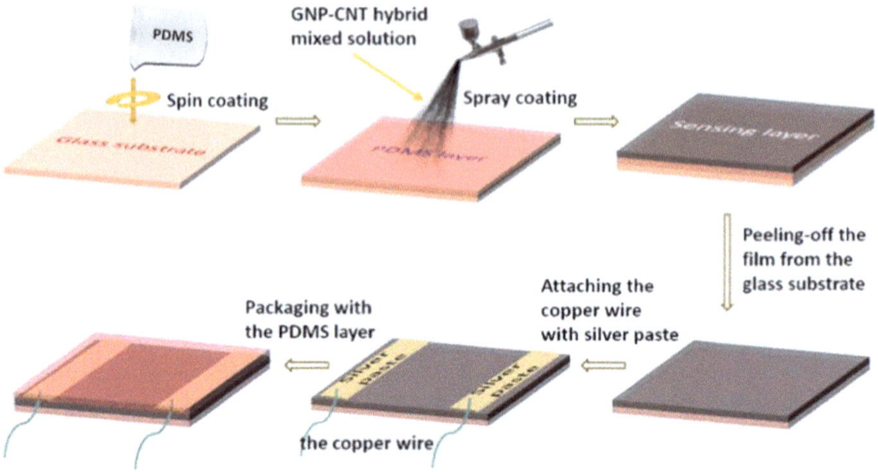

Fig. 16 fabrication process of strain sensors using carbon nanotubes and graphene nanoplatelets hybrid

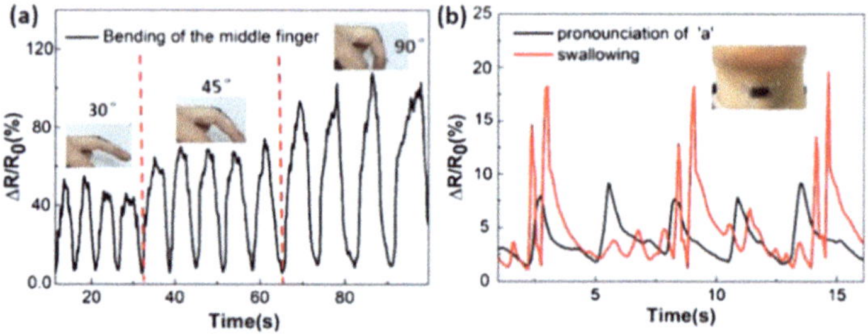

Fig. 17 Performance of strain sensors for **a** finger bending. **b** swallowing of the throat

wearable devices, natural fiber-reinforced polymer composites, and environmental sensing devices for measuring the humidity. The Fig. 18 shows the block diagram of the device measuring the humidity in an environmental chamber [57].

6 Conclusions

Graphene with its high sensing ability, it can be used as a recognition element even in non-enzymatic manner. Since graphene can be produced at low cost and in large scale, it can be a suitable candidate to replace gold nanoparticles in the field of sensors. Because of its plasmonic properties, graphene is suitable for making optical sensors. Since graphene possess high electrical conductivity, electrical- based FET transistors can be constructed for sensing. Graphene exhibits large surface area and porous nature

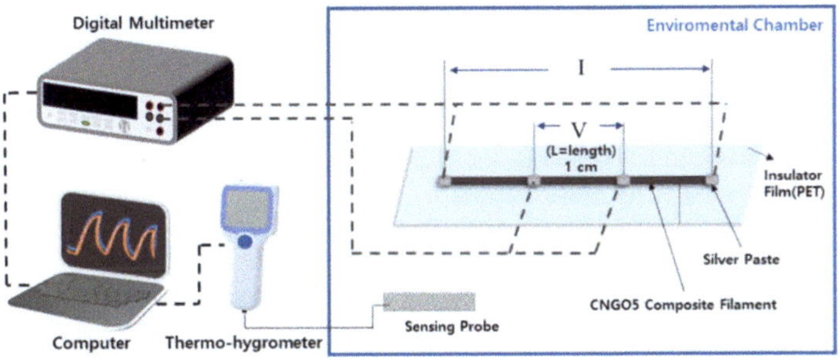

Fig. 18 Block diagram of a humidity sensor using Cellulose nanofiber/graphene oxide hybrid filament

that are suitable for fabricating portable electrochemical sensors. Graphene can be hybridized with other nanomaterials and hence their synergistic effect becomes useful for the detection of low concentration of analytes like in gas sensors and for early detection of diseases in medical field since graphene provides high selectivity in addition to high sensitivity.

References

1. Jana, A., Scheer, E., Polarz, S.: Synthesis of graphene–transition metal oxide hybrid nanoparticles and their application in various fields. Beilstein J. Nanotechnol. **8**(1), 688–714 (2017)
2. Kavitha, C., Bramhaiah, K., John, N.S.: Reduced graphene oxide/nanoparticle hybrid structures: a new generation smart materials for optical sensors Mater. Today: Proc. **5**(1), 2609–2618 (2018)
3. Choi, Y., Bae, H.S., Seo, E., Jang, S., Park, K.H., Kim, B.-S.: Hybrid gold nanoparticle-reduced graphene oxide nanosheets as active catalysts for highly efficient reduction of nitroarenes. J. Mater. Chem. **21**(39), 15431–15436 (2011)
4. Piszter, G., Kertész, K., Molnár, G., Pálinkás, A., Deák, A., Osváth, Z.: Vapour sensing properties of graphene-covered gold nanoparticles. Nanoscale Adv. **1**(6), 2408–2415 (2019)
5. Rattan, S., Kumar, S., Goswamy, J.K.: Gold nanoparticle decorated graphene for efficient sensing of NO_2 gas. Sens. Int. **3**, 100147 (2022)
6. Walters, F., Rozhko, S., Buckley, D., Ahmadi, E.D., Ali, M., Tehrani, Z., Mitchell, J., et al.: Real-time detection of hepatitis B surface antigen using a hybrid graphene-gold nanoparticle biosensor. 2D Mater. **7**(2), 024009 (2020)
7. Mevold, A.H., Hsu, W.W., Hardiansyah, A., Huang, L.Y., Yang, M.C., Liu, T.Y., Chan, T.Y., et al.: Fabrication of gold nanoparticles/graphene-PDDA nanohybrids for bio-detection by SERS nanotechnology. Nanoscale Res. Lett. **10**, 1–7 (2015)
8. Leem, J., Wang, M.C., Kang, P., Nam, S.W.: Mechanically self-assembled, three-dimensional graphene–gold hybrid nanostructures for advanced nanoplasmonic sensors. Nano Lett. **15**(11), 7684–7690 (2015)
9. Medhat, A., Salah, D., Boichuk, N., Hassan, I., Vitusevich, S., Kasry, A.: Graphene nanoplatelet–au nanoparticle hybrid as a capacitive-metal–oxide–semiconductor pH sensor. ACS Appl. Electron. Mater. **3**(1), 430–436 (2020)

10. Lin, K.L., Yang, T., Zhang, F.F., Lei, G., Zou, H.Y., Li, Y.F., Huang, C.Z.: Luminol and gold nanoparticle-co-precipitated reduced graphene oxide hybrids with long-persistent chemiluminescence for cholesterol detection. J. Mater. Chem. B **5**(35), 7335–7341 (2017)
11. Akere, T.H., de Medeiros, A.M., Martinez, D.S.T., Ibrahim, B., Ali-Boucetta, H., Valsami-Jones, E.: Synthesis and characterisation of a graphene oxide-gold nanohybrid for use as test material. Nanomaterials **13**(1), 33 (2022)
12. Wang, X., Qiaoling, X., Xiaoye, H., Han, F., Zhu, C.: Silver-nanoparticles/graphene hybrids for effective enrichment and sensitive SERS detection of polycyclic aromatic hydrocarbons. Spectrochim. Acta Part A Mol. Biomol. Spectrosc. **228**, 117783 (2020)
13. Lozovskis, P., Jankauskaitė, V., Guobienė, A., Kareivienė, V., Vitkauskienė, A.: Effect of graphene oxide and silver nanoparticles hybrid composite on P. aeruginosa strains with acquired resistance genes. Int. J. Nanomed. 5147–5163 (2020)
14. Vo, T.T., Lee, H.-J., Kim, S.-Y., Suk, J.W.: Synergistic effect of graphene/silver nanowire hybrid fillers on highly stretchable strain sensors based on spandex composites. Nanomaterials **10**(10), 2063 (2020)
15. Htwe, Y.Z.N., Mariatti, M.: Performance of water-based printed hybrid graphene/silver nanoparticle conductive inks for flexible strain sensor applications. Synth. Met. **300**, 117495 (2023)
16. Zhang, M., Zheng, Z., Liu, H., Wang, D., Chen, T., Liu, J., Yucheng, W.: Rationally designed graphene/bilayer silver/Cu hybrid structure with improved sensitivity and stability for highly efficient SERS sensing. ACS Omega **3**(5), 5761–5770 (2018)
17. Cui, S., Mao, S., Wen, Z., Chang, J., Zhang, Y., Chen, J.: Controllable synthesis of silver nanoparticle-decorated reduced graphene oxide hybrids for ammonia detection. Analyst **138**(10), 2877–2882 (2013)
18. Fan, W., Lee, Y.H., Pedireddy, S., Zhang, Q., Liu, T., Ling, X.Y.: Graphene oxide and shape-controlled silver nanoparticle hybrids for ultrasensitive single-particle surface-enhanced Raman scattering (SERS) sensing. Nanoscale **6**(9), 4843–4851 (2014)
19. Chettri, P., Singh, M.K., Tripathi, A., Pathak, A.P., Mandal, R.K., Tiwari, A.: Eriochrome Black T sensing using silver nanoparticle-reduced graphene oxide composite via luminescent "turn-off" mechanism and its biosorption on guava (Psidium guajava) leaf powder. Graphene Technol. **4**, 41–51 (2019)
20. Pardeshi, N.M., Ghuge, R.S., Birla, P.N., Chauhan, R., Bhalekar, S.P., Shinde, M.D., Sivalingam, Y., Kale, R.D., Rane, S.B.: Reduced Graphene Oxide@ Bimodal TiO_2 nanocomposites as an efficacious console for room temperature n-butanol gas sensing. ACS Appl. Electron. Mater. (2024)
21. Sagadevan, S., Lett, J.A., Weldegebrieal, G.K., Oh, W.C., Alshahateet, S.F., Fatimah, I., Mohammad, F., et al.: Enhanced gas sensing and photocatalytic activity of reduced graphene oxide loaded TiO_2 nanoparticles. Chem. Phys. Lett. **780**, 138897 (2021)
22. Dhulkefl, A.J., Atacan, K., Bas, S.Z., Ozmen, M.: An Ag–TiO_2–reduced graphene oxide hybrid film for electrochemical detection of 8-hydroxy-2′-deoxyguanosine as an oxidative DNA damage biomarker. Anal. Methods **12**(4), 499–506 (2020)
23. Tobaldi, D.M., Dvoranová, D., Lajaunie, L., Rozman, N., Figueiredo, B., Seabra, M.P., Škapin, A.S., Calvino, J.J., Brezová, V., Labrincha, J.A.: Graphene-TiO_2 hybrids for photocatalytic aided removal of VOCs and nitrogen oxides from outdoor environment. Chem. Eng. J. **405**, 126651 (2021)
24. Baragau, I.A., Buckeridge, J., Nguyen, K.G., Heil, T., Sajjad, M.T., Thomson, S.A., Rennie, A., et al.: Outstanding visible light photocatalysis using nano-TiO_2 hybrids with nitrogen-doped carbon quantum dots and/or reduced graphene oxide. J. Mater. Chem. A **11**(18), 9791–9806 (2023)
25. Sordello, F., Zeb, G., Kaiwen, H., Calza, P., Minero, C., Szkopek, T., Cerruti, M.: Tuning TiO_2 nanoparticle morphology in graphene–TiO_2 hybrids by graphene surface modification. Nanoscale **6**(12), 6710–6719 (2014)
26. Tian, J., Yang, G., Jiang, D., Su, F., Zhang, Z.: A hybrid material consisting of bulk-reduced TiO_2, graphene oxide and polyaniline for resistance-based sensing of gaseous ammonia at room temperature. Microchimica Acta **183**, 2871–2878 (2016)

27. Galstyan, V., Ponzoni, A., Kholmanov, I., Natile, M.M., Comini, E., Nematov, S., Sberveglieri, G.: Investigation of reduced graphene oxide and a Nb-doped TiO_2 nanotube hybrid structure to improve the gas-sensing response and selectivity. ACS Sens. **4**(8), 2094–2100 (2019)
28. Zhu, M., Murtaza, I., Xie, W., Li, S., Zou, C., Xiang, Y.: Reduced graphene oxide/MXene-derived TiO_2 hybrid interface layer for the improvement of zinc oxide nanorod growth and their applications in glutathione sensing. Nano **17**(08), 2250063 (2022)
29. Galstyan, V., Comini, E., Kholmanov, I., Ponzoni, A., Sberveglieri, V., Poli, N., Faglia, G., Sberveglieri, G.: Graphene-zinc oxide-based nanomaterials for gas sensing devices. Procedia Eng. **168**, 1172–1175 (2016)
30. Dong, X., Ma, Y., Zhu, G., Huang, Y., Wang, J., Chan-Park, M.B., Wang, L., Huang, W., Chen, P.: Synthesis of graphene–carbon nanotube hybrid foam and its use as a novel three-dimensional electrode for electrochemical sensing. J. Mater. Chem. **22**(33), 17044–17048 (2012)
31. Photaram, W., Liangruksa, M., Aiempanakit, M., Suwanchawalit, C., Wisitsoraat, A., Sukunta, J., Laksee, S., Siriwong, C.: Design and fabrication of zinc oxide-graphene nanocomposite for gas sensing applications. Appl. Surf. Sci. **595**, 153510 (2022)
32. Pal, P., Yadav, A., Chauhan, P.S., Parida, P.K., Gupta, A.: Reduced graphene oxide-based hybrid functionalized films for hydrogen detection: Theoretical and experimental studies. Sens. Int. **2**, 100072 (2021)
33. Li, W., Chen, R., Qi, W., Cai, L., Sun, Y., Sun, M., Li, C., et al.: Reduced graphene oxide/mesoporous ZnO NSs hybrid fibers for flexible, stretchable, twisted, and wearable NO_2 E-textile gas sensor. ACS Sens. **4**(10), 2809–2818 (2019)
34. Pitiphattharabun, S., Auewattanapun, K., Htet, T.L., Thu, M.M., Panomsuwan, G., Techapiesan-charoenkij, R., Ohta, J., Jongprateep, O.: Reduced graphene oxide/zinc oxide composite as an electrochemical sensor for acetylcholine detection. Sci. Rep. **14** (2024)
35. Vatandost, E., Ghorbani-Hasan Saraei, A., Chekin, F., Raeisi, S.N., Shahidi, S.A.: Electrochemical sensor based on magnetic Fe_3O_4–reduced graphene oxide hybrid for sensitive detection of binaphthol. Russian J. Electrochem. **57**(5), 490–498 (2021)
36. Bhattacharyya, R., Prakash, O., Roy, S., Singh, A.P., Bhattacharya, T.K., Maiti, P., Bhattacharyya, S., Das, S.: Graphene oxide-ferrite hybrid framework as enhanced broadband absorption in gigahertz frequencies. Sci. Rep. 9(1), 12111 (2019)
37. Lukose, R., Zurauskiene, N., Balevicius, S., Stankevic, V., Keršulis, S., Plausinaitiene, V., Navickas, R.: Hybrid graphene-manganite thin film structure for magnetoresistive sensor application. Nanotechnology **30**(35), 355503 (2019)
38. Verma, M., Bahuguna, G., Shukla, S., Gupta, R.: SnO_2 Nanoparticle-reduced graphene oxide hybrids for highly selective and sensitive NO_2 sensors fabricated using a component combinatorial approach. ACS Appl. Nano Mater. **5**(12), 19053–19061 (2022)
39. Yusof, N.M., Ibrahim, S., Rozali, S.: Synthesis of graphene oxide-cobalt oxide hybrid materials for gas sensors application. Mater. Today: Proc. **66**, 2680–2684 (2022)
40. Chu, J., Li, X., Qi, J.: Hydrothermal synthesis of CdS microparticles–graphene hybrid and its optical properties. CrystEngComm **14**(5), 1881–1884 (2012)
41. Luan, X., Dai, H., Li, Q., Fugui, X., Mai, Y.: A hybrid photocatalyst composed of CdS nanoparticles and graphene nanoribbons for visible-light-driven hydrogen production. ACS Appl. Energy Mater. **5**(7), 8621–8628 (2022)
42. Chan, Y., Dahua, Z., Jun, Y., Linlong, T., Chongqian, L., Jun, S.: Fabrication of hybrid graphene/CdS quantum dots film with the flexible photo-detecting performance. Physica E **124**, 114216 (2020)
43. Alhammadi, S., Reddy, V.R.M., Gedi, S., Park, H., Sayed, M.S., Shim, J.-J., Kim, W.K.: Performance of graphene–cds hybrid nanocomposite thin film for applications in Cu (In, Ga) Se2 solar cell and h2 production. Nanomaterials **10**(2), 245 (2020)
44. Zhang, X., Xu, F., Zhao, B., Ji, X., Yao, Y., Wu, D., Gao, Z., Jiang, K.: Synthesis of CdS quantum dots decorated graphene nanosheets and non-enzymatic photoelectrochemical detection of glucose. Electrochimica Acta **133**, 615–622 (2014)
45. Paimard, G., Shamsipur, M., Gholivand, M.B.: A three-dimensional hybrid of CdS quantum dots/chitosan/reduced graphene oxide-based sensor for the amperometric detection of ceftazidime. Chemi. Papers **77**(1), 437–449 (2023)

46. Li, H., Xia, X., Guo, C., Ge, L., Li, F.: Laser-induced nano-bismuth decorated CdS–graphene hybrid for plasmon-enhanced photoelectrochemical analysis. Chem. Commun. **56**(89), 13784–13787 (2020)
47. Yue, N., Weicheng, J., Rongguo, W., Guomin, D., Yifan, H.: Hybrid nanostructures combining graphene–MoS_2 quantum dots for gas sensing. J. Mater. Chem. A **4**(21), 8198–8203 (2016)
48. Hu, J., Zou, C., Su, Y., Li, M., Hu, N., Ni, H., Yang, Z., Zhang, Y.: Enhanced NO_2 sensing performance of reduced graphene oxide by in situ anchoring carbon dots. J. Mater. Chem. C **5**(27), 6862–6871 (2017)
49. Van Tam, T., Hur, S.H., Chung, J.S., Choi, W.M.: Ultraviolet light sensor based on graphene quantum dots/reduced graphene oxide hybrid film. Sens. Actuators A: Phys. **233**, 368–373 (2015)
50. Li, L., Liu, D., Shi, A., You, T.: Simultaneous stripping determination of cadmium and lead ions based on the N-doped carbon quantum dots-graphene oxide hybrid. Sens. Actuators, B Chem. **255**, 1762–1770 (2018)
51. Yang, C., Hu, L.-W., Zhu, H.-Y., Ling, Y., Tao, J.-H., Xu, C.-X.: rGO quantum dots/ZnO hybrid nanofibers fabricated using electrospun polymer templates and applications in drug screening involving an intracellular H_2O_2 sensor. J. Mater. Chem. B **3**(13), 2651–2659 (2015)
52. Dong, X., Cao, Y., Wang, J., Chan-Park, M.B., Wang, L., Huang, W., Chen, P.: Hybrid structure of zinc oxide nanorods and three-dimensional graphene foam for supercapacitor and electrochemical sensor applications. RSC Adv. **2**(10), 4364–4369 (2012)
53. Li, Y., Ai, Q., Mao, L., Guo, J., Gong, T., Lin, Y., Guitai, W., Huang, W., Zhang, X.: Hybrid strategy of graphene/carbon nanotube hierarchical networks for highly sensitive, flexible wearable strain sensors. Sci. Rep. **11**(1), 21006 (2021)
54. Hong, Z., Zheng, Z., Kong, L., Zhao, L., Liu, S., Li, W., Shi, J.: Welded carbon nanotube-graphene hybrids with tunable strain sensing behavior for wide-range bio-signal monitoring. Polymers **16**(2), 238 (2024)
55. Wu, Q., Liu, W., Xiangrui, B., Haiyang, W., Wang, C., Li, X., Wang, X.: Graphene oxide/graphene hybrid film with ultrahigh ammonia sensing performance. Nanotechnology **32**(11), 115501 (2020)
56. Liu, K., Hu, J., Kong, Z., Hu, J., Tian, Z., Hou, J., Qin, J., et al.: High-yield, high-conductive graphene/nanocellulose hybrids prepared by Co-exfoliation of low-oxidized expanded graphite and microfibrillated cellulose. Compos. Part B: Eng. **225**, 109250 (2021)
57. Kim, H.C., Panicker, P.S., Kim, D., Adil, S., Kim, J.: High-strength cellulose nanofiber/graphene oxide hybrid filament made by continuous processing and its humidity monitoring. Sci. Rep. **11**(1), 13611 (2021)
58. Zhang, X., Liu, Q., Shi, X., Asiri, A.M., Luo, Y., Sun, X., Li, T.: TiO_2 nanoparticles–reduced graphene oxide hybrid: an efficient and durable electrocatalyst toward artificial N_2 fixation to NH_3 under ambient conditions. J. Mater. Chem. A **6**(36), 17303–17306 (2018)
59. Liu, S., Wang, Z., Zhang, Y., Dong, Z., Zhang, T.: Preparation of zinc oxide nanoparticle–reduced graphene oxide–gold nanoparticle hybrids for detection of NO_2. RSC Adv. **5**(111), 91760-91765 (2015)
60. High-performance gas sensors have been developed using ternary hybrids as zinc oxide nanoparticle–reduced graphene oxide–gold nanoparticle hybrids for the detection of NO_2 gas. The sensor shows quick response and high selectivity
61. Ling, L., Ruiyi, L., Guangli, W., Zhiguo, G., Zaijun, L.: Graphene quantum dots-NaYF4: Yb, Er hybrid with significant enhancement of upconversion emission for fluorescent detection of carcinoembryonic antigen with exonuclease III-aided target recycling amplification. Sens. Actuators B: Chem. **285**, 453–461 (2019)

Zero to Multi-dimensional Nanohybrid Materials: Synthesis and Characterization

Tijjani Adam and Subash C. B. Gopinath

1 Introduction

As research progresses on two-dimensional (2D) materials, there is a growing interest in identifying practical applications for them [1]. This interest is particularly significant in the energy sector, given the urgent need to transition to more sustainable practices and reduce carbon emissions. To optimize a material for a specific application, it's common practice to tailor its properties and characteristics to align with desired parameters. Historically, this has involved techniques like alloying metals or doping semiconductors, enabling precise control over material properties [2]. For 2D materials, creating heterostructures or hybrids with other nanomaterials emerges as a promising avenue for such controlled manipulation. The category of zero-dimensional materials (0D) primarily includes fullerenes, organic molecules, quantum dots, and atomic clusters. Hybrid structures combining 0D and 2D materials have been successfully implemented in various devices, often exhibiting enhanced performance, stability, or chemical properties [3]. However, creating such mixed-dimensional heterostructures demands meticulous control at the atomic level due to

T. Adam (✉)
Faculty of Electronic Engineering & Technology, Universiti Malaysia Perlis (UniMAP), 02600 Arau, Perlis, Malaysia
e-mail: tijjani@unimap.edu.my

Institute of Nano Electronic Engineering, Universiti Malaysia Perlis (UniMAP), 01000 Kangar, Perlis, Malaysia

Micro System Technology, Centre of Excellence (CoE), Universiti Malaysia Perlis (UniMAP), Pauh Campus, 02600 Arau, Perlis, Malaysia

S. C. B. Gopinath
Center for Global Health Research, Saveetha Medical College & Hospital, Saveetha Institute of Medical and Technical Sciences (SIMATS), Thandalam, Chennai 602 105, Tamil Nadu, India

Faculty of Chemical Engineering & Technology, Universiti Malaysia Perlis (UniMAP), 02600 Arau, Perlis, Malaysia

S. C. B. Gopinath and M. M. Ramli (eds.), *Hybrid-Nanomaterials*, Engineering Materials, https://doi.org/10.1007/978-981-97-9022-7_5

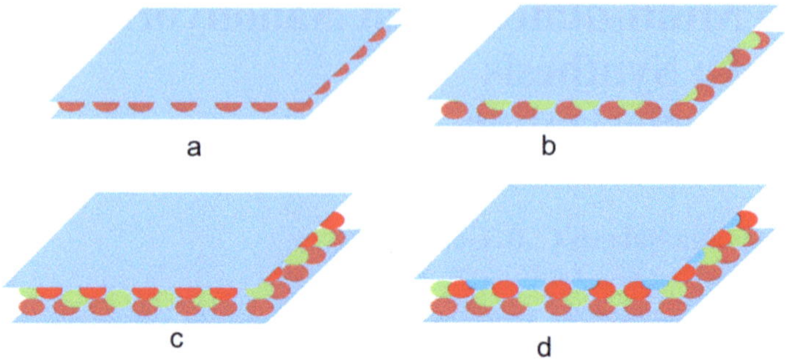

Fig. 1 Zero-Dimensional (0D) Hybrid Materials, One-Dimensional (1D) Hybrid Materials, Two-Dimensional (2D) Hybrid Materials and Three-Dimensional (3D) Hybrid Materials

differences in crystal structure and thermal stability, as well as the need for a high-quality interface. Consequently, 0D-2D materials pose unique challenges and opportunities for developing innovative synthesis methods. In recent years, researchers have synthesized these heterostructures using diverse approaches such as wet chemistry, thermal deposition, and chemical reaction pathways [4]. Typically, the more stable 2D material is synthesized first, followed by the introduction of the 0D material. The experimental parameters, including temperature, atmosphere, and pressure, vary depending on the specific 0D material and the desired concentration and uniformity on the 2D host. Unlocking the capabilities of a 0D-2D composite necessitates designing a system that maximizes the unique strengths of each component [5]. Ideally, this results in a combined effect or interaction that surpasses what each part can achieve individually. The structure and manner in which the composite components interact are largely influenced by the synthesis method utilized, a topic that will be discussed in the following section [6]. The chapter of the book discussed zero to multi-dimensional nanohybrid materials, including Zero-Dimensional (0D) Hybrid Materials, One-Dimensional (1D) Hybrid Materials, Two-Dimensional (2D) Hybrid Materials and Three-Dimensional (3D) Hybrid Materials as shown in Fig. 1.

2 Surface Characteristics and Interactions in Zero to Multi-dimensional Nanohybrid Materials

Zero- to multidimensional nanohybrid materials are complex structures that blend multiple dimensional nanomaterials (0D, 1D, 2D, and 3D) to take advantage of their distinct features for improved performance in a variety of applications [7]. Understanding the surface properties and interactions of these nanohybrids is critical for improving their functionality. Here's a summary of these topics, along with an example involving TEM imaging and chemical formulas. Surface Characteristics, Surface Functionalization Surface functionalization includes adding chemical groups to the surface of nanomaterials to increase their stability, dispersibility, and compatibility with other materials [8]. For example, Silicon nanoparticles (SiNPs) can be functionalized with hydroxide groups (OH) to connect to organic or inorganic substrates, and Surface Charge as illustrated in Fig. 2. The surface charge, generally represented by the zeta potential, influences the colloidal stability and interactions of nanomaterials. For example, negatively charged surfaces can reject particles with similar charges, avoiding aggregation. Nanomaterials have a high surface area-to-volume ratio, which provides more active places for interaction [9]. This is especially significant in catalysis and sensing applications. Surface Defects and Vacancies, Defects and vacancies on the surface can improve reactivity and catalysis. For example, oxygen vacancies on SiO_2 nanoparticles improve photocatalytic activity [10].

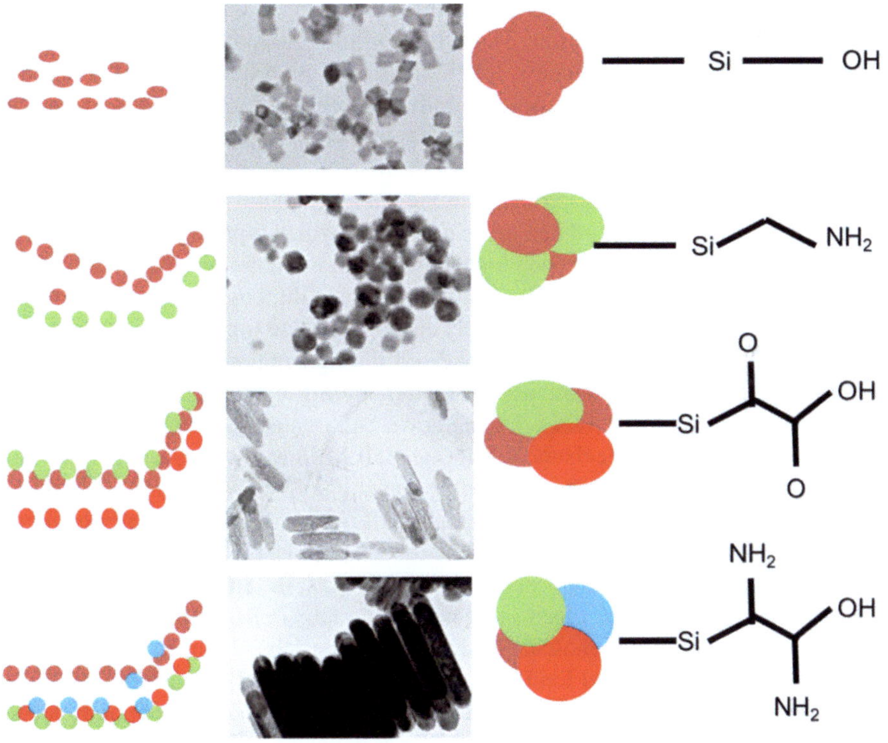

Fig. 2 Surface characteristics and interactions in Zero to multi-dimensional nanohybrid materials

2.1 Zero-Dimensional (0D) Hybrid Materials

Materials with all dimensions under the nanometer range are classified as zero-dimensional (0D) nanomaterials. Examples of 0D nanomaterials include nanoparticles, quantum dots, carbon nanodots, and fullerenes. Over the past decades, significant progress has been made in the field of 0D nanomaterials [11]. These materials possess inherent structural properties, such as a high surface-to-volume ratio and ultra-small size, which provide a high number of active sites per unit mass. Common examples include carbon quantum dots (CQDs), graphene quantum dots (GQDs), fullerenes, inorganic quantum dots (IQDs), polymer dots (Pdots), and noble metal nanoparticles [12]. The reduction in dimensions grants these nanomaterials unique properties not seen in their bulk counterparts, typically forming in spherical or quasi-spherical shapes with diameters less than 100 nm. Carbon-based nanomaterials are particularly studied due to their low cost, low inherent toxicity, and flexible surface functionalization. These 0D carbon nanomaterials offer enhanced electrical and optical performance, low toxicity, and high quantum yield. Carbon-based quantum dots exhibit enhanced chemiluminescence, fluorescence, and electro-chemiluminescence properties, making CQDs useful for applications in bioimaging and targeted drug delivery [13].

Zero-dimensional (0D) materials, also known as quantum dots or nanoparticles, have all three dimensions confined to the nanoscale, which typically ranges from 1 to 10 nm. Quantum confinement phenomena provide these materials distinct electrical, optical, and chemical features. The synthesis of zero-dimensional materials incorporates numerous methodologies, broadly categorized into top-down and bottom-up processes. The synthesis process are realised through major two approaches: top down and bottom up; The synthesis or manufacturing of these materials is either done top-down, which involves breaking down bulk materials into nanoscale particles [14]. They include Mechanical Milling: Ball Milling: A high-energy ball mill grinds bulk materials into nanoparticles by causing collisions and friction with the substance. Cryomilling is a low-temperature ball milling technique that uses liquid nitrogen to prevent particle aggregation and oxidation. Laser Ablation: A high-power laser beam vaporizes a bulk substance in a liquid or gas, forcing it to condensate into nanoparticles. Lithography: Electron Beam Lithography (EBL): This technique uses an electron beam to form nanoscale designs on a substrate, which are then etched to make nanoparticles. Focused Ion Beam (FIB): Similar to EBL, but using ion beams. In the other hand, Bottom-up approaches create nanoparticles one atom or molecule at a time [15, 15]. This includes: Chemical vapor deposition (CVD): Precursor gases react with a substrate to produce nanoparticles. Particle size and distribution are controlled by parameters such as temperature, pressure, and gas flow rate.

Sol–gel Synthesis: Metal alkoxides or metal salts undergo hydrolysis and polycondensation processes in solution, resulting in the production of a colloidal suspension (sol) that eventually becomes a gel. Drying and calcination produce nanoparticles. Hydrothermal and Solvothermal Techniques: Reactions conducted in a sealed vessel at high temperatures and pressures aid in the nucleation and development of nanoparticles. Solvothermal procedures employ organic solvents, whereas hydrothermal processes use water. Interactions in NanoHybrid Materials for chemical bonding. Covalent, ionic, or metallic connections between the various components of nanohybrids enable strong contact and stability [17]. For example, covalent bonding between graphene oxide and polymer chains in a nanocomposite. Furthermore, weak van der Waals forces can play an important role in nanohybrid assembly and interaction, such as the stacking of 2D materials like graphene. electrostatic Interactions: Electrostatic interactions between charged nanomaterials influence the creation and stability of nanohybrids. Positively charged gold nanoparticles can interact with negatively charged DNA strands for biosensing purposes. Hydrogen bonding Hydrogen bonding can help to stabilize hybrid structures, particularly in bio-nanohybrids, which combine biological molecules and nanomaterials. Characteristics: Zero-dimensional hybrid materials are typically composed of nanoscale particles or clusters that are confined in all three dimensions [18]. The typica nano hybrid is graphene Oxide-Gold Nanoparticle Hybrid, Chemical Formula: Graphene oxide (GO) functionalized with gold nanoparticles (AuNPs). The TEM image would typically show gold nanoparticles (small, dark spots) evenly distributed on the graphene oxide sheet (large, lighter area with a wrinkled appearance). Quantum dots, nanospheres, core–shell nanoparticles. properties: They exhibit quantum confinement effects, size-dependent properties, and high surface-to-volume ratios. Their properties can be tuned by varying the size and composition of the nanoparticles [19].

2.2 One-Dimensional (1D) Hybrid Materials

Characteristics: One-dimensional hybrid materials have a nanoscale structure in one dimension while being extended in the other two dimensions. Examples: Nanowires, nanorods, nanotubes. Properties: They have high aspect ratios, good elongated-dimension electrical and thermal conductivity, and prospective uses in energy storage, sensors, and electronics. One-dimensional (1D) nanomaterials' electrical, mechanical, chemical, and magnetic properties are influenced by their large length-to-diameter ratio. Furthermore, 1D nanomaterials have benefits over their bulk equivalents due to their frequent high crystallinity, superior homogeneity and dispersion, and simple production. The enormous surface area, volume change adaptability, and porous or hollow architectures of 1D nanomaterials make them especially well suited for applications involving hydrogen storage [20].

Porous nanostructures can be found in single-wall carbon nanotubes, which are made up of sp2 hybridized atoms. Because of their larger specific surface area and lower mass density, carbon-based 1D nanomaterials such as nanotubes, rods, and wires have received a lot of attention from researchers for hydrogen storage. Researchers have reported that nanotubes built from inorganic materials are promising for reversible hydrogen storage because pure carbon nanotubes have low hydrogen sorption capability. Hydrogen can be held in nanotubes in both the internal and external cavities, as well as the spaces between them. Carbon nanotubes' hydrogen storage capacity can be improved by increasing their activity, resulting in larger porosity and specific surface area. Nanorods are one-dimensional nanomaterials that provide a guided channel for charge carrier transfer [21]. Furthermore, because of the quantum size effect, the bandgap of nanorods can be adjusted by varying their radius. Due to the fact that doping facilitates better sorption and desorption processes, studies have demonstrated that doped nanorods can increase hydrogen storage capacity. Because of its exceptional electrochemical activity, mechanical and chemical stability, and suitability for use in energy storage and transfer devices, nanorods [22].

2.3 *Two-Dimensional (2D) Hybrid Materials*

Two-dimensional hybrid materials are stretched in the third dimension while retaining a nanoscale structure in the two dimensions. Examples include graphene oxide, graphene, and transition metal dichalcogenides (such as MoS2 and WS2). Characteristics: They possess distinct mechanical, optical, and electronic qualities, including strong electron mobility, mechanical pliability, and optical transparency. Applications in photonics, electronics, and energy storage seem promising. Since 2004, 2D graphene sheets have made significant advancements, making them an important class of 2D nanomaterials. The remarkable electrical and thermal conductivities of a single layer of carbon atoms organized in a honeycomb lattice were shown by researchers Andre Geim and Konstantin Novoselov [23].

At higher temperatures (over 100 K), graphene's charge carrier dynamics exhibit unusual features such as quantum oscillations due to its ballistic transport over extended distances, which is governed by a linear energy–momentum connection. The Nobel Prize in Physics given in 2010 for pioneering work with 2D graphene greatly increased interest in nanomaterials and nanotechnology among the general public and scientists. 2D nanomaterials are made up of planar structures that are atomic layer thick, joined to one another by weak van der Waals forces. Their small atomic thickness and strong in-plane covalent connections account for their exceptional mechanical strength and flexibility. Because of their malleability, these layered materials can be easily shaped into 2D nanosheets and nanoplates. Ultrathin 2D nanomaterials have sheet-like structures with thicknesses of about 5 nm and transverse dimensions of a few hundred nanometers [24]. They are very beneficial for a variety of applications due to their unique topography, increased surface area, and anisotropic chemical and physical characteristics. 2D nanomaterials are especially helpful in situations where surface activity is important because of their huge flat surface areas. The uses of 2D nanomaterials in sensors, optoelectronics, electronics, and energy storage and conversion have increased recently. Two-dimensional nanomaterials possess distinct physicochemical features due to the one-dimensional quantum confinement of electrons within them. These materials allow the production of traditional two-dimensional electron systems, which are widely employed in high electron mobility transistors, by confining electrons within their ultrathin areas. The enhanced transport characteristics and distinct attributes of quantum transport offer novel prospects for the development of next-generation electronics and optoelectronics devices [25].

2.4 Three-Dimensional (3D) Hybrid Materials:

Among the 3D new material is graphene, since 2004, 2D graphene sheets have seen tremendous progress, making them essential elements in the field of 2D nanomaterials. The remarkable electrical and thermal conductivities of a single layer of carbon atoms organized in a honeycomb lattice were shown by researchers Andre Geim and Konstantin Novoselov. Graphene's charge carrier dynamics are characterized by a linear energy–momentum connection. Additionally, the material's ballistic transport across large distances gives rise to distinct transport characteristics, including quantum oscillations at temperatures higher than 100 K. The Nobel Prize in Physics given in 2010 for these pioneering experiments greatly increased interest in nanomaterials and nanotechnology among professionals and the general public. 2D nanomaterials are made up of planar structures that are atomic layer thick, joined to one another by weak van der Waals forces. Their thinnest atomic layer and strong in-plane covalent bonding account for their exceptional mechanical strength and flexibility [25].

This is because of their malleability, these layered materials can be easily shaped into 2D nanosheets and nanoplates. Ultrathin 2D nanomaterials have sheet-like structures with thicknesses of about 5 nm and transverse dimensions of a few hundred nanometers. They are very beneficial for a variety of applications due to their unique topography, increased surface area, and anisotropic chemical and physical characteristics. 2D nanomaterials are especially helpful in situations where surface activity is important because of their huge flat surface areas. Advancements in 2D nanomaterials have led to an expansion of their applications in energy storage and conversion, sensors, optoelectronics, electronics, and other fields. Two-dimensional nanomaterials possess distinct physicochemical features due to the one-dimensional quantum confinement of electrons within them. The ultrathin regions of these materials enable the creation of classical two-dimensional electron systems, which are widely used in high electron density applications. Next-generation electronics and optoelectronics devices have new possibilities thanks to quantum transport's exceptional transport capabilities and special characteristics. Three-dimensional hybrid materials consist of a three-dimensionally extended nanoscale structure. Aerogels, covalent organic frameworks (COFs), and metal–organic frameworks (MOFs) are a few examples. Features: They incorporate the high surface area, porosity, structural stability, and adjustable chemical functionality of their component materials. Applications for them include sensing, catalysis, and gas storage [26].

2.5 *Synthesis of Zero to Multi-dimensional Nanohybrid Materials*

The synthesis of zero- to multidimensional nanohybrid materials entails combining various types of nanomaterials, each with its own set of properties, to produce composites with improved or novel functions. These materials can include zero-dimensional (0D) quantum dots and nanoparticles, one-dimensional (1D) nanowires and nanotubes, two-dimensional (2D) nanosheets, and three-dimensional (3D) bulk nanostructures. The synthesis process usually consists of multiple steps to ensure effective integration and interaction between various components. A typical methodology for fabricating nanostructured materials can be divided into two main categories: (i) the top-down approach and (ii) the bottom-up approach, as illustrated in Fig. 2. The top-down approach begins with bulk material at the microscale level and uses various techniques to break it down into nanoscale particles or structures. Conversely, the bottom-up approach starts with the smallest possible units and builds up to the desired nanoscale particles or structures. Both approaches have their own advantages and disadvantages [27].

2.6 *Characterization of Zero to Multi-dimensional Nanohybrid Materials*

0D Nanoparticles: These include nanoparticles like quantum dots and fullerenes, which have all dimensions in the nanoscale. They exhibit high surface-to-volume ratios and unique optical properties, making them useful in imaging and drug, while, 1D Nanoparticles. This category includes nanowires and nanotubes, which have one dimension outside the nanoscale. Their high aspect ratios improve electrical and thermal conductivities, and they are used in electronics and energy storage, 2D Nanoparticles: Graphene and MXenes are prime examples. These materials have exceptional electrical, thermal, and mechanical properties due to their planar structures [28]. They are pivotal in electronics, sensors, and energy applications. 3D Nanoparticles: These include bulk nanostructured materials, where all three dimensions are in the nanoscale. They often have complex architectures providing high surface areas and are used in catalysis and structural applications. Each type of nanomaterial offers distinct advantages due to its dimensional properties, impacting its application in various technological fields. Each type of nanomaterial offers distinct advantages due to its dimensional properties, impacting its application in various technological fields [29].

0D Nanoparticles: These include quantum dots and fullerenes, characterized by their nanoscale dimensions in all directions. Their high surface-to-volume ratios and unique optical properties make them suitable for applications in imaging and drug delivery [30]. However, their small size can limit their structural stability and make them prone to aggregation. 1D Nanoparticles: Nanowires and nanotubes fall into this category, with one dimension outside the nanoscale. They possess high aspect ratios that improve electrical and thermal conductivities, making them ideal for electronics and energy storage. Nonetheless, their fabrication can be challenging, and they may suffer from issues related to alignment and integration into devices. 2D Nanoparticles: Examples like graphene and MXenes exhibit exceptional electrical, thermal, and mechanical properties due to their planar structures. These materials are pivotal in electronics, sensors, and energy applications, offering large surface areas and robust mechanical properties. However, their synthesis and scalability can be complex, and maintaining their unique properties in practical applications can be difficult. 3D Nanoparticles: Bulk nanostructured materials, where all three dimensions are in the nanoscale, often have complex architectures providing high surface areas. These are used in catalysis and structural applications. While they offer excellent mechanical strength and chemical activity, their fabrication processes are often more complex and expensive, and ensuring uniformity can be challenging [31].

3 Type and Application of Zero to Multi-dimensional Nanohybrid Materials

Type and application of zero to multi-dimensional nanohybrid materials, much emphasis has been paid to the new prospects that nanotechnology will provide in the next generation of technological developments. The application of modified nanomaterials for the selective detection of biomolecules has accelerated progress in this research, especially in the diagnostic domain. The wide range of applications of nanoengineering technology, particularly in biosensors, demonstrates the significant impact it has had [32]. Traditionally, the number of dimensions that fall within the nanoscale range is used to classify nanomaterials. Dimension, morphology, size, form, structure, and phase of matter are therefore regarded as defining characteristics for the classification of nanomaterials. The dimensionality-based classification of nanomaterials is shown in Fig. 3. As an example, nanoparticles are nanomaterials having all three dimensions in the range of the nanoscale, with the shortest and longest axes having the least amount of difference. Nanofibers, on the other hand, are single-dimensional objects that fall inside the nanoscale range. They can also have hollow or solid structures, such as nanotubes or nanorods. In the nanoscale range, nanoplates and nanoribbons likewise have one dimension. Nanomaterials can be roughly classified into two categories based on their morphology: consolidated materials and nano-dispersions [33]. Nanopowders and nanoparticles are two instances of one-dimensional nanodispersive systems. The latter group is further separated into supermolecules, nanoclusters, nanocrystals, and nanotubes. Solid materials with a minimum of one zone with nanoscale dimensions and at least one distinct chemical or physical region are called nanocomposites. A liquid or solid matrix containing a gaseous phase, at least one of which has nanoscale dimensions, makes up a nanofoam.

Nanocrystalline materials are made of nanoscale crystal grains, whereas nanoporous materials are solids containing cavities of nanoscale size [34]. There are examples of nanomaterials that fall into the categories of one-dimensional (1D), two-dimensional (2D), three-dimensional (3D), and zero-dimensional (0D). The several nanomaterials used so far for point-of-care (POC) applications are shown in Table 1 below. This is a summary of newly created biosensors that use 0D, 1D, 2D, and 3D nanomaterials to detect a range of disease biomarkers and associated analytes. According to Table 1 analysis, researchers prefer to use gold and graphene nanomaterials for POC biosensing devices because of their improved optical and electrical properties, among other benefits [35] (Figs. 4 and 5).

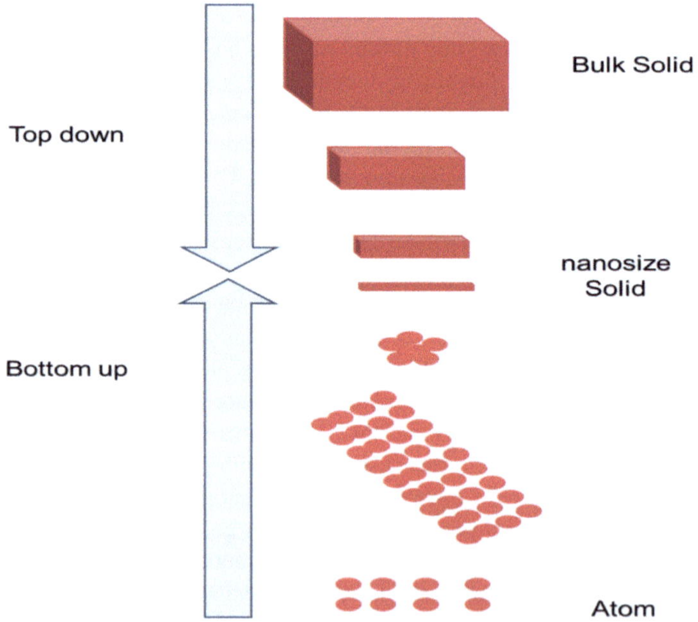

Fig. 3 Bottom-up and top-down process for realizing zero to multi-dimensional hybrid nanomaterial

Table 1 Characteristics comparison of 0D, 1D, 2D and 3D

Nanomaterials and dimension	Characteristics
Zero-dimensional	Excellent in applications requiring high surface reactivity and unique optical properties but face stability issues
One-dimensional	Offer superior conductive properties ideal for electronic applications but require precise fabrication techniques
Two-dimensional	Combine large surface areas with exceptional mechanical and electrical properties, suitable for advanced sensors and energy storage, though their production can be challenging
Three-dimensional	Provide robust structural integrity and high catalytic activity but are often more complex and costly to produce

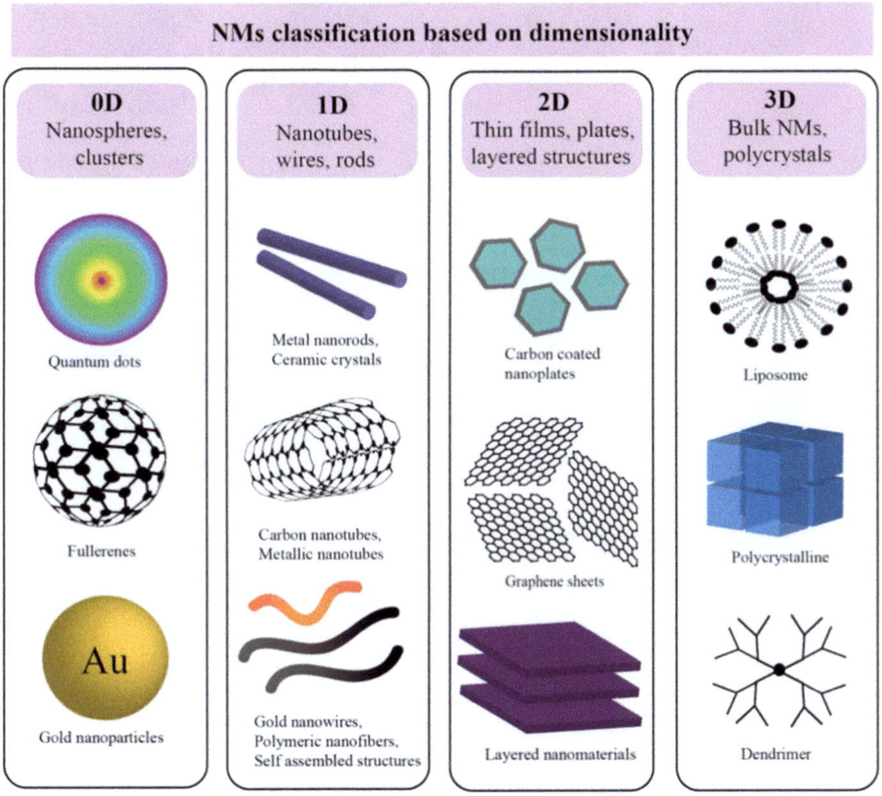

Fig. 4 Characteristics comparison of 0D, 1D, 2D and 3D

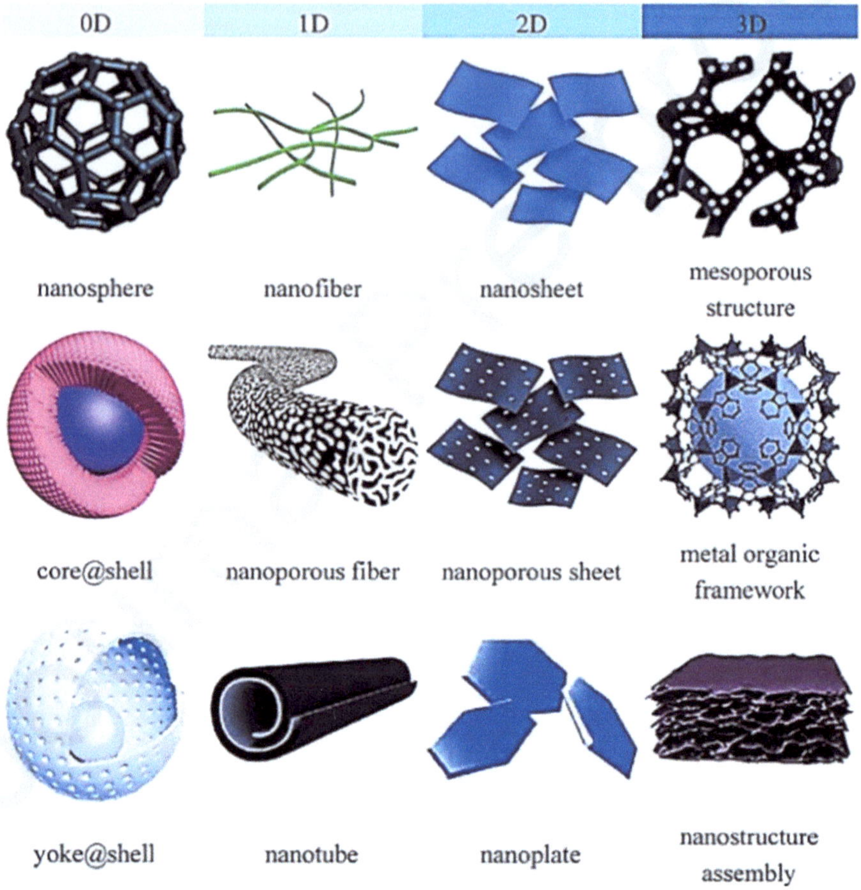

Fig. 5 Types of 0D, 1D, 2D and 3D and zero dimension to multidimensional hybrid nanomaterials

4 Conclusion

In summary, the book chapter from zero-dimensional to multi-dimensional nanohybrid materials represents a significant advancement in nanotechnology. Through innovative synthesis techniques and meticulous characterization methods, researchers have expanded the realm of possibilities for nanomaterials in various applications. The synthesis of zero-dimensional nanohybrids, such as nanoparticles and nanoclusters, has laid the foundation for exploring their unique properties and applications. These nanomaterials exhibit remarkable characteristics, including size-dependent properties, high surface area-to-volume ratios, and tunable surface chemistry, making them suitable for a wide range of applications in catalysis, sensing, drug delivery, and more. Moving beyond zero-dimensional structures, the exploration of one-dimensional nanohybrids, such as nanowires and nanotubes, has opened new avenues for engineering materials with enhanced mechanical, electrical, and thermal properties. These elongated nanostructures offer opportunities for applications in nanoelectronics, energy storage, and reinforcement in composite materials. The advancement into two-dimensional nanohybrids, exemplified by nanosheets and nanoplates, has further expanded the horizon of nanomaterials. These ultrathin materials possess unique properties, including high aspect ratios, large surface areas, and excellent mechanical flexibility, making them promising candidates for applications in flexible electronics, membrane technologies, and optoelectronic devices. Finally, the synthesis and characterization of three-dimensional nanohybrids have propelled nanotechnology into the realm of advanced materials with intricate architectures and tailored functionalities. These complex nanostructures, such as hierarchical assemblies and nanofoams, exhibit synergistic properties derived from their multi-dimensional architectures, paving the way for transformative applications in tissue engineering, environmental remediation, and beyond. Overall, the synthesis and characterization of zero to multi-dimensional nanohybrid materials represent a paradigm shift in materials science, offering unprecedented control over the structure–property relationships at the nanoscale. As researchers continue to push the boundaries of nanotechnology, these innovative materials hold the promise of revolutionizing various fields and addressing complex societal challenges in the years to come.

References

1. Poh, T.Y., et al.: Inhaled nanomaterials and the respiratory microbiome. Particle Fibre Toxicol. **15**(1), 1–16 (2018) [Online]. Available: https://doi.org/10.1186/s12989-018-0282-0%0A, https://doi.org/10.1186/s12989-018-0282-0
2. Yang, M., et al.: Zero→two-dimensional metal nanostructures: An overview on methods of preparation, characterization, properties, and applications. Nanomaterials **11**(8) (2021). https://doi.org/10.3390/nano11081895
3. Iswanti, F.C., Nurulita, I., Djauzi, S., Sadikin, M., Witarto, A.B., Yamazaki, T.: Preparation, characterization, and evaluation of chitosan-based nanoparticles as CpG ODN carriers. Biotechnol. Biotechnol. Equip. **33**(1), 390–396 (2019). https://doi.org/10.1080/13102818.2019.1578690
4. Kang, E., Kim, D., Suhito, I.R., Lee, W., Song, I., Kim, T.: Platforms to control mesenchymal stem cell differentiation. Biomater. Res. **22**, 1–12 (2018)
5. Zheng, J., et al.: 2D titanoniobate-titaniumcarbide nanohybrid anodes for ultrafast lithium-ion batteries. J. Power Sour. **512**, 230523 (2021). https://doi.org/10.1016/j.jpowsour.2021.230523
6. Alzate, M., Gamba, O., Daza, C., Santamaria, A., Gallego, J.: Iron/multiwalled carbon nanotube (Fe/MWCNT) hybrid materials characterization: thermogravimetric analysis as a powerful characterization technique. J. Therm. Anal. Calorim. **147**(22), 12355–12363 (2022). https://doi.org/10.1007/s10973-022-11446-w
7. Baig, N.: Two-dimensional nanomaterials: a critical review of recent progress, properties, applications, and future directions. Compos. Part A: Appl. Sci. Manuf. **165**, 107362 (2023). https://doi.org/10.1016/j.compositesa.2022.107362
8. Sahay, R., Reddy, V.J., Ramakrishna, S.: Synthesis and applications of multifunctional composite nanomaterials. Int. J. Mech. Mater. Eng. **9**(1), 1–13 (2014). https://doi.org/10.1186/s40712-014-0025-4

9. Chang, F., Davies, G.L.: From 0D to 2D: Synthesis and bio-application of anisotropic magnetic iron oxide nanomaterials. Progress Mater. Sci. **144**, 101267 (2024). https://doi.org/10.1016/j.pmatsci.2024.101267
10. Pareek, V., Bhargava, A., Panwar, J.: Biomimetic approach for multifarious synthesis of nanoparticles using metal tolerant fungi: a mechanistic perspective. Mater. Sci. Eng. B **262**, 114771 (2020). https://doi.org/10.1016/j.mseb.2020.114771
11. Xing, C., et al.: Constructing and regulating nanochannels in two-dimensional-material-based membranes for specified separation applications. Microstructures **3**(4) (2023). https://doi.org/10.20517/microstructures.2023.11
12. Tiwari, J.N., Tiwari, R.N., Kim, K.S.: Zero-dimensional, one-dimensional, two-dimensional and three-dimensional nanostructured materials for advanced electrochemical energy devices. Prog. Mater. Sci. **57**(4), 724–803 (2012). https://doi.org/10.1016/j.pmatsci.2011.08.003
13. Paras, et al.: A review on low-dimensional nanomaterials: nanofabrication, characterization and applications. Nanomaterials **13**(1), 1–44 (2023). https://doi.org/10.3390/nano13010160
14. B. Bera and D. Banerjee, "Preparation and Characterization of Amorphous Carbon Nanotube-MoS2 Nanohybrid," no. 8, pp. 5–8, 2019.
15. Chazapis, N., et al.: Iridium-based nanohybrids: synthesis, characterization, optical limiting, and nonlinear optical properties. Nanomaterials **13**(14) (2023). https://doi.org/10.3390/nano13142131
16. Alkhafaji, E.N., Oda, N.A., Ahmad, L.M.: Characterization of silver nanohybrid with layers double hydroxide and demonstration inhibition of antibiotic-resistance staphylococcus aureus. Egypt. J. Chem. **65**(6), 143–157 (2022). https://doi.org/10.21608/EJCHEM.2022.97856.4564
17. Wang, Z., Hu, T., Liang, R., Wei, M.: Application of zero-dimensional nanomaterials in biosensing. Front. Chem. **8**(April), 1–19 (2020). https://doi.org/10.3389/fchem.2020.00320
18. ChemElectroChem–2024–Breczko–Zero-Dimensional Carbon Nanomaterials for Electrochemical Energy Storage.pdf
19. Yeşildağ, N., Ünsal, Ö.F., Gömeç, R., Bedeloğlu, A.Ç.: Detecting acetone from breath using a $PrFeO_3$-doped PANi/TiO_2-coated PAN nanofiber sensor for non-invasive diabetic diagnosis. J. Innovat. Eng. Nat. Sci. **2**(2), 153–166 (2023). https://doi.org/10.29228/jiens.70059
20. Dawadi, S., et al.: Current research on silver nanoparticles: synthesis, characterization, and applications. J. Nanomater. **2021** (2021). https://doi.org/10.1155/2021/6687290
21. "Zero-dimensional-carbon-nanomaterials-a-carbon-dots-and-b-fullerene-One-dimensional
22. South, F., Usf, T., Graduate, D., Usf, G., Theses, D., Alamro, T.S.: Synthesis and characterization of molybdenum disulfide/synthesis and characterization of molybdenum disulfide/conducting polymer nanocomposite materials for supercapacitor conducting polymer nanocomposite materials for supercapacitor applications application (2019) [Online]. Available: https://digitalcommons.usf.edu/etd
23. Majumder, P., Gangopadhyay, R.: Evolution of graphene oxide (GO)-based nanohybrid materials with diverse compositions: an overview. R. Soc. Chem. **12**(9) (2022)
24. Falola, T.O.: Nanoparticles modified electrodes: synthesis, modification, and characterization—a review. World J. Nano Sci. Eng. **12**(03), 29–62 (2022). https://doi.org/10.4236/wjnse.2022.123003
25. Maity, D., Sahoo, S.R., Saha, S.: Synthesis and characterization of nanomaterials for electrochemical sensors. ACS Symp. Ser. **1437**, 193–222 (2023). https://doi.org/10.1021/bk-2023-1437.ch009
26. Stefan, M., et al.: Synthesis and characterization of Fe_3O_4 @ZnS and Fe_3O_4 @Au@ZnS core-shell nanoparticles. Appl. Surf. Sci. **288**, 180–192 (2014). https://doi.org/10.1016/j.apsusc.2013.10.005
27. Al-Sabagh, A.M., Betiha, M.A., Osman, D.I., Mahmoud, T.: Synthesis and characterization of nanohybrid of poly(Octadecylacrylates derivatives)/montmorillonite as pour point depressants and flow improver for waxy crude oil. J. Appl. Polym. Sci. **136**(17), 1–15 (2019). https://doi.org/10.1002/app.47333
28. Hasan, M., et al.: Biological entities as chemical reactors for synthesis of nanomaterials: progress, challenges and future perspective. Mater. Today Chem. **8**, 13–28 (2018). https://doi.org/10.1016/j.mtchem.2018.02.003

29. Byakodi, M., et al.: Emerging 0D, 1D, 2D, and 3D nanostructures for efficient point-of-care biosensing. Biosens. Bioelectron. X **12**, 100284 (2022). https://doi.org/10.1016/j.biosx.2022.100284
30. Vahdati, M., Tohidi Moghadam, T.: Synthesis and characterization of selenium nanoparticles-lysozyme nanohybrid system with synergistic antibacterial properties. Sci. Rep. **10**(1), 1–10 (2020). https://doi.org/10.1038/s41598-019-57333-7
31. Syed Ibrahim, G.P., Isloor, A.M., Ismail, A.F., Farnood, R.: One-step synthesis of zwitterionic graphene oxide nanohybrid: Application to polysulfone tight ultrafiltration hollow fiber membrane. Sci. Rep. **10**(1), 1–13 (2020). https://doi.org/10.1038/s41598-020-63356-2
32. Cobianu, C., et al.: Sonochemically synthetized ZnO-graphene nanohybrids and its characterization. Rev. Adv. Mater. Sci. **59**(1), 176–187 (2020). https://doi.org/10.1515/rams-2020-0013
33. Dkhar, D.S., Kumari, R., Chandra, P.: Chemically engineered unzipped multiwalled carbon nanotube and rGO nanohybrid for ultrasensitive picloram detection in rice water and soil samples. Sci. Rep. **13**(1), 1–14 (2023). https://doi.org/10.1038/s41598-023-34536-7
34. Gunjakar, J.L., Kim, T.W., Kim, I.Y., Lee, J.M., Hwang, S.J.: Highly efficient visible light-induced O_2 generation by self-assembled nanohybrids of inorganic nanosheets and polyoxometalate nanoclusters. Sci. Rep. **3**, 1–8 (2013). https://doi.org/10.1038/srep02080
35. Crisci, M., et al.: Nanostructured 2D WS2@PANI nanohybrids for electrochemical energy storage. Front. Chem. **10**, 1–11 (2022). https://doi.org/10.3389/fchem.2022.1000910

Graphene-Based Hybrid Materials and Their Applications

Ismariza Ismail, Muhammad Mahyiddin Ramli, and Norizah Abd Karim

1 Introduction to Graphene

Graphene, a revolutionary material, was initially discovered in 2004 by Andre Geim and Konstantin Novoselov. Their pioneering technique involved using Scotch tape on a block of graphite to isolate single layers of graphene. This breakthrough earned them the Nobel Prize in Physics in 2010. Graphene has since captivated the scientific world due to its extraordinary properties; it is stronger than diamond, more conductive than copper, and incredibly thin, being a million times thinner than paper. Graphene is a two-dimensional material comprising a single sheet of carbon atoms arranged in a honeycomb structure. It serves as a fundamental building block for various carbon allotropes. Each carbon atom in graphene forms covalent bonds with three neighboring carbon atoms, resulting in an atom-thick sheet that is considered a single molecule and acts as the foundational unit for other carbon structures. The unique structure of graphene is depicted in Fig. 1.

The honeycomb structure of graphene serves as the structural basis for several carbon allotropes. Stacking multiple layers of graphene yields graphite, a soft black solid used in applications such as pencils, lubricants, and batteries. Rolling graphene sheets produce carbon nanotubes, which exhibit exceptional strength and lightness,

I. Ismail
Faculty of Chemical Engineering & Technology, University Malaysia Perlis, Kawasan Perindustrian Jejawi (UniMAP), Kompleks Pusat Pengajian Jejawi 3, 02600 Arau, Perlis, Malaysia

I. Ismail · M. M. Ramli · N. Abd Karim (✉)
Institute of Nano Electronic Engineering, Universiti Malaysia Perlis (UniMAP), 01000 Kangar, Perlis, Malaysia
e-mail: norizah@unimap.edu.my

N. Abd Karim
Faculty of Mechanical Engineering Technology, University Malaysia Perlis, Pauh Putra Campus, 02600 Arau, Perlis, Malaysia

S. C. B. Gopinath and M. M. Ramli (eds.), *Hybrid-Nanomaterials*, Engineering Materials, https://doi.org/10.1007/978-981-97-9022-7_6

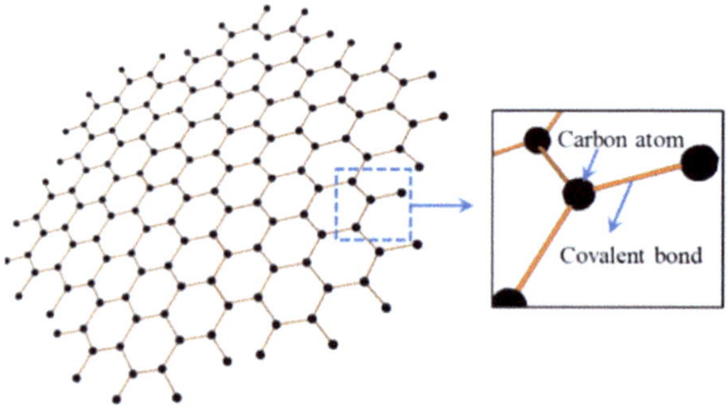

Fig. 1 Basic structure of graphene

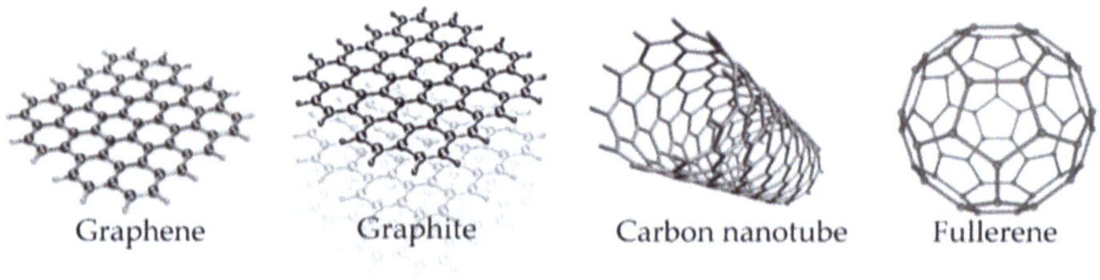

Fig. 2 The structures of graphene, graphite, carbon nanotubes, and fullerenes

making them ideal for reinforcing materials. Carbon nanotubes can exist as single-walled structures, comprising a single layer of carbon atoms, or multi-walled structures, consisting of multiple concentric layers of carbon atoms. Conversely, wrapping graphene sheets into closed spherical or cage-like structures results in the formation of fullerenes, also known as buckyballs or buckytubes, which find diverse applications, particularly in medicine. The structures of graphene, graphite, carbon nanotubes, and fullerenes are depicted in Fig. 2 for comparison.

Graphene's exceptional properties extend beyond its structural versatility. It exhibits outstanding electrical conductivity at room temperature due to the sp2 hybridization of its carbon atoms, allowing electrons to move freely. This property makes graphene well-suited for advanced transistors and high-speed electronic applications. Additionally, graphene is an excellent heat conductor, making it suitable for heat dissipation systems in electronics and thermal management in buildings [1]. Theoretical calculations suggest that graphene could be up to 200 times stronger than steel per unit weight, making it ideal for aerospace components and sports equipment. Furthermore, a single layer of graphene is nearly transparent, allowing it to transmit all light through it. This property, combined with its excellent conductivity, makes graphene a suitable candidate for transparent electrodes in solar cells and touchscreens [2, 3]. Graphene also serves as an excellent gas barrier material due to its tightly packed atomic structure. Its impermeability to most gases makes it highly effective in applications requiring gas separation membranes or ultra-thin barrier films [4]. These outstanding properties position graphene as a highly promising material with broad applicability across diverse fields, including electronics, energy storage, aerospace, and medicine.

2 Synthesis Methods of Graphene

Graphene synthesis employs two primary strategies: the top-down approach and the bottom-up approach. The top-down method disassembles larger materials, like graphite, into graphene sheets. This process involves exfoliating graphite, where graphene is peeled layer by layer from the bulk graphite, overcoming the Van der Waals attractions between adjacent graphene layers. Conversely, the bottom-up approach constructs graphene atom by atom or molecule by molecule until the graphene structure is achieved.

2.1 The Top-Down Approach

2.1.1 Mechanical Exfoliation

In this method, graphene is produced by manually or mechanically peeling off layers from a graphite crystal. This technique was first introduced by Andre Geim and Konstantin Novoselov [4] in their discovery of graphene, known as the "Scotch tape method." It involves using adhesive tape to extract graphene layers from graphite. The tape is pressed firmly onto the surface, and upon removal, it pulls away the topmost graphene layers. This process is repeated iteratively, progressively thinning the remaining graphite until isolated graphene sheets are achieved. This method exploits the weak Van der Waals forces between graphite layers, allowing individual layers to be peeled off. Notably, this method was used by the winners of the 2010 Nobel Prize in Physics [5]. Despite revealing many exceptional properties of graphene, this method is labor-intensive and time-consuming, limiting its applicability to laboratory research and making industrial-scale production seem impractical. Another mechanical exfoliation method is the sonication technique, which is highly effective for exfoliating graphene and producing monolayer or few-layer graphene in high concentrations. This technique utilizes ultrasonic waves to induce physical or chemical changes in systems by generating cavitation bubbles. As the waves propagate, they create areas of high and low pressure, leading molecules to be pushed and pulled. During low-pressure phases, microbubbles form and grow until they collapse, producing powerful shockwaves [6].

Sonication can be performed using two primary types: bath sonication (BS) and tip sonication (TS). These methods have been used together or separately to exfoliate graphene and produce monolayer or few-layer graphene sheets [7]. Sonication-assisted liquid-phase exfoliation (LPE) typically involves three main steps: preparing a dispersion of graphite in a solvent, exfoliating the dispersion through sonication, and purifying the resulting graphene. The effectiveness of sonication depends on various factors such as sonication power, the liquid medium used for dispersing graphene nanosheets, and the centrifugation rate [8].

Additionally, ball milling, a common technique in powder production technology, can also be utilized to exfoliate graphite into graphene flakes. In the ball milling process, graphite flakes are enclosed in a container with hard balls, typically made of materials like stainless steel. The container undergoes rotation, leading the balls to grind against the graphite flakes, resulting in the exfoliation of graphene layers [9]. Ball milling can be classified into wet ball milling and dry ball milling. Wet ball milling involves dispersing graphite in solvents with a matched surface energy to overcome the Van der Waals forces between adjacent graphene flakes, allowing for better dispersal of graphite flakes in the liquid medium and potentially higher-quality graphene. Dry ball milling, on the other hand, is conducted in the absence of a liquid medium, making it a simpler process but may result in less uniform exfoliation and higher defect density in the graphene [10]. The choice between the two methods depends on the specific requirements of the graphene application and the desired balance between quality and production efficiency.

2.1.2 Chemical Exfoliation

Chemical exfoliation produces graphene by breaking down larger graphite structures into graphene layers using chemical reactions. There are three major steps involved in this method which are intercalation, exfoliation, and reduction. Intercalation is the initial step where a chemical agent is introduced between the graphite layers to disrupt the weak van der Waals forces holding the layers together, making them more susceptible to separation. Various intercalating agents, such as strong oxidizing agents, Lewis's acids, organic solvents, and molten salts, can be used for this purpose [11–13]. For example, in the Hummers' method [14], intercalation is achieved by treating graphite with a mixture of sulfuric acid (H_2SO_4) and an oxidizing agent, such as potassium permanganate ($KMnO_4$). The sulfuric acid acts as an intercalant, penetrating the interlayer spaces of the graphite and expanding the interlayer distance. The oxidizing agent helps to introduce defects and functional groups on the graphene layers, further facilitating exfoliation.

Once the intercalating agent has loosened the interlayer bonding, various techniques are used to break down the expanded graphite into individual graphene sheets. This exfoliation step involves separating the graphene oxide (GO) produced in the previous step into individual graphene sheets. This can be achieved through sonication, chemical etching, and shearing [15, 16]. Sonication uses high-frequency sound waves to create shear forces that separate the layers, while chemical etching employs strong acids or bases to further break down the intercalated graphite. Shearing applies mechanical forces to disperse the graphite into smaller flakes.

The reduction process aims to partially reverse the effects of oxidation and restore some of the desirable properties of pristine graphene. To reduce graphene oxide

(GO) back to graphene or reduced graphene oxide (rGO), which exhibits physical and chemical properties close to graphene, both thermal annealing and chemical reduction methods are commonly employed. Thermal annealing involves heating the GO to high temperatures (typically above 600 °C) in an inert atmosphere, such as argon or nitrogen, to drive off the oxygen-containing functional groups through thermal decomposition, thereby reducing GO to rGO [17]. On the other hand, the chemical reduction method involves the application of chemical-reducing agents like hydrazine (N_2H_4) and hydroiodic acid (HI) [18]. These agents target the oxygen functional groups, such as hydroxyl (–OH), epoxy (–O–), and carboxyl (–COOH) groups, breaking the bonds and removing the oxygen atoms. This action helps restore some of the sp2 carbon-carbon bonds, partially returning the graphene structure to its original form. The reduction process is crucial for restoring the electrical conductivity and other desirable properties of graphene, making it more similar to pristine graphene and suitable for various applications.

2.1.3 Plasma Exfoliation

Plasma exfoliation presents a promising method for graphene's efficient and scalable production. This technique utilizes highly reactive plasma, a state of matter consisting of charged particles (ions and electrons) and neutral species, generated by ionizing a gas. The plasma is then directed towards a graphite target, typically in the form of foil or powder. The high-energy ions in the plasma bombard the surface of the graphite, breaking the weak van der Waals forces holding the graphene layers together [19]. This process effectively exfoliates the graphite, leading to the production of individual or few-layer graphene sheets. The utilization of plasma for graphene synthesis presents significant challenges, as it requires the simultaneous delivery of small precursor amounts, and rapid and uniform nucleation while avoiding any damage due to ion bombardment [20]. Plasma exfoliation offers several advantages compared to traditional methods for graphene production. Unlike mechanical exfoliation using the Scotch tape method, which is a time-consuming and low-yield process, plasma exfoliation has the potential for large-scale, continuous production [21]. Additionally, compared to chemical exfoliation methods that often involve harsh chemicals and introduce defects into the graphene structure, plasma exfoliation offers a cleaner and potentially more controlled approach.

2.2 *The Bottom-Up Approach*

2.2.1 Chemical Vapor Deposition

Chemical vapor deposition (CVD) is regarded as the most promising approach for the synthesis of high-quality graphene. During the CVD process, a carbon-rich gas like methane (CH_4) or ethylene (C_2H_4) is introduced into a heated chamber containing a substrate, often made of copper, nickel, or other metals. This setup, heated to temperatures typically between 800 and 1000 °C, triggers the breakdown of carbon atoms in the gas, leading to the formation of a graphene layer on the substrate. The metal substrate serves a dual role as a catalyst, reducing the energy barrier of the reaction, and as a determinant of the graphene deposition mechanism, ultimately influencing the quality of the graphene produced [22]. Post-synthesis, graphene can be transferred to insulating substrates for further characterization and applications.

Copper is highly regarded as the preferred substrate for graphene growth due to its ability to exclusively support the formation of monolayer graphene. In contrast, nickel substrates are conducive to the controlled growth of few-layer graphene. Despite investigations into other transition metals like ruthenium, iridium, palladium, platinum, rhodium, rhenium, or gold as potential CVD substrates, none have significantly surpassed copper, nickel, or cobalt. This is attributed to factors such as substrate cost, the quality and transferability of graphene, and the feasibility of extending graphene sheets to larger dimensions [23].

The precursor gas in graphene formation can decompose through two main mechanisms: homogeneous gas-phase reactions and heterogeneous chemical reactions [24]. Homogeneous gas-phase reactions involve the direct decomposition of the precursor gas in the gas phase before reaching the substrate. This process, triggered by heat or other factors, leads to the formation of carbon-containing molecules or atoms that can adhere to the substrate, aiding in graphene growth. On the other hand, heterogeneous chemical reactions occur when the precursor gas reaches the heated substrate surface before decomposing. In this reaction, the high temperature of the substrate surface facilitates the decomposition of the precursor gas, resulting in the formation of carbon atoms or molecules directly on or near the substrate surface. These carbon species then diffuse on the substrate surface and assemble into graphene layers [25]. Both processes play crucial roles in graphene growth through CVD, with the choice of precursor gas and process parameters influencing which reaction pathway dominates. Understanding these mechanisms is essential for controlling the growth kinetics, quality, and properties of the graphene film produced using CVD. CVD provides precise control over graphene thickness and the capability to produce high-quality, uniform films. Nevertheless, the process is intricate, demanding high temperatures and specialized equipment. Furthermore, achieving large-scale production while maintaining quality and uniformity can be challenging.

2.2.2 Epitaxial Growth

Epitaxial growth offers a meticulous approach for fabricating high-quality graphene films. The process commences with the selection of a suitable crystalline substrate, most commonly silicon carbide (SiC). This substrate undergoes rigorous cleaning and polishing protocols to achieve an atomically pristine surface. Subsequently, the substrate is subjected to high temperatures within a vacuum or controlled atmosphere. This high-temperature treatment facilitates the desorption of surface contaminants, ensuring a contaminant-free environment conducive to graphene growth [26]. The introduction of a carbon-containing precursor gas, such as methane (CH_4), initiates the growth process. Carbon atoms from the decomposed gas absorb onto the substrate and subsequently diffuse across the surface. Crucially, these carbon atoms arrange themselves in a hexagonal lattice structure that mirrors the underlying substrate's crystal lattice [27].

This epitaxial relationship between the graphene film and the substrate is paramount for achieving high-quality graphene with superior electrical and structural properties. Following the formation of the desired graphene layer, the substrate undergoes a controlled cooling process. This gradual decrease in temperature stabilizes the newly formed film, with careful optimization of the cooling rate and final temperature minimizing defect formation and preserving the integrity of the graphene layer [28, 29]. In addition to SiC, metals such as Cu, Ni, and Pt, as well as insulating substrates like quartz or sapphire, can serve as alternative materials for the epitaxial growth of graphene [30–32]. Each substrate type presents distinct advantages, such as strong adhesion for metals and a pristine surface for insulating substrates. However, they also have drawbacks, including possible impurities for metals and the necessity for supplementary processing steps for insulating substrates. The selection of a substrate hinges on the specific demands of the graphene application and the targeted properties of the graphene film.

2.2.3 Molecular Beam Epitaxy

Molecular Beam Epitaxy (MBE) presents a unique method for producing high-quality, atomically precise graphene films. Unlike Chemical Vapor Deposition (CVD), which relies on gaseous precursors, MBE operates within an ultra-high vacuum chamber, allowing for the precise construction of graphene layers by depositing one carbon atom at a time. This process involves generating a beam of carbon atoms through the sublimation of a solid carbon source, often a graphite rod, at elevated temperatures [33]. Silicon carbide (SiC) reigns supreme as a substrate for high-quality graphene production due to its well-matched crystal structure with graphene. Apart from SiC, MBE can utilize other single-crystal materials as well. These include gallium nitride (GaN), certain transition metals such as nickel and copper, and even some insulating materials such as mica and hexagonal boron nitride (h-BN) [34–36]. The substrate's temperature is crucial in the process, typically ranging from 1000 to 1400 °C. This heating facilitates the diffusion of carbon atoms across the surface, promoting efficient bonding and the formation of graphene layers [37].

The standout feature of MBE is its precision. By controlling the carbon beam flux, the growth rate can be precisely adjusted, enabling the production of films with remarkable uniformity and thickness at the atomic level [33]. This level of precision has the potential to result in high-quality graphene with minimal defects, which is particularly advantageous for specific applications. Moreover, MBE provides some flexibility in substrate selection, allowing for integration with various device platforms. For example, using gallium nitride (GaN) as a substrate could enable integration with GaN-based electronics, while hexagonal boron nitride (h-BN), a well-known insulator, could be used to create graphene-based heterostructures with unique properties [38, 39]. Despite its advantages, MBE comes with its own set of challenges. The meticulous, atom-by-atom construction leads to slower growth rates compared to CVD, which hampers large-scale production. Moreover, the complex equipment and ultra-high vacuum requirements make MBE a more costly technique. Additionally, some substrates may not withstand the high temperatures required for effective MBE graphene growth, thus limiting the available material options.

3 Characterization Techniques for Graphene

Graphene, a single layer of carbon atoms arranged in a honeycomb lattice, is a wonder material with unique properties. Due to its exceptional electrical, mechanical, and thermal characteristics, graphene has garnered significant interest in various fields of research and technology. However, characterizing graphene is crucial to ensure its quality, number of layers, and defects. Several characterization techniques are employed to assess the properties of graphene.

3.1 Raman Spectroscopy

Raman spectroscopy is widely used to characterize graphene. It provides information about the number of layers, defects, and strain in graphene by analyzing the frequency shifts and intensities of Raman-active modes. This technique analyses the vibrational modes of the carbon atoms in graphene. The characteristic Raman spectrum of graphene allows researchers to determine the number of layers (single, few-layer, or multilayer), presence of defects, and level of doping. The G and 2D peaks are characteristic features of graphene in Raman spectra as shown in Fig. 3 [40, 41].

3.2 Scanning Electron Microscopy (SEM)

SEM provides high-resolution images of the graphene surface, revealing its morphology, size, and distribution. It also provides information about the structure, such as layer thickness and uniformity at high resolution. Example of SEM images of graphene nanosheets can be found in Fig. 4.

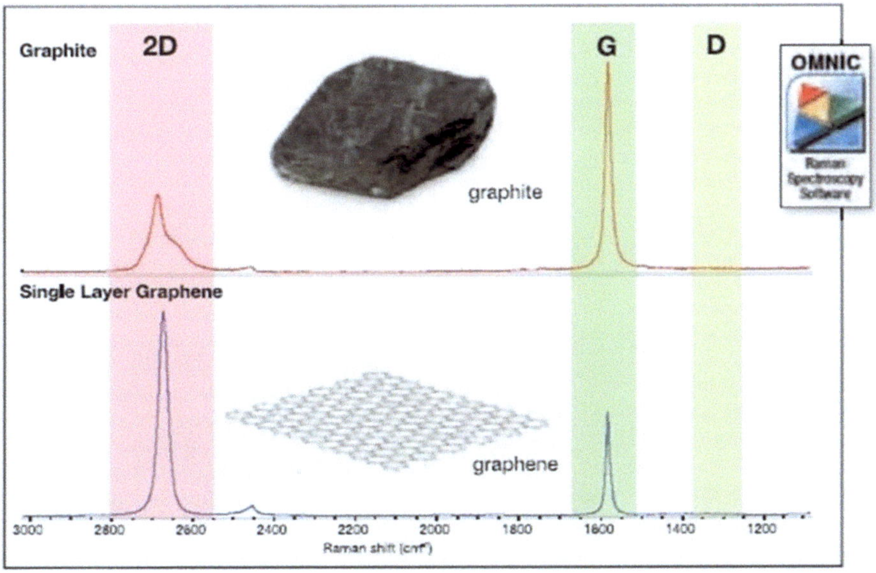

Fig. 3 The Raman spectra of graphite and angle layer graphene, collected with 532 nm excitation [42]

3.3 Transmission Electron Microscopy (TEM)

TEM offers direct imaging of individual graphene sheets at the atomic level resolution, enabling visualization of the graphene lattice structure, number of layers, and presence of defects or contaminants and grain boundaries. The sample of transmission electron microscopy images for graphene can be shown in Fig. 5.

3.4 X-ray Diffraction (XRD)

XRD provides useful information on phase analysis and vital tool to characterize the crystal structure and orientation of graphene samples. It can provide information about the stacking order of graphene layers and the presence of interlayer spacing. XRD also can be used to study the crystallographic orientation and alignment of graphene layers in multilayered graphene samples. Single sharp peak of graphene, due to its single atomic layer structure, produces a single, sharp peak in the XRD pattern. This peak is typically located around 26.5° (2θ) for Cu Kα radiation (wavelength of 1.5406 Å). The position of the peak (2θ) corresponds to the interlayer spacing between the graphene sheets. In pristine graphene, this spacing reflects the distance between the stacked carbon layers. A strong and sharp peak indicates good crystallinity and a high degree of order in the graphene structure. Conversely, a broad

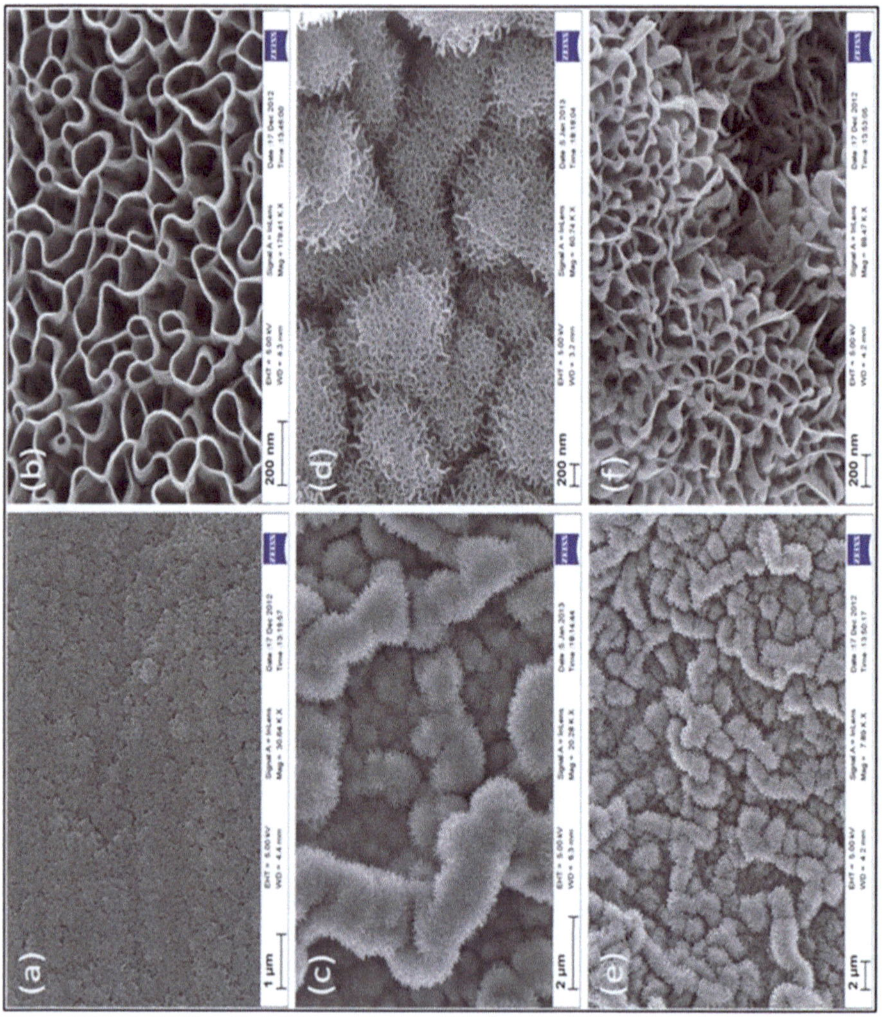

Fig. 4 Scanning electron microscopy (SEM) images of graphene nanosheets on **a** and **b** copper, **c** and **d** nickel and **e** and **f** platinum substrates [43]

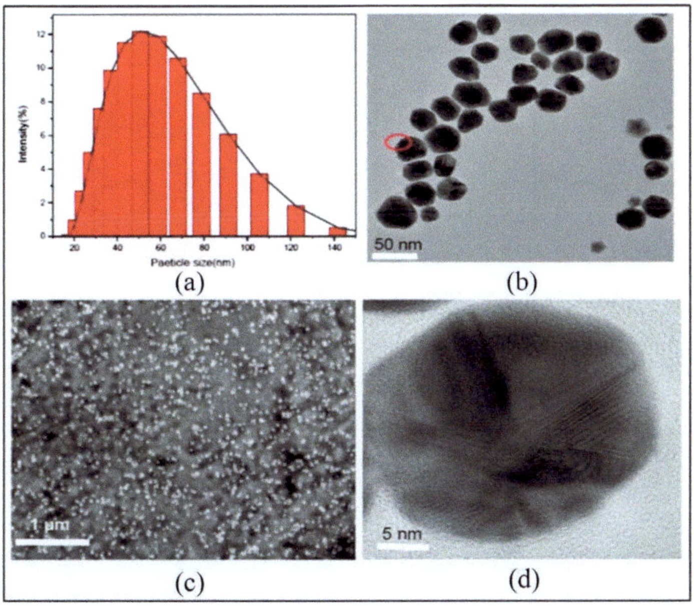

Fig. 5 Electron microscopy characterization of CAO-AuNPs: **a** particle size distribution graph; **b** transmission electron microscope (TEM) image of CAO-AuNPs; **c** scanning electron microscope (SEM) image of CAO-AuNPs; **d** high-resolution TEM (HRTEM) images of CAO–AuNPs [44]

or weak peak suggests lower crystallinity or the presence of defects. The difference pattern for XRD peaks for graphite, graphene and graphene oxide can be shown in Fig. 6. It is crucial to distinguish the XRD pattern of graphene from its parent material, graphite. Graphite, with its layered structure, shows a similar peak at a slightly lower angle (around 26.0°) due to the larger interlayer spacing caused by the presence of weak van der Waals forces between the graphene layers [45–47].

Fig. 6 XRD patterns of graphite, graphene oxide and graphene nano sheets [48]

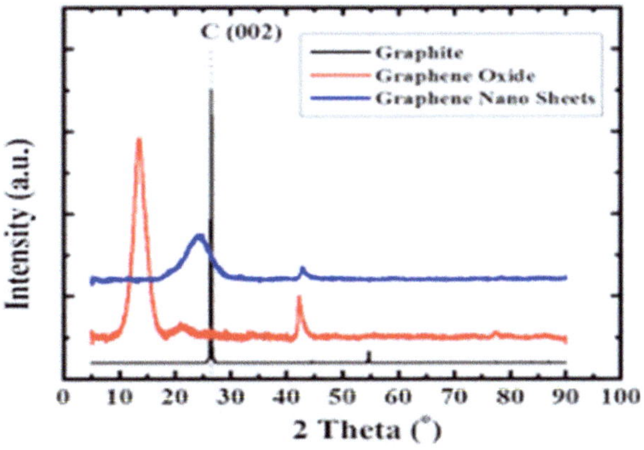

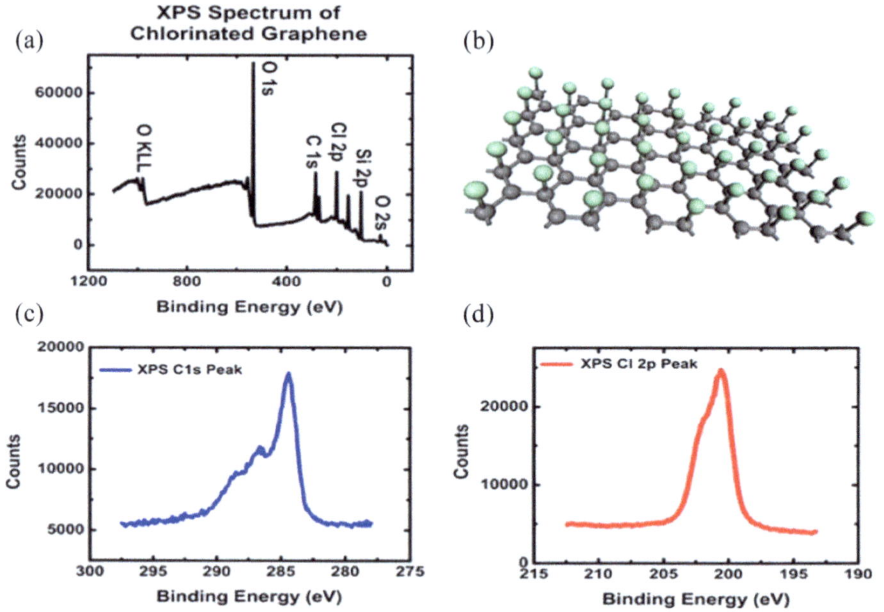

Fig. 7 **a–c** X-ray photoelectron spectroscopy (XPS) characterization of the chlorinated graphene. The graphene samples were treated in Cl2 plasma with dc bias of 8 V for 30 s. **a** Full XPS spectrum of the sample. **b** Schematic representation of the atomic structure of chlorinated graphene with a C:Cl ratio of 2:1. **c** Zoom-in view of the C 1 s peak of (**a**). **d** Zoom-in view of the Cl 2p peak of (**a**) [49]

3.5 X-ray Photoelectron Spectroscopy (XPS)

XPS provides information about the chemical composition and electronic state of the graphene surface. It can identify the presence of impurities or functional groups on the graphene surface. The chemical composition sample of graphene can be viewed in Fig. 7.

3.6 Atomic Force Microscopy (AFM)

AFM measures the surface topography of graphene with high precision, allowing determination of thickness, roughness, and presence of wrinkles or folds. It can also be used to probe mechanical properties such as stiffness and elasticity. AFM images of graphene can be shown in Fig. 8.

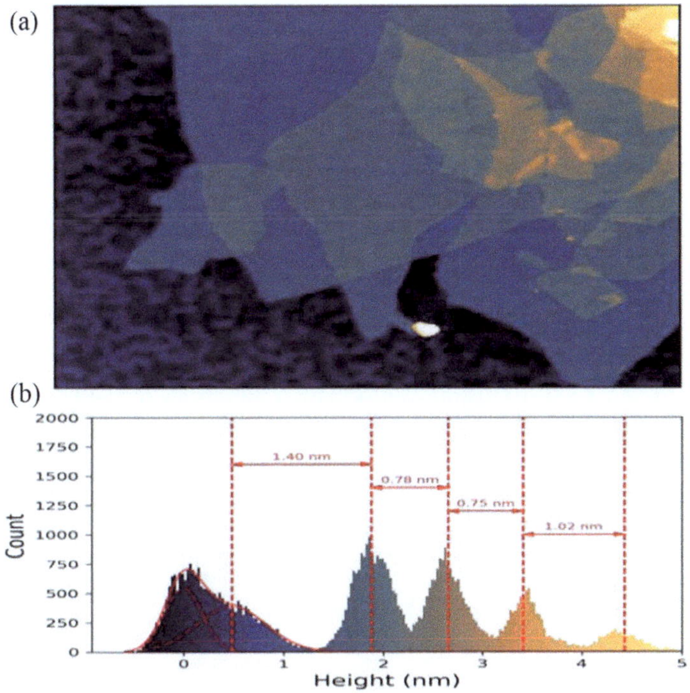

Fig. 8 Measuring the thickness of multilayer graphene. **a** AFM topographical image of graphene oxide with lateral dimensions of 5.11 × 5.11 μm^2. **b** Histogram of the heights in (**a**) showing the thickness of the first 4 layers [50]

3.7 Fourier Transform Infrared Spectroscopy (FTIR)

FTIR is used to analyse the functional groups and chemical bonding in graphene-based materials. It can provide information about the presence of oxygen-containing functional groups, such as hydroxyl and epoxy groups, which are commonly found in graphene oxide. Pristine graphene, composed solely of carbon atoms in a honeycomb lattice, has minimal functional groups. Therefore, the FTIR spectrum of graphene is relatively featureless, with a weak peak around 1580 cm^{-1} corresponding to the in-plane vibrations of the aromatic C = C bonds. Whereas graphene oxide (GO) is heavily decorated with oxygen-containing functional groups like hydroxyl (OH), carbonyl (C = O), and epoxy groups. These functional groups introduce distinct peaks in the FTIR spectrum of GO as shown in Fig. 9 [46, 51]:

i. Broad peak around 3400 cm^{-1}: stretching vibration of O–H groups
ii. Peak around 1720 cm^{-1}: stretching vibration of C = O groups in carboxylic acids

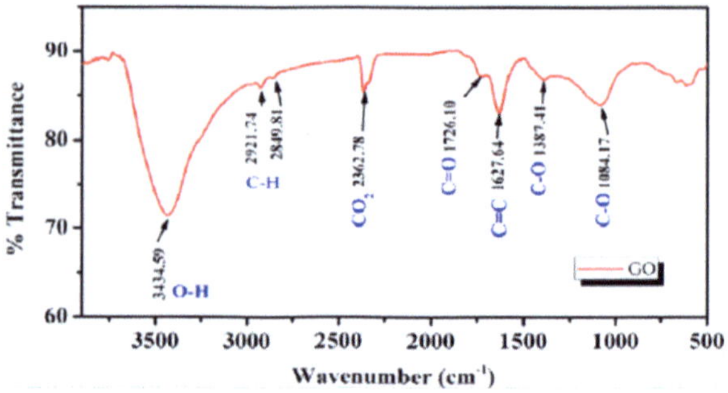

Fig. 9 FTIR spectrum of graphene oxide [52]

iii. Peaks around 1200–1050 cm^{-1}: deformation vibrations of C–O bonds in epoxy and alkoxy groups.

4 Graphene-Based Hybrid Materials

4.1 Introduction to Hybrid Materials

Hybrid materials are composed of two or more distinct components, often with differing chemical or physical properties, combined at nanometer, molecular, or macroscopic scales to achieve synergistic properties or capabilities that surpass those of the individual components. Graphene-based hybrid materials combine graphene, a two-dimensional sheet of carbon atoms arranged in a hexagonal lattice, with other materials, representing a class of materials with enhanced attributes and functionalities. Graphene's exceptional mechanical strength, high electrical conductivity, and superior thermal properties make it an appealing candidate for improving the performance of various materials and devices when incorporated into hybrids.

The exceptional mechanical strength of graphene is a key advantage in its integration into hybrid materials. Graphene, with a tensile strength of about 130 GPa [53], ranks among the strongest known materials. Its inclusion significantly enhances the mechanical properties of the hybrid material, improving stiffness and toughness, thus imparting greater durability and resistance to mechanical stress. Moreover, graphene's high conductivity is a crucial advantage in hybrid material fabrication. Its high electron mobility, exceeding 200,000 cm^2/Vs [54], makes it ideal for enhancing the electrical conductivity of insulating materials or establishing conductive pathways in electronic devices. Moreover, graphene's outstanding thermal conductivity is a considerable advantage in hybrid materials. With a thermal conductivity of up to 5000 W/mK at room temperature [55], graphene can significantly improve heat dissipation and thermal stability in hybrid materials, which is crucial for applications in thermal management and electronics. Additionally, graphene's substantial surface area per unit mass provides abundant active sites for chemical reactions or adsorption processes, rendering it particularly valuable in catalytic and sensing applications

[56]. Graphene's flexibility and transparency further expand its applications, particularly in flexible electronics or transparent conductive coatings. Harnessing the unique properties of graphene allows researchers and engineers to design hybrid materials with customized properties for a wide array of applications, including structural materials, electronics, energy storage, and biomedical devices.

4.2 *Types of Graphene-Based Hybrids and Their Synthesis Methods*

Graphene-based hybrids represent a diverse class of composite materials that combine graphene with metals, polymers, oxides, and other carbon-based materials. This combination results in materials with improved properties or novel functionalities. These hybrid materials demonstrate great potential in various applications due to their unique combination of properties.

4.2.1 Graphene-Metal Hybrids

Hybrids or composites of graphene and metals combine graphene with metal nanoparticles or substrates, resulting in materials with distinct properties. These hybrids have garnered considerable interest for their synergistic effects, which enhance mechanical, electrical, and catalytic properties. Graphene-metal nanoparticle composites are formed by incorporating metal nanoparticles, typically ranging in size from a few nanometers to several hundred nanometers, onto sheets of graphene, graphene oxide (GO), or reduced graphene oxide (rGO). These composites are synthesized using both in situ methods, where nanoparticles are grown directly on the graphene surface, and ex-situ methods, where pre-made nanoparticles are attached to the graphene. GO and rGO serve as effective templates for this purpose due to their surfaces containing defects and oxygen functional groups. These features facilitate the nucleation, growth, and attachment of various metal nanoparticles, such as Au, Ag, and Pt [57–59].

Graphene-metal nanoparticle composites are typically synthesized by reducing metallic salts using chemical agents such as ethylene glycol, sodium citrate, and sodium borohydride. The negatively charged functional groups on the surface of graphene oxide (GO) facilitate the nucleation of positively charged metallic salts, leading to the growth of metal nanoparticles on the GO surface. Alternatively, chemical linkers with a strong affinity for graphene's surface can be used to attach metal nanoparticles through π–π stacking [60]. A novel, nontoxic method for synthesizing graphene composites decorated with various nanoparticles (Au, Ag, Pt, Pd, and latex beads) was introduced by Liu and colleagues [61]. They utilized bovine serum albumin (BSA) not only for reducing GO to graphene but also for attaching nanoparticles onto the graphene surface. Similarly, Li et al. developed a green synthesis method for Ag–GO nanocomposites using glucose as a reducing and stabilizing agent for Ag nanoparticles, eliminating the need for toxic agents [62].

In addition to chemical reduction methods, electrochemical techniques can also be used to fabricate graphene-nanoparticle composite structures. Yang et al. [63] demonstrated the electrochemical deposition of graphene-Au/Pd nanoparticle composites onto electrodes by applying a potential of 0.2 V in the presence of metallic salts. Claussen et al. [64] employed electrochemical methods to fabricate platinum nanoparticle-decorated multilayered graphene nanosheets, with the size and morphology of the Pt nanoparticles controlled by adjusting the intensity of the pulse current. These findings indicate that electrochemical techniques offer precise control over the fabrication of graphene-nanoparticle composites. The various methods available for synthesizing graphene-metal hybrids highlight their versatility and the potential to tailor the properties of graphene-metal nanoparticle composites for diverse applications.

4.2.2 Graphene-Polymer Hybrids

Graphene-polymer hybrids are composite materials in which graphene is embedded within a polymer matrix. These composites can be classified into three types based on the 3D arrangement and the interaction between graphene sheets and polymers: graphene-filled polymer composites, layered graphene-polymer films, and polymer-functionalized graphene nanosheets. Graphene-filled polymer composites are a promising alternative to conventional carbon-based fillers like carbon nanotubes (CNTs) for enhancing the properties of polymer matrices. While CNTs are effective, they are also costly. Graphene-based fillers offer a potential replacement or supplement to CNTs. Achieving optimal properties in these composites requires attention to factors such as the dispersity of the graphene filler and its bonding with the polymer matrix. These composites are typically prepared using methods like solution mixing, melt blending, and in situ polymerization [65–67]. Solution mixing involves mixing graphene oxide (GO) with water-soluble polymers like poly (vinyl alcohol) (PVA) to create stable dispersions of graphene derivatives in polymer solutions, facilitated by sonication [68]. On the other hand, melt compounding blends graphene fillers and polymer matrices using high-shear forces and high-temperature melting, suitable for composites like polylactide (PLA)-exfoliated graphite (EG) and PET-rGO graphene composites [69]. In-situ polymerization polymerizes monomers in the presence of graphene fillers, such as in graphene-epoxy composites, where a curing agent initiates the polymerization process, or in graphene-polyaniline (PANi) composites, where oxidative agents facilitate polymerization [70]. Each method offers a distinct approach to incorporating graphene into polymer matrices, tailored for various applications.

In contrast to the conventional dispersion of graphene fillers within polymer matrices, graphene derivatives can also be integrated into polymers in layered structures tailored for specific applications such as directional load-bearing membranes and thin films for photovoltaic devices. A layer-by-layer (LbL) assembly technique, employing the Langmuir-Blodgett (LB) method, has been utilized to deposit graphene oxide (GO) sheets onto films composed of polyelectrolytes like poly (allylamine hydrochloride) (PAH) and poly (sodium 4-styrene sulfonate) (PSS) multi-layers [71]. This approach has led to a significant enhancement in the directional elastic modulus, increasing from 1.5 to 20 GPa with an 8 vol% loading of graphene. Similar strategies have been applied to fabricate multi-layer films comprising polyvinyl alcohol (PVA) and GO, which have demonstrated improved elastic modulus and hardness [72]. Layered graphene-polymer composites are also employed in polymer-based photovoltaic devices, where composite films are fabricated through sequential spin-coating of functional components.

Polymer-functionalized graphene nanosheets offer a unique approach to enhancing the properties of polymers. Instead of being used merely as fillers, graphene derivatives can serve as templates for polymer decoration through covalent and non-covalent functionalization. This process not only improves the solubility of the graphene derivatives but also imparts additional functionality to the resulting hybrid nanosheets. Covalent functionalization involves the reaction between polymer functional groups and oxygenated groups on graphene oxide (GO) or reduced graphene oxide (rGO) surfaces. For instance, the esterification of carboxylic groups in GO with hydroxyl groups in polyvinyl alcohol (PVA) produces GO-PVA composite sheets [73]. On the other hand, non-covalent functionalization relies on van der Waals forces or electrostatic interactions, making it simpler and preserving the chemical structure of the graphene sheets [74]. This approach allows for the tailoring of electronic/optical properties and solubility, offering versatile possibilities for various applications.

4.2.3 Graphene-Ceramic Hybrids

Graphene-ceramic composites represent a class of advanced materials that combine the exceptional properties of graphene with the desirable characteristics of ceramics. Ceramics are renowned for their high stiffness, strength, and thermal stability, yet they often suffer from brittleness and poor electrical conductivity. In contrast, graphene offers remarkable mechanical strength, excellent electrical and thermal conductivity, and a large surface area. The types of ceramics that have been combined with graphene, such as alumina, silicon carbide, titanium dioxide, and zirconia, are well-documented in the literature [75, 76]. By incorporating graphene into ceramic matrices, these composites aim to improve the overall properties of ceramics. A variety of methods are utilized for processing ceramics containing graphene-type fillers, a critical yet sometimes overlooked aspect with significant implications for ceramic performance, particularly in composites where improper processing can yield detrimental results. The main processing routes encompass powder processing, colloidal methods, sol-gel and salt precursor techniques, preceramic polymers, and in-situ graphene synthesis.

The powder processing route, the most prevalent method, entails dispersing submicron matrix powders and graphene nanofillers in a solvent. This approach employs various strategies, such as separate dispersion of the components followed by mixing, or simultaneous milling of the matrix and graphene. The choice of dispersant, whether water or organic solvent, depends on the type of nanofillers. Solvent removal methods, such as spray drying or freeze drying, are utilized to maintain a uniform dispersion state. Powder processing is extensively utilized for non-oxide ceramic composites, as well as for systems involving Al_2O_3 and SiO_2 [77, 78]. Colloidal methods are utilized for simpler systems, such as fine Al_2O_3 powders and graphene oxide (GO) fillers. The process often involves hetero coagulation of GO and Al_2O_3 suspensions to precipitate the mixture [79]. Sol–Gel and Salt Precursor Methods offer a refined approach to mixing matrix and nanofillers in ceramic processing. For instance, the sol-gel process can incorporate an aluminum nitrate salt mixed with a graphene oxide (GO)/water dispersion, followed by thermal decomposition [80]. Another technique involves a reactive titanium oxycarbide (TiOC) precursor infiltrating a 3D graphene network, leading to the formation of TiC-based graphene composites post-thermal treatment [81].

On the other hand, the processing of preceramic polymers with graphene fillers is still in its nascent stages. In one study, a polycarbosilane was mixed with graphene nanoplatelets (GNP) in a solvent, followed by subsequent steps of crosslinking, pyrolysis, and crystallization [82]. In-situ Graphene Synthesis involves epitaxial graphene growth on Silicon Carbide (SiC) grains during the sintering SiC powders in a Spark Plasma Sintering (SPS) furnace, or the nucleation of a graphene network in preceramic polymers treated at high temperatures in SPS [83]. Densification for the above-described processing methods is typically achieved through Spark Plasma Sintering (SPS), offering benefits like higher densities and reduced damage to graphene nanostructures. These methods play a crucial role in enhancing the performance of ceramics, particularly in demanding applications requiring superior properties [84, 85].

Graphene fillers have demonstrated remarkable success in reinforcing ceramic matrices, leading to enhanced electrical functionality and promising advancements in structural ceramics. These composites also exhibit superior performance under friction and wear conditions, a key advantage given ceramics' broad applications in these areas. While not all composites show increased thermal conductivity, graphene fillers provide high directionality, offering potential benefits for thermal management and protection applications. The adaptability of graphene nanostructures to various ceramic processing methods has enabled innovative developments, including layered composites, coatings, and 3D assemblies. Despite ongoing challenges related to specific reinforcement models, thermal stability, directional properties, and nanostructure dispersibility, the current research indicates a positive outlook for the future of these materials.

4.2.4 Other Types of Graphene-Based Hybrids

In addition to the aforementioned hybrids, several other types of graphene-based hybrids have been investigated for their unique properties and potential applications. Among these, graphene-carbon nanotube hybrids and graphene-quantum dot hybrids have emerged as particularly promising materials. Graphene-carbon nanotube hybrids represent a compelling combination of two advanced carbon-based materials, each renowned for their exceptional properties. Graphene, with its 2D structure and high surface area, boasts remarkable electrical and thermal conductivity as well as exceptional mechanical strength. In contrast, carbon nanotubes exhibit unique one-dimensional structures with outstanding mechanical, thermal, and electrical properties. The synergistic integration of these materials in hybrids results in enhanced mechanical strength, electrical conductivity, and thermal properties, making them highly attractive for a range of applications [86–88]. These include lightweight and high-strength composites, flexible electronics, and energy storage devices.

Graphene-quantum dot hybrids, meanwhile, blend the exceptional properties of graphene with those of semiconductor quantum dots, which are nanoscale semiconductor particles with tunable electronic properties due to quantum confinement effects. The strategic incorporation of quantum dots onto the graphene surface enables precise control over their interactions, leading to tailored electronic properties in the hybrid material. This tunability renders hybrids suitable for applications in optoelectronics, photovoltaics, and quantum computing. Furthermore, the integration of quantum dots enhances the light absorption capabilities of graphene, making these hybrids promising candidates for use in solar cells and other photonic devices [89–91].

5 Applications of Graphene-Based Hybrid Materials

Graphene-based hybrid materials combine the extraordinary properties of graphene with other materials, resulting in synergistic effects that can lead to enhanced performance and novel functionalities. These hybrid materials have garnered significant interest across various fields due to their potential applications in electronics and optoelectronics, energy storage and conversion, composite materials, biomedicals applications and environmental applications.

5.1 Electronics and Optoelectronics

Graphene's exceptional electronic and optoelectronic properties make it a promising material for various applications in these fields. It is a highly attractive material for next-generation electronic and optoelectronic devices.

5.1.1 Electronic Devices

i. **Transistor**: Graphene-based transistors have been explored for their high carrier mobility, which can lead to faster switching speeds and lower power consumption compared to traditional silicon transistors. Graphene's high carrier mobility allows for the fabrication of high-frequency transistors suitable for radio-frequency (RF) and microwave applications [92].
ii. **Transparent Conducting Electrodes (TCEs)**: Graphene's transparency and conductivity make it a superior alternative to traditional indium tin oxide (ITO) electrodes used in touchscreens, solar cells, and organic light-emitting diodes (OLEDs). Graphene-based TCEs offer better flexibility, higher conductivity, and improved light transmission [93].
iiii. **Radio Frequency (RF) Electronics**: Graphene's high carrier mobility and thermal stability make it suitable for high-frequency electronics used in radiofrequency identification (RFID) tags, microwave circuits, and antennas [94].

5.1.2 Optoelectronic Devices

i. **Photodetectors**. Graphene's broad light absorption spectrum and high-speed response make it ideal for photodetectors that can operate across a wide range of wavelengths, from ultraviolet to infrared. This has applications in optical communication, imaging, and sensing [95].
ii. **Light Modulators**. By manipulating the conductivity of graphene with an electric field or light, researchers can create efficient light modulators. These devices are essential components in optical communication systems for controlling the intensity or phase of light signals [96].
iii. **Solar Cells**. Graphene's transparency and ability to collect light across a broad spectrum make it a potential material for next-generation solar cells. Researchers are exploring ways to integrate graphene with other materials to improve light absorption and energy conversion efficiency [97].

5.2 Energy Storage and Conversion (E.g., Batteries, Supercapacitors)

Graphene's unique properties make it highly promising for applications in energy storage and conversion, particularly in batteries and supercapacitors. Its unique properties, exceptional electrical conductivity, high surface area, and remarkable mechanical strength make it a perfect candidate for next-generation batteries and supercapacitors.

5.2.1 Batteries

i. **Lithium-Ion Batteries (LIBs)**. Graphene can be used as an additive or electrode material in lithium-ion batteries to enhance their performance. Graphene-based electrodes offer high electrical conductivity, large surface area, and mechanical flexibility, leading to improved charge/discharge rates, higher energy density, and longer cycle life [98, 99].
ii. **Sodium-Ion Batteries (SIBs)**. Graphene-based electrodes have also been investigated for use in sodium-ion batteries, which are considered a low-cost alternative to lithium-ion batteries. Graphene's ability to accommodate large volume changes during charge/discharge cycles improves the stability and cycling performance of sodium-ion batteries [100–102].

5.2.2 Supercapacitors

i. **Electrochemical Capacitors**. Graphene-based supercapacitors, also known as electrochemical capacitors or ultracapacitors, store energy through the electrostatic adsorption of ions on the surface of graphene electrodes. Graphene's high surface area, electrical conductivity, and mechanical strength enable the fabrication of supercapacitors with high energy density, power density, and cycling stability [103, 104].
ii. **Hybrid Supercapacitors**. Graphene can be combined with other materials, such as metal oxides or conductive polymers, to form hybrid supercapacitor electrodes. These hybrid materials leverage the synergistic effects between graphene and other components to further enhance energy storage performance, such as increasing capacitance, improving charge transfer kinetics, and widening the operating voltage window [105, 106].

5.2.3 Other Energy Storage and Conversion Devices

i. **Fuel Cells**. Graphene-based materials have shown potential as catalyst supports or electrode materials in fuel cells for hydrogen or methanol oxidation reactions. Graphene's high surface area and electrical conductivity enhance the catalytic activity and stability of fuel cell electrodes [107, 108].
ii. **Solar Cells**. Graphene-based materials can be incorporated into solar cell devices, such as dye-sensitized solar cells (DSSCs) or perovskite solar cells, as transparent conductive electrodes or electron transport layers. Graphene's high electrical conductivity and transparency improve the efficiency and stability of solar cells [97, 109].

5.3 Composite Materials

Graphene is widely used as a reinforcing agent in composite materials due to its exceptional mechanical, electrical, thermal, and barrier properties. By combining graphene with other materials, scientists create entirely new classes of materials with properties exceeding those of the individual components [110].

5.3.1 Polymer Composites

Polymer Matrix: Graphene is often dispersed within polymer matrices such as epoxy, polyethylene, polypropylene, and polystyrene to form graphene-polymer composites. The high aspect ratio and surface area of graphene sheets provide effective reinforcement, improving mechanical properties such as tensile strength, stiffness, and toughness. Graphene-polymer composites find applications in automotive components, aerospace structures, sporting goods, and construction materials, where lightweight, high-strength materials are required [111].

5.3.2 Metal Matrix Composites (MMCs)

Metal Matrix: Graphene can be incorporated into metal matrix such as aluminum, copper, and magnesium to form metal matrix composites. Graphene enhances the mechanical properties and thermal conductivity of metal matrices, leading to improved strength, stiffness, and heat dissipation. Graphene-metal composites are used in automotive parts, electronic packaging, heat sinks, and aerospace components, where high thermal conductivity and mechanical strength are critical [112, 113].

5.3.3 Ceramic Composites

Ceramic Matrix: Graphene can be integrated into ceramic matrix such as silicon carbide, alumina, and zirconia to form ceramic composites. Graphene enhances the fracture toughness, thermal shock resistance, and electrical conductivity of ceramic matrices, improving their mechanical and thermal properties. Graphene-ceramic composites are used in cutting tools, armor materials, aerospace components, and thermal management systems, where high hardness, wear resistance, and thermal stability are required [114, 115].

5.3.4 Carbon-Based Composites

Carbon Matrix: Graphene can be combined with carbon matrix such as carbon nanotubes (CNTs) or carbon fibers to form carbon-based composites. Graphene serves as a nucleation site and interfacial bonding agent, improving the dispersion and alignment of carbon nanofillers, leading to enhanced electrical and mechanical properties. Graphene-carbon composites find applications in conductive additives, electromagnetic interference (EMI) shielding, and energy storage devices such as supercapacitors and lithium-ion batteries [116, 117].

5.4 Biomedical Applications

Graphene, with its unique properties like high surface area, conductivity, and biocompatibility (depending on form and modification), is making waves in the field of biomedicine. Its unique properties also make it highly promising for various biomedical applications, including drug delivery, biosensing, tissue engineering, imaging, and cancer therapy.

5.4.1 Drug Delivery

Graphene Nanocarriers: Graphene-based nanocarriers can efficiently encapsulate and deliver therapeutic agents such as drugs, genes, proteins, and peptides to target sites in the body. The large surface area and functional groups of graphene enable high drug loading capacity and controlled release kinetics, improving therapeutic efficacy and minimizing side effects. Graphene-based drug delivery systems are used for cancer therapy, regenerative medicine, and targeted drug delivery to specific organs or tissues [118, 119].

5.4.2 Biosensing

Graphene Biosensors: Graphene-based biosensors offer high sensitivity, selectivity, and rapid response times for detecting biomolecules such as DNA, proteins, enzymes, and viruses. Functionalized graphene surfaces can immobilize biomolecular recognition elements and transduce biomolecular interactions into measurable signals. Graphene biosensors are used for early disease diagnosis, point-of-care testing, environmental monitoring, food safety, and drug screening [120, 121].

5.4.3 Tissue Engineering

Graphene Scaffolds: Graphene-based scaffolds provide a three-dimensional (3D) substrate for cell growth, proliferation, and differentiation in tissue engineering applications. Graphene's mechanical strength, electrical conductivity, and biocompatibility support tissue regeneration and functional integration. Graphene scaffolds are used for bone regeneration, neural tissue engineering, cardiac repair, and skin tissue engineering [122, 123].

5.4.4 Imaging

Graphene Nanoprobes: Graphene-based nanoprobes can be functionalized with imaging agents such as fluorescent dyes, quantum dots, or magnetic nanoparticles for various imaging modalities, including fluorescence imaging, magnetic resonance imaging (MRI), and photoacoustic imaging. Graphene-based imaging probes enable high-resolution imaging of biological structures, molecular interactions, and disease processes in vivo and in vitro [124, 125].

5.4.5 Cancer Therapy

Graphene-Based Theranostics: Graphene-based materials can serve as multifunctional platforms for cancer theranostics, combining diagnostic imaging and therapeutic capabilities in a single system. Functionalized graphene nanomaterials can target cancer cells, deliver therapeutic agents, and monitor treatment response. Graphene-based theranostic agents are used for cancer diagnosis, imaging-guided drug delivery, photothermal therapy, photodynamic therapy, and radiotherapy sensitization [126–128].

5.5 Environmental Applications

In the fight for a cleaner environment, graphene emerges as a promising warrior. This wonder material, a single layer of carbon atoms arranged in a honeycomb lattice, boasts unique properties that can be harnessed for various environmental applications Graphene's remarkable properties make it highly promising for various environmental applications, including pollution remediation, water purification, environmental monitoring, and renewable energy generation.

5.5.1 Pollution Remediation

Graphene-based Adsorbents: Graphene and graphene oxide (GO) can absorb a wide range of pollutants, including heavy metals, organic contaminants, and dyes, from air, water, and soil. The large surface area, high adsorption capacity, and tunable surface chemistry of graphene enable efficient pollutant removal. Graphene-based adsorbents are used for wastewater treatment, air purification, soil remediation, and industrial pollution control [129, 130].

5.5.2 Water Purification

Graphene Membranes: Graphene-based membranes offer high water permeability, selectivity, and fouling resistance for water purification applications. Graphene oxide membranes can selectively separate ions, molecules, and nanoparticles based on size, charge, and hydrophilicity. Graphene membranes are used for desalination, nanofiltration, reverse osmosis, and wastewater recycling in municipal, industrial, and domestic water treatment plants [131–133].

5.5.3 Environmental Monitoring

Graphene Sensors: Graphene-based sensors offer high sensitivity, selectivity, and rapid response times for detecting environmental pollutants, gases, and biomolecules. Functionalized graphene surfaces can immobilize specific receptors and transduce analyte interactions into measurable signals. Graphene sensors are used for air quality monitoring, water quality monitoring, environmental surveillance, and detection of hazardous substances in soil and groundwater [134, 135].

5.5.4 Renewable Energy Generation

Graphene-Based Catalysts: Graphene and graphene-based materials can serve as efficient catalysts for various energy conversion and storage processes, including electrocatalysis, photocatalysis, and solar energy harvesting. Functionalized graphene surfaces can enhance catalytic activity, charge transfer kinetics, and stability. Graphene-based catalysts are used for hydrogen production, oxygen reduction reaction (ORR), water splitting, carbon dioxide conversion, and dye-sensitized solar cells (DSSCs) [136, 137].

5.5.5 Carbon Capture and Storage (CCS)

Graphene Membranes: Graphene-based membranes can selectively capture carbon dioxide (CO_2) from flue gas emissions and industrial exhaust streams. The high surface area and tunable porosity of graphene enable efficient CO_2 adsorption and separation. Graphene membranes are used for carbon capture and storage in power plants, cement factories, and other industrial facilities to mitigate greenhouse gas emissions and combat climate change [138].

6 Conclusions and Outlook

Graphene-based hybrid materials stand at the forefront of materials science, showcasing a remarkable synergy of properties that make them highly versatile and valuable across a broad spectrum of applications. Their unique combination of mechanical strength, thermal conductivity, electrical properties, and chemical stability has revolutionized the development of functional composites. These materials offer a pathway to address pressing challenges in energy storage, conversion, electronics, and beyond. The versatility of graphene-based hybrids is further enhanced by the ability to fine-tune their properties through precise control of the hybrid structure. This level of control enables the design of materials with tailored characteristics, leading to enhanced performance and efficiency in a wide range of applications.

Looking forward, the field of graphene-based hybrid materials is poised for continued growth and innovation. Ongoing research is expected to drive advancements in fabrication techniques, leading to more efficient and scalable production methods. Additionally, the exploration of novel hybrid architectures and the integration of graphene-based materials into practical devices are expected to pave the way for new applications and discoveries. Despite the existing challenges associated with large-scale production and cost-effective synthesis, the notable progress achieved thus far underscores the promising prospects of graphene-based hybrid materials. Continued research endeavors and technological advancements hold the promise of revolutionizing industries and catalyzing innovation across diverse sectors, including renewable energy, electronics, and healthcare.

References

1. Jauregui, L.A., Yue, Y., Sidorov, A.N., Hu, J., Yu, Q., Lopez, G., Chen, Y.P.: Thermal transport in graphene nanostructures: experiments and simulations. ECS Trans. **28**(5), 73 (2010)
2. Lee, C., Wei, X., Kysar, J.W., Hone, J.: Measurement of the elastic properties and intrinsic strength of monolayer graphene. Science **321**(5887), 385–388 (2008)
3. Zhu, Y., Murali, S., Cai, W., Li, X., Suk, J.W., Potts, J.R., Ruoff, R.S.: Graphene and graphene oxide: synthesis, properties, and applications. Adv. Mater. **22**(35), 3906–3924 (2010)
4. Berry, V.: Impermeability of graphene and its applications. Carbon **62**, 1–10 (2013)
5. Novoselov, K.S., Geim, A.K., Morozov, S.V., Jiang, D.E., Zhang, Y., Dubonos, S.V., et al.: Electric field effect in atomically thin carbon films. Science **306**(5696), 666–669 (2004)
6. Ciesielski, A., Samorì, P.: Graphene via sonication assisted liquid-phase exfoliation. Chem. Soc. Rev. **43**(1), 381–398 (2014)

7. Lotya, M., King, P.J., Khan, U., De, S., Coleman, J.N.: High-concentration, surfactant-stabilized graphene dispersions. ACS Nano **4**(6), 3155–3162 (2010)
8. Han, J.T., Jang, J.I., Kim, H., Hwang, J.Y., Yoo, H.K., Woo, J.S., Lee, G.W.: Extremely efficient liquid exfoliation and dispersion of layered materials by unusual acoustic cavitation. Sci. Rep. **4**(1), 5133 (2014)
9. Huang, G., Jiang, P., Zhang, X., Zhou, C., Zhou, J., Huang, Y., Hao, Y.: Efficient preparation and characterization of graphene based on ball milling. Diam. Relat. Mater. **130**, 109438 (2022)
10. Yi, M., Shen, Z.: A review on mechanical exfoliation for the scalable production of graphene. J. Mater. Chem. A **3**(22), 11700–11715 (2015)
11. Kovtyukhova, N.I., Wang, Y., Berkdemir, A., Cruz-Silva, R., Terrones, M., Crespi, V.H., Mallouk, T.E.: Non-oxidative intercalation and exfoliation of graphite by Brønsted acids. Nat. Chem. **6**(11), 957–963 (2014)
12. Lavi, A., Pyrikov, M., Ohayon-Lavi, A., Tadmor, R., Shachar-Michaely, G., Leibovitch, Y., et al.: Total exfoliation of graphite in molten salts. Phys. Chem. Chem. Phys. **25**(3), 2618–2628 (2023)
13. Ou, E., Xie, Y., Peng, C., Song, Y., Peng, H., Xiong, Y., Xu, W.: High concentration and stable few-layer graphene dispersions prepared by the exfoliation of graphite in different organic solvents. RSC Adv. **3**(24), 9490–9499 (2013)
14. Hummers, W.S., Jr., Offeman, R.E.: Preparation of graphitic oxide. J. Am. Chem. Soc. **80**(6), 1339–1339 (1958)
15. Viinikanoja, A., Kauppila, J., Damlin, P., Mäkilä, E., Leiro, J., Ääritalo, T., Lukkari, J.: Interactions between graphene sheets and ionic molecules used for the shear-assisted exfoliation of natural graphite. Carbon **68**, 195–209 (2014)
16. Lavin-Lopez, M., Valverde, J.L., Sanchez-Silva, L., Romero, A.: Solvent-based exfoliation via sonication of graphitic materials for graphene manufacture. Ind. Eng. Chem. Res. **55**(4), 845–855 (2016)
17. Wang, X., Zhi, L., Müllen, K.: Transparent, conductive graphene electrodes for dye-sensitized solar cells. Nano Lett. **8**(1), 323–327 (2008)
18. Park, S., Ruoff, R.S.: Chemical methods for the production of graphenes. Nat. Nanotechnol. **4**(4), 217–224 (2009)
19. Meyer-Plath, A., Beckert, F., Tölle, F.J., Sturm, H., Mülhaupt, R.: Stable aqueous dispersions of functionalized multi-layer graphene by pulsed underwater plasma exfoliation of graphite. J. Phys. D Appl. Phys. **49**(4), 045301 (2015)
20. Lee, H., Bratescu, M.A., Ueno, T., Saito, N.: Solution plasma exfoliation of graphene flakes from graphite electrodes. RSC Adv. **4**(93), 51758–51765 (2014)
21. Segundo, E.H., Fontana, L.C., Recco, A.A., Scholtz, J.S., Vomstein, M.A.N., Becker, D.: Graphene nanosheets obtained through graphite powder exfoliation in pulsed underwater electrical discharge. Mater. Chem. Phys. **217**, 1–4 (2018)
22. Zhang, Y.I., Zhang, L., Zhou, C.: Review of chemical vapor deposition of graphene and related applications. Acc. Chem. Res. **46**(10), 2329–2339 (2013)
23. Plutnar, J., Pumera, M., Sofer, Z.: The chemistry of CVD graphene. J. Mater. Chem. C **6**(23), 6082–6101 (2018)
24. Munoz, R., Gómez-Aleixandre, C.: Review of CVD synthesis of graphene. Chem. Vapor Deposition **19**(10–11–12), 297–322 (2013)
25. Deng, B., Liu, Z., Peng, H.: Toward mass production of CVD graphene films. Adv. Mater. **31**(9), 1800996 (2019)
26. Yu, X.Z., Hwang, C.G., Jozwiak, C.M., Köhl, A., Schmid, A.K., Lanzara, A.: New synthesis method for the growth of epitaxial graphene. J. Electron Spectrosc. Relat. Phenom. **184**(3–6), 100–106 (2011)
27. Tetlow, H., De Boer, J.P., Ford, I.J., Vvedensky, D.D., Coraux, J., Kantorovich, L.: Growth of epitaxial graphene: theory and experiment. Phys. Rep. **542**(3), 195–295 (2014)
28. De Heer, W.A., Berger, C., Wu, X., First, P.N., Conrad, E.H., Li, X., et al.: Epitaxial graphene. Solid State Commun. **143**(1–2), 92–100 (2007)

29. Shan, J., Sun, J., Liu, Z.: Chemical vapor deposition synthesis of graphene over sapphire substrates. ChemNanoMat **7**(5), 515–525 (2021)
30. Gao, M., Pan, Y., Huang, L., Hu, H., Zhang, L.Z., Guo, H.M., et al.: Epitaxial growth and structural property of graphene on Pt (111). Appl. Phys. Lett. **98**(3) (2011)
31. Salzmann, I., Moser, A., Oehzelt, M., Breuer, T., Feng, X., Juang, Z.Y., et al.: Epitaxial growth of π-stacked perfluoropentacene on graphene-coated quartz. ACS Nano **6**(12), 10874–10883 (2012)
32. Yang, W., Chen, G., Shi, Z., Liu, C.C., Zhang, L., Xie, G., et al.: Epitaxial growth of single-domain graphene on hexagonal boron nitride. Nat. Mater. **12**(9), 792–797 (2013)
33. Moreau, E., Ferrer, F.J., Vignaud, D., Godey, S., Wallart, X.: Graphene growth by molecular beam epitaxy using a solid carbon source. Physica Status Solidi (a) **207**(2), 300–303 (2010)
34. Wofford, J.M., Oliveira, M.H., Schumann, T., Jenichen, B., Ramsteiner, M., Jahn, U., Riechert, H.: Molecular beam epitaxy of graphene on ultra-smooth nickel: growth mode and substrate interactions. New J. Phys. **16**(9), 093055 (2014)
35. Garcia, J.M., Wurstbauer, U., Levy, A., Pfeiffer, L.N., Pinczuk, A., Plaut, A.S., Taniguchi, T.: Graphene growth on h-BN by molecular beam epitaxy. Solid State Commun. **152**(12), 975–978 (2012)
36. Lippert, G., Dabrowski, J., Yamamoto, Y., Herziger, F., Maultzsch, J., Baringhaus, J., Lupina, G.: Molecular beam epitaxy of graphene on mica. Physica Status Solidi (b) **249**(12), 2507–2510 (2010)
37. Hussain, A., Mehdi, S.M., Abbas, N., Hussain, M., Naqvi, R.A.: Synthesis of graphene from solid carbon sources: a focused review. Mater. Chem. Phys. **248**, 122924 (2020)
38. Fernández-Garrido, S., Ramsteiner, M., Gao, G., Galves, L.A., Sharma, B., Corfdir, P., Geelhaar, L.: Molecular beam epitaxy of GaN nanowires on epitaxial graphene. Nano Lett. **17**(9), 5213–5221 (2017)
39. Zuo, Z., Xu, Z., Zheng, R., Khanaki, A., Zheng, J.G., Liu, J.: In-situ epitaxial growth of graphene/h-BN van der Waals heterostructures by molecular beam epitaxy. Sci. Rep. **5**(1), 14760 (2015)
40. Ferrari, A.C., et al.: Raman spectrum of graphene and graphene layers. Phys. Rev. Lett. **97**(18), 1–4 (2006). https://doi.org/10.1103/PhysRevLett.97.187401
41. Nanda, S.S., Kim, M.J., Yeom, K.S., An, S.S.A., Ju, H., Yi, D.K.: Raman spectrum of graphene with its versatile future perspectives. TrAC—Trends Anal. Chem. **80**, 125–131 (2016). https://doi.org/10.1016/j.trac.2016.02.024
42. Dresselhaus, M.S., Jorio, A., Souza Filho, A.G., Saito, R.: Defect characterization in graphene and carbon nanotubes using Raman spectroscopy. Philos. Trans. R. Soc. A Math. Phys. Eng. Sci. **368**(1932), 5355–5377 (2010). https://doi.org/10.1098/rsta.2010.0213
43. Thomas, R., Rao, G.M.: Synthesis of 3-dimensional porous graphene nanosheets using electron cyclotron resonance plasma enhanced chemical vapour deposition. RSC Adv. **5**(103), 84927–84935 (2015). https://doi.org/10.1039/c5ra09087c
44. Chen, X., Zhao, X., Gao, Y., Yin, J., Bai, M., Wang, F.: Green synthesis of gold nanoparticles using carrageenan oligosaccharide and their in vitro antitumor activity. Mar. Drugs **16**(8) (2018). https://doi.org/10.3390/md16080277
45. Krishnamoorthy, K., Veerapandian, M., Yun, K., Kim, S.J.: The chemical and structural analysis of graphene oxide with different degrees of oxidation. Carbon N.Y. **53**, 38–49 (2013). https://doi.org/10.1016/j.carbon.2012.10.013
46. Surekha, G., Krishnaiah, K.V., Ravi, N., Padma Suvarna, R.: FTIR, Raman and XRD analysis of graphene oxide films prepared by modified Hummer's method. J. Phys. Conf. Ser. **1495**(1) (2020). https://doi.org/10.1088/1742-6596/1495/1/012012
47. Sabzevari, M., Cree, D., Wilson, L.: Preparation and characterization of graphene oxide cross-linked composites. **2014** (2018). https://doi.org/10.25071/10315/35427
48. Siburian, R., Sihotang, H., Lumban Raja, S., Supeno, M., Simanjuntak, C.: New route to synthesize of graphene nano sheets. Orient. J. Chem. **34**(1), 182–187 (2018). https://doi.org/10.13005/ojc/340120

49. Zhang, X., et al.: Impact of chlorine functionalization on high-mobility chemical vapor deposition grown graphene. ACS Nano **7**(8), 7262–7270 (2013). https://doi.org/10.1021/nn4026756
50. Teklu, A., et al.: Mechanical characterization of reduced graphene oxide using AFM. Adv. Condens. Matter Phys. **2019** (2019). https://doi.org/10.1155/2019/8713965
51. Hidayah, N.M.S., et al.: Comparison on graphite, graphene oxide and reduced graphene oxide: Synthesis and characterization. AIP Conf. Proc. **1892** (2017). https://doi.org/10.1063/1.5005764
52. Manoratne, C.H., Rosa, S.R.D., Kottegoda, I.R.M.: XRD-HTA, UV visible, FTIR and SEM interpretation of reduced graphene oxide synthesize purity vein graphite. Mater. Sci. Res. India **14**(1), 19–30 (2017). https://doi.org/10.13005/msri/140104
53. Kochnev, A.S., Ovid'Ko, I.A., Semenov, B.N.: Tensile strength of graphene containing 5-8-5 defects. Rev. Adv. Mater. Sci. **37** (2014)
54. Bolotin, K.I., Sikes, K.J., Jiang, Z., Klima, M., Fudenberg, G., Hone, J., Stormer, H.L.: Ultrahigh electron mobility in suspended graphene. Solid State Commun. **146**(9–10), 351–355 (2008)
55. Nika, D.L., Ghosh, S., Pokatilov, E.P., Balandin, A.A.: Lattice thermal conductivity of graphene flakes: comparison with bulk graphite. Appl. Phys. Lett. **94**(20) (2009)
56. Korkut, S., Roy-Mayhew, J.D., Dabbs, D.M., Milius, D.L., Aksay, I.A.: High surface area tapes produced with functionalized graphene. ACS Nano **5**(6), 5214–5222 (2011)
57. Leem, J., Wang, M.C., Kang, P., Nam, S.: Mechanically self-assembled, three-dimensional graphene–gold hybrid nanostructures for advanced nanoplasmonic sensors. Nano Lett. **15**(11), 7684–7690 (2015)
58. Xu, L.Y., Yang, G.Y., Jing, H.Y., Wei, J., Han, Y.D.: Ag–graphene hybrid conductive ink for writing electronics. Nanotechnology **25**(5), 055201 (2014)
59. Giovanni, M., Poh, H.L., Ambrosi, A., Zhao, G., Sofer, Z., Šaněk, F., et al.: Noble metal (Pd, Ru, Rh, Pt, Au, Ag) doped graphene hybrids for electrocatalysis. Nanoscale **4**(16), 5002–5008 (2012)
60. Huang, J., Zhang, L., Chen, B., Ji, N., Chen, F., Zhang, Y., Zhang, Z.: Nanocomposites of size-controlled gold nanoparticles and graphene oxide: formation and applications in SERS and catalysis. Nanoscale **2**(12), 2733–2738 (2010)
61. Liu, J., Fu, S., Yuan, B., Li, Y., Deng, Z.: Toward a universal "adhesive nanosheet" for the assembly of multiple nanoparticles based on a protein-induced reduction/decoration of graphene oxide. J. Am. Chem. Soc. **132**(21), 7279–7281 (2010)
62. Li, J., Kuang, D., Feng, Y., Zhang, F., Xu, Z., Liu, M., Wang, D.: Green synthesis of silver nanoparticles–graphene oxide nanocomposite and its application in electrochemical sensing oftryptophan. Biosens. Bioelectron. **42**, 198–206 (2013)
63. Yang, J., Deng, S., Lei, J., Ju, H., Gunasekaran, S.: Electrochemical synthesis of reduced graphene sheet–AuPd alloy nanoparticle composites for enzymatic biosensing. Biosens. Bioelectron. **29**(1), 159–166 (2011)
64. Claussen, J.C., Kumar, A., Jaroch, D.B., Khawaja, M.H., Hibbard, A.B., Porterfield, D.M., Fisher, T.S.: Nanostructuring platinum nanoparticles on multilayered graphene petal nanosheets for electrochemical biosensing. Adv. Func. Mater. **22**(16), 3399–3405 (2012)
65. Cai, D., Yusoh, K., Song, M.: The mechanical properties and morphology of a graphite oxide nanoplatelet/polyurethane composite. Nanotechnology **20**(8), 085712 (2009)
66. Zhang, H.B., Zheng, W.G., Yan, Q., Yang, Y., Wang, J.W., Lu, Z.H., Yu, Z.Z.: Electrically conductive polyethylene terephthalate/graphene nanocomposites prepared by melt compounding. Polymer **51**(5), 1191–1196 (2010)
67. Wei, T., Luo, G., Fan, Z., Zheng, C., Yan, J., Yao, C., Zhang, C.: Preparation of graphene nanosheet/polymer composites using in situ reduction–extractive dispersion. Carbon **47**(9), 2296–2299 (2009)
68. Xu, Y., Hong, W., Bai, H., Li, C., Shi, G.: Strong and ductile poly (vinyl alcohol)/graphene oxide composite films with a layered structure. Carbon **47**(15), 3538–3543 (2009)

69. Kim, I.H., Jeong, Y.G.: Polylactide/exfoliated graphite nanocomposites with enhanced thermal stability, mechanical modulus, and electrical conductivity. J. Polym. Sci. Part B: Polym. Phys. **48**(8), 850–858 (2010)
70. Yan, J., Wei, T., Shao, B., Fan, Z., Qian, W., Zhang, M., Wei, F.: Preparation of a graphene nanosheet/polyaniline composite with high specific capacitance. Carbon **48**(2), 487–493 (2010)
71. Kulkarni, D.D., Choi, I., Singamaneni, S.S., Tsukruk, V.V.: Graphene oxide–polyelectrolyte nanomembranes. ACS Nano **4**(8), 4667–4676 (2010)
72. Zhao, X., Zhang, Q., Hao, Y., Li, Y., Fang, Y., Chen, D.: Alternate multilayer films of poly (vinyl alcohol) and exfoliated graphene oxide fabricated via a facial layer-by-layer assembly. Macromolecules **43**(22), 9411–9416 (2010)
73. Salavagione, H.J., Gomez, M.A., Martínez, G.: Polymeric modification of graphene through esterification of graphite oxide and poly (vinyl alcohol). Macromolecules **42**(17), 6331–6334 (2009)
74. Björk, J., Hanke, F., Palma, C.A., Samori, P., Cecchini, M., Persson, M.: Adsorption of aromatic and anti-aromatic systems on graphene through $\pi-\pi$ stacking. J. Phys. Chem. Lett. **1**(23), 3407–3412 (2010)
75. Tang, Y., Zhang, Z., Li, L., Guo, J., Yang, P.: Thermal transport enhancement resolution for graphene/Si and graphene/SiC interfaces. Int. J. Therm. Sci. **171**, 107231 (2022)
76. Bhanvase, B.A., Shende, T.P., Sonawane, S.H.: A review on graphene–TiO_2 and doped graphene–TiO_2 nanocomposite photocatalyst for water and wastewater treatment. Environ. Technol. Rev. **6**(1), 1–14 (2017)
77. Zeng, Z., Liu, Y., Chen, W., Li, X., Zheng, Q., Li, K., Guo, R.: Fabrication and properties of in situ reduced graphene oxide-toughened zirconia composite ceramics. J. Am. Ceram. Soc. **101**(8), 3498–3507 (2018)
78. Porwal, H., Tatarko, P., Grasso, S., Khaliq, J., Dlouhý, I., Reece, M.J.: Graphene reinforced alumina nano-composites. Carbon **64**, 359–369 (2013)
79. Fan, Y., Wang, L., Li, J., Li, J., Sun, S., Chen, F., et al.: Preparation and electrical properties of graphene nanosheet/Al_2O_3 composites. Carbon **48**(6), 1743–1749 (2010)
80. Fan, Y., Jiang, W., Kawasaki, A.: Highly conductive few-layer graphene/Al_2O_3 nanocomposites with tunable charge carrier type. Adv. Func. Mater. **22**(18), 3882–3889 (2012)
81. Lee, B., Koo, M.Y., Jin, S.H., Kim, K.T., Hong, S.H.: Simultaneous strengthening and toughening of reduced graphene oxide/alumina composites fabricated by molecular-level mixing process. Carbon **78**, 212–219 (2014)
82. Wang, X., Lu, M., Qiu, L., Huang, H., Li, D., Wang, H., Cheng, Y.B.: Graphene/titanium carbide composites prepared by sol–gel infiltration and spark plasma sintering. Ceram. Int. **42**(1), 122–131 (2016)
83. Rahman, A., Singh, A., Karumuri, S., Harimkar, S.P., Kalkan, K.A., Singh, R.P.: Graphene reinforced silicon carbide nanocomposites: processing and properties. In: Composite, Hybrid, and Multifunctional Materials, Volume 4: Proceedings of the 2014 Annual Conference on Experimental and Applied Mechanics, pp. 165–176. Springer International Publishing (2015)
84. Miranzo, P., Ramírez, C., Román-Manso, B., Garzón, L., Gutiérrez, H.R., Terrones, M., Belmonte, M.: In situ processing of electrically conducting graphene/SiC nanocomposites. J. Eur. Ceram. Soc. **33**(10), 1665–1674 (2013)
85. Walker, L.S., Marotto, V.R., Rafiee, M.A., Koratkar, N., Corral, E.L.: Toughening in graphene ceramic composites. ACS Nano **5**(4), 3182–3190 (2011)
86. Al-Saleh, M.H.: Electrical and mechanical properties of graphene/carbon nanotube hybrid nanocomposites. Synth. Met. **209**, 41–46 (2015)
87. Jyoti, J., Singh, B.P.: A review on 3D graphene–carbon nanotube hybrid polymer nanocomposites. J. Mater. Sci. **56**(31), 17411–17456 (2021)
88. Tung, V.C., Chen, L.M., Allen, M.J., Wassei, J.K., Nelson, K., Kaner, R.B., Yang, Y.: Low-temperature solution processing of graphene–carbon nanotube hybrid materials for high-performance transparent conductors. Nano Lett. **9**(5), 1949–1955 (2009)

89. Konstantatos, G., Badioli, M., Gaudreau, L., Osmond, J., Bernechea, M., De Arquer, F.P.G., Koppens, F.H.: Hybrid graphene–quantum dot phototransistors with ultrahigh gain. Nat. Nanotechnol. **7**(6), 363–368 (2012)
90. Liu, Y., Wu, P.: Graphene quantum dot hybrids as efficient metal-free electrocatalyst for the oxygen reduction reaction. ACS Appl. Mater. Interfaces **5**(8), 3362–3369 (2013)
91. Chiang, C.W., Haider, G., Tan, W.C., Liou, Y.R., Lai, Y.C., Ravindranath, R., Chen, Y.F.: Highly stretchable and sensitive photodetectors based on hybrid graphene and graphene quantum dots. ACS Appl. Mater. Interfaces **8**(1), 466–471 (2016)
92. Anichini, C., Samorì, P.: Graphene-based hybrid functional materials. Small **17**(33), 1–48 (2021). https://doi.org/10.1002/smll.202100514
93. Sang, M., Shin, J., Kim, K., Yu, K.J.: Electronic and thermal properties of graphene and recent advances in graphene based electronics applications. Nanomaterials **9**(3), 1–33 (2019). https://doi.org/10.3390/nano9030374
94. Liao, L., Duan, X.: Graphene for radio frequency electronics. Mater. Today **15**(7–8), 328–338 (2012). https://doi.org/10.1016/S1369-7021(12)70138-4
95. Liu, J., Liang, Q., Zhao, R., Lei, S., Hu, W.: Application of organic-graphene hybrids in high performance photodetectors. Mater. Chem. Front. **4**(2), 354–368 (2020). https://doi.org/10.1039/c9qm00517j
96. Yu, R., Pruneri, V., García De Abajo, F.J.: Resonant visible light modulation with graphene. ACS Photon. **2**(4), 550–558 (2015). https://doi.org/10.1021/ph5004829
97. Velasco Davoise, L., Díez-Pascual, A.M., Peña Capilla, R.: Application of graphene-related materials in organic solar cells. Materials (Basel) **15**(3) (2022). https://doi.org/10.3390/ma15031171
98. Joy, R., et al.: Graphene: chemistry and applications for lithium-ion batteries. Electrochem **3**(1), 143–183 (2022). https://doi.org/10.3390/electrochem3010010
99. Wen, Y., Liu, H., Jiang, X.: Preparation of graphene by exfoliation and its application in lithium-ion batteries. J. Alloys Compd. **961**, 170885 (2023). https://doi.org/10.1016/j.jallcom.2023.170885
100. Cai, X., Lai, L., Shen, Z., Lin, J.: Graphene and graphene-based composites as Li-ion battery electrode materials and their application in full cells. J. Mater. Chem. A **5**(30), 15423–15446 (2017). https://doi.org/10.1039/c7ta04354f
101. Wasalathilake, K.C., Li, H., Xu, L., Yan, C.: Recent advances in graphene based materials as anode materials in sodium-ion batteries. J. Energy Chem. **42**, 91–107 (2020). https://doi.org/10.1016/j.jechem.2019.06.016
102. Lu, Y., Lu, Y., Niu, Z., Chen, J.: Graphene-based nanomaterials for sodium-ion batteries. Adv. Energy Mater. **8**(17), 1–21 (2018). https://doi.org/10.1002/aenm.201702469
103. Oyedotun, K.O., Manyala, N.: Graphene foam–based electrochemical capacitors. Curr. Open. Electrochem. **21**, 125–131 (2020). https://doi.org/10.1016/j.coelec.2019.12.010
104. Wu, H., et al.: Graphene based architectures for electrochemical capacitors. Energy Storage Mater. **5**, 8–32 (2016). https://doi.org/10.1016/j.ensm.2016.05.003
105. Zhang, H., Yang, D., Lau, A., Ma, T., Lin, H., Jia, B.: Hybridized graphene for supercapacitors: beyond the limitation of pure graphene. Small **17**(12), 1–22 (2021). https://doi.org/10.1002/smll.202007311
106. Ma, Y., Chang, H., Zhang, M., Chen, Y.: Graphene-based materials for lithium-ion hybrid supercapacitors. Adv. Mater. **27**(36), 5296–5308 (2015). https://doi.org/10.1002/adma.201501622
107. Osman, S.H., Kamarudin, S.K., Karim, N.A., Basri, S.: Application of graphene in low temperature fuel cell technology: an overview. Int. J. Energy Res. **45**(13), 18318–18336 (2021). https://doi.org/10.1002/er.6969
108. Su, H., Hu, Y.H.: Recent advances in graphene-based materials for fuel cell applicating. Sci. Eng. **9**(7), 958–983 (2021). https://doi.org/10.1002/ese3.833
109. Mahmoudi, T., Wang, Y., Hahn, Y.B.: Graphene and its derivatives for solar cells application. Nano Energy **47**(January), 51–65 (2018). https://doi.org/10.1016/j.nanoen.2018.02.047

110. Yusuf, M., Kumar, M., Khan, M.A., Sillanpääillanpa, M., Arafat, H.: A review on exfoliation, characterization, environmental and energy applications of graphene and graphene-based composites. Adv. Colloid Interface Sci. **273**, 1–23 (2019). https://doi.org/10.1016/j.cis.2019.102036
111. Phiri, J., Gane, P., Maloney, T.C.: General overview of graphene: Production, properties and application in polymer composites. Mater. Sci. Eng. B **215**, 9–28 (2017). https://doi.org/10.1016/j.mseb.2016.10.004
112. Guler, O., Bagci, N.: A short review on mechanical properties of graphene reinforced metal matrix composites. J. Mater. Res. Technol. **9**(3), 6808–6833 (2020). https://doi.org/10.1016/j.jmrt.2020.01.077
113. Zhao, Z., et al.: An overview of graphene and its derivatives reinforced metal matrix composites: preparation, properties and applications. Carbon N. Y. **170**, 302–326 (2020). https://doi.org/10.1016/j.carbon.2020.08.040
114. Ramírez, C., Belmonte, M., Miranzo, P., Osendi, M.I.: Applications of ceramic/graphene composites and hybrids. Materials (Basel) **14**(8) (2021). https://doi.org/10.3390/ma14082071
115. Markandan, K., Chin, J.K., Tan, M.T.T.: Recent progress in graphene based ceramic composites: a review. J. Mater. Res. **32**(1), 84–106 (2017). https://doi.org/10.1557/jmr.2016.390
116. Dang, V.T., et al.: Recent trends in preparation and application of carbon nanotube-graphene hybrid thin films. Adv. Nat. Sci. Nanosci. Nanotechnol. **7**(3) (2016). https://doi.org/10.1088/2043-6262/7/3/033002
117. Maarouf, A.A., Kasry, A., Chandra, B., Martyna, G.J.: A graphene-carbon nanotube hybrid material for photovoltaic applications. Carbon N. Y. **102**, 74–80 (2016). https://doi.org/10.1016/j.carbon.2016.02.024
118. Hashemzadeh, H., Raissi, H.: Understanding loading, diffusion and releasing of Doxorubicin and Paclitaxel dual delivery in graphene and graphene oxide carriers as highly efficient drug delivery systems. Appl. Surf. Sci. **500**, 144220 (2020). https://doi.org/10.1016/j.apsusc.2019.144220
119. Ye, Y., Mao, X., Xu, J., Kong, J., Hu, X.: Functional graphene oxide nanocarriers for drug delivery. Int. J. Polym. Sci. **2019** (2019). https://doi.org/10.1155/2019/8453493
120. Suvarnaphaet, P., Pechprasarn, S.: Graphene-based materials for biosensors: a review. Sensors (Switzerland) **17**(10) (2017). https://doi.org/10.3390/s17102161
121. Jiang, Z., Feng, B., Xu, J., Qing, T., Zhang, P., Qing, Z.: Graphene biosensors for bacterial and viral pathogens. Biosens. Bioelectron. **166**, 112471 (2020) https://doi.org/10.1016/j.bios.2020.112471
122. Wang, W., et al.: Engineered 3D printed poly(ε-caprolactone)/graphene scaffolds for bone tissue engineering. Mater. Sci. Eng. C **100**(March), 759–770 (2019). https://doi.org/10.1016/j.msec.2019.03.047
123. Jakus, A.E., Secor, E.B., Rutz, A.L., Jordan, S.W., Hersam, M.C., Shah, R.N.: Three-dimensional printing of high-content graphene scaffolds for electronic and biomedical applications. ACS Nano **9**(4), 4636–4648 (2015). https://doi.org/10.1021/acsnano.5b01179
124. Hui, F., et al.: Graphene coated nanoprobes: a review. Crystals **7**(9) (2017) https://doi.org/10.3390/cryst7090269
125. Hui, F., et al.: Moving graphene devices from lab to market: advanced graphene-coated nanoprobes. Nanoscale **8**(16), 8466–8473 (2016). https://doi.org/10.1039/c5nr06235g
126. Lage, T., Rodrigues, R.O., Catarino, S., Gallo, J., Bañobre-López, M., Minas, G.: Graphene-based magnetic nanoparticles for theranostics: an overview for their potential in clinical application. Nanomaterials **11**(5) (2021). https://doi.org/10.3390/nano11051073
127. Viseu, T., Lopes, C.M., Fernandes, E., Real Oliveira, M.E.C.D., Lúcio, M.: A systematic review and critical analysis of the role of graphene-based nanomaterials in cancer theranostics. Pharmaceutics **10**(4) (2018). https://doi.org/10.3390/pharmaceutics10040282
128. Alhazmi, H.A., et al.: Graphene-based biosensors for disease theranostics: development, applications, and recent advancements. Nanotechnol. Rev. **11**(1), 96–116 (2021). https://doi.org/10.1515/ntrev-2022-0009

129. Ali, I., et al.: Graphene based adsorbents for remediation of noxious pollutants from wastewater. Environ. Int. **127**(March), 160–180 (2019). https://doi.org/10.1016/j.envint.2019.03.029
130. Baig, N., Ihsanullah, M.S., Saleh, T.A.: Graphene-based adsorbents for the removal of toxic organic pollutants: a review. J. Environ. Manag. **244**, 370–382 (2019). https://doi.org/10.1016/j.jenvman.2019.05.047
131. Sun, M., Li, J.: Graphene oxide membranes: Functional structures, preparation environmental applications. Nano Today **20**, 121–137 (2018). https://doi.org/10.1016/j.nantod.2018.04.007
132. Sun, P., Wang, K., Zhu, H.: Recent developments in graphene-based membranes: structure, mass-transport mechanism and potential applications. Adv. Mater. **28**(12), 2287–2310 (2016). https://doi.org/10.1002/adma.201502595
133. Homaeigohar, S., Elbahri, M.: Graphene membranes for water desalination. NPG Asia Mater. **9**(8), 1–16 (2017). https://doi.org/10.1038/AM.2017.135
134. Parvizi, R., Azad, S., Dashtian, K., Ghaedi, M., Heidari, H.: Natural source-based graphene as sensitising agents for air quality monitoring. Sci. Rep. **9**(1), 1–15 (2019). https://doi.org/10.1038/s41598-019-40433-9
135. Karim, N.A., Ghazali, C.M.R., Ramli, M.M., Sandu, A.V., Chirila, E., Abdullah, M.M.A.: Optimization of processing parameters for graphitization of oil palm trunk waste at lower heating temperature. IOP Conf Ser Mater Sci Eng **572**(1) (2019). https://doi.org/10.1088/1757-899X/572/1/012060
136. Karim, N.A., et al.: Graphene oxide production via exfoliation process of low-temperature synthetic graphite from oil palm trunk waste. Acta Phys. Pol. A **142**(1), 81–83 (2022). https://doi.org/10.12693/APhysPolA.142.81
137. Karim, N.A., Ramli, M.M., Ghazali, C.M.R., Osman, N.H., Denesh, G., Nadia, C.A.I.: Synthesis of graphene oxide from synthetic graphite utilizing oil palm trunk waste via modification of hummers' method. AIP Conf. Proc. **2339** (2021). https://doi.org/10.1063/5.0044867
138. Taheri Najafabadi, A.: Emerging applications of graphene and its derivatives in carbon capture and conversion: Current status and future prospects. Renew. Sustain. Energy Rev. **41**, 1515–1545 (2015). https://doi.org/10.1016/j.rser.2014.09.022

Hybrid Nanomaterials in Biosensors for Advanced Point-of-Care Diagnostics: DNA-, Enzyme-, or Immunosensor-Based, for Healthcare

Adilah Ayoib

1 Nanomaterials Discovery: From Ancient Advances to Modern Innovations

Throughout history, nanoscale materials have quietly exerted significant influence. In the 4th century, the resourceful Romans utilized minuscule gold and silver particles, approximately 10–100 nm in size, to produce glass that exhibited iridescent hues: when lit from the outside, the glass appeared slate green, yet it emitted a red glow when lit from within [1]. While their comprehension of the scientific underpinnings behind this phenomenon may have been poor, their practical application of nanotechnology was unquestionably inventive. The renowned Lycurgus Cup (Fig. 1) serves as a noteworthy example displaying the significant use of nanomaterials in ancient times. It was only later in 1990 that transmission electron microscopy (TEM) analysis revealed that dichroism in the cup resulted from the dispersion of 66.2% silver, 31.2% gold, and 2.6% copper nanoparticles, typically 50–100 nm in diameter [2]. The unique light absorption properties of these nanoparticles led to the cup showing both red and green colors.

Similar to ancient Roman practices, European medieval artisans in the 14th century also understood that incorporating insignificant amounts of gold and silver into stained-glass windows facilitated the production of vibrant reds and yellows [12], exemplifying the premodern use of nanomaterials. In fact, the silver stain technique, initially discovered in Islamic lands (specifically Egypt and Syria) from the 7th

A. Ayoib (✉)
Faculty of Chemical Engineering & Technology, Universiti Malaysia Perlis (UniMAP), 02100 Padang Besar, Perlis, Malaysia
e-mail: adilahayoib@unimap.edu.my

Institute of Nano Electronic Engineering, Universiti Malaysia Perlis (UniMAP), 01000 Kangar, Perlis, Malaysia

Carbon Sustainability Nexus (CaSNex), Special Interest Group (SIG), Universiti Malaysia Perlis (UniMAP), 02600 Arau, Perlis, Malaysia

S. C. B. Gopinath and M. M. Ramli (eds.), *Hybrid-Nanomaterials*, Engineering Materials, https://doi.org/10.1007/978-981-97-9022-7_7

Fig. 1 Historical examples of nanomaterials in art and artifacts **a** The dichroic glass of the Lycurgus cup (4th century) demonstrates color changes from red-purple to yellow-green due to the unique light absorption properties of silver, gold, and copper nanoparticles. ©The British Museum [3]. **b** Stained glass of the South rose window of Chartres Cathedral, France (1194–1220 CE), utilizing nanoparticle plasmonic resonances to produce colorful pictures. ©dvdbramhall [4]. **c** Stained glass window of Al-Aqsa Masjid, Palestine (685–715 CE), embedded within nanometer-scale particles of gold. ©Elazhar [5]. **d** Gold and silver nanoparticles exhibit varied color properties based on their sizes and shapes, due to light interacting with their electrons, causing resonance at various frequencies, resulting in the different colors observed in window glasses [6, 7]. **e** Egyptian bowl (9th century) painted with a metallic stain of copper and silver nanoparticles, which produces red or brown, and yellow colors, respectively. ©The Corning Museum of Glass [8]. **f** Meissen porcelain teapot containing gold nanoparticles creating iridescence (1723–1724). ©The Art Institute of Chicago [9]. Chinese painted enameled porcelains and metalware from the Qing dynasty: **g** R1175 tray containing gold, silver, and copper nanoparticles (1735), **h** R975 bottle containing copper nanoparticles (1730–1796). ©Musée du Louvre [10]. **i** Iranian saber (shamshir) of Damascus steel (1777–1778) containing carbon nanotubes encapsulated with iron-carbide (cementite) nanowires. ©The MET Museum [11]

through the 9th century, involved metallic silver nanoparticles (1–30 nm) dispersed in the glass for stained glasses, and was later adapted with variations for luster ceramics [13, 14]. Recent evidence has verified that red glass, colored by copper metal nanoparticles, dated back to the European Bronze Age [15]. Concurrently, artists from various regions, including China, western Asia, Europe, and the Middle East, also utilized silver and copper nanoparticles in ceramics, particularly in pottery glazes, to bestow a distinctive luster to items such as tiles and bowls [16–19]. Another historical finding can also be depicted through modern TEM analysis of a 17th-century sample of Damascus saber steel that revealed the presence of carbon nanotubes and cementite nanowires [20]. This steel, rooted in "wootz" steel originating from India in 900 CE [21], sheds light on the exceptional properties and distinctive banding pattern present in this ultrahigh-carbon steel, derived from a lost ancient recipe.

The 19th century heralded a more informed exploration of these minute materials when scientist Michael Faraday undertook rigorous research to elucidate the mechanisms by which gold nanoparticles generate solutions of diverse colors under specific conditions. His investigation of nanosized gold colloidal suspensions led to a detailed understanding of their distinctive optical and electronic properties. This intersection of art and science showcases the enduring impact of nanoscale materials throughout history. Fast forward to 1959, and physicist Richard Feynman, who then, in his lecture, asked if we could "write the entire 24 volumes of the Encyclopedia Britannica on the head of a pin?" proved his hypotheses correct with the concept of nanotechnology and became the father of modern nanotechnology [22], though term "nanotechnology" was only coined later by scientist Norio Taniguch in a 1974 paper [23].

In the late 1970 and 1980s, engineer Eric Drexler played a key role in popularizing the potential of molecular nanotechnology, while the inception of modern nanotechnology occurred in 1981 through the creation of the scanning tunneling microscope, enabling observation and manipulation of individual atoms. Subsequently, the atomic force microscope (AFM), considered fundamental to nanotechnology, was developed by Binnig, Quate, and Gerber in 1986 [24], further boosting the development of nanotechnology. The invention of AFM significantly expanded the capabilities of nanoscience, allowed for the imaging of insulating materials at atomic resolution, and enabled direct measurement and manipulation at the atomic scale. Biochemist Ned Seeman played a key role in realizing nanoscale machines through deoxyribonucleic acid (DNA) self-assembly, which led to the emergence of the field of DNA nanotechnology [25], occurring almost thirty years after the DNA structure was first revealed by James Watson and Francis Crick in 1953 [26].

The 1990s saw substantial government funding for nanoscience as a separate discipline, leading to impactful technological developments and interdisciplinary collaboration across various scientific fields. By the early 2000s, the first consumer products incorporating nanotechnology emerged on the market. These included various innovations such as lightweight and resilient automobile bumpers, golf balls for improved flight, stiffer tennis rackets enhancing ball rebound, more flexible and high-performance baseball bats, antibacterial nano-silver socks, clothing resistant to wrinkles and stains, cosmetics with deep-penetrating properties, scratch-resistant glass coatings, quicker-recharging batteries for cordless tools, and advanced displays for electronics [27].

In recent years, scientists and engineers have shown increasing interest in developing hybrid nanomaterials by merging diverse types of nanomaterials to tailor their physicochemical properties, as the amalgamation of different nanomaterials has exhibited exceptional physicochemical properties in contrast to individual nanoparticles. It is widely acknowledged that a single nanomaterial typically lacks all the superior properties necessary for specific applications, rendering them less competitive for practical use.

2 Overview of Hybrid Nanomaterials in Biosensors

Since its conceptualization in 1959, nanotechnology has significantly impacted various fields and has garnered widespread attention in scientific and translational research. Ideal nanoparticles usually fall between 1 and 100 nm and offer unique structures, materials, and devices [12]. By manipulating materials and devices at the nanoscale level, nanotechnology has revolutionized diverse industries, including electronics, medicine, energy, and environmental science. On a scale of measurements, a nanometer (nm) is equivalent to one-billionth of a meter (10^{-9} m) and is incredibly small in size, the scale at which we measure atoms and the molecules they make. For instance, human DNA strands are typically around 2.5 nm long, while a single strand of human hair can be approximately from 80,000 to 100,000 nm wide, and the diameter of a standard ping pong ball is approximately 40,000,000 nm [28] (Fig. 2). The distinct properties and high surface-area/volume ratio of nanoparticles are a consequence of their diminutive size.

Nanomaterials play a crucial role in the realm of nanotechnology, with hybrid nanomaterials representing a novel material category that be widely applied across multiple sectors, notably in biosensors. The capacity of these materials to combine distinct features of various constituents on a nanoscale level has rendered them a favorable option for developing high-performance biosensors, particularly in sophisticated healthcare point-of-care (POC) diagnostics. In a broader sense, biosensors are devices that transform biological occurrences into digital output signals [26]. The

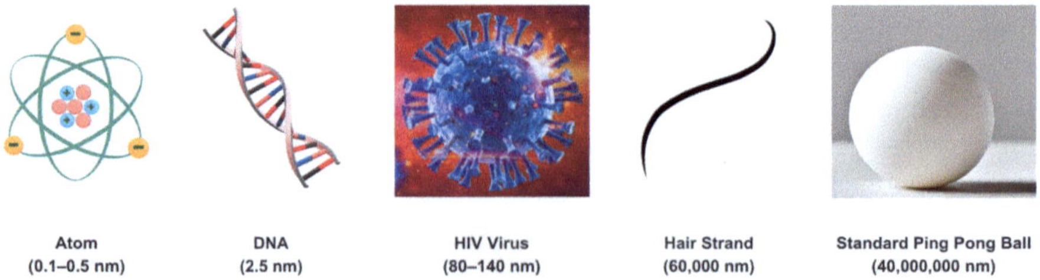

Fig. 2 Size comparison of matters in nanometer dimensions

categorization of biosensors can rely on the transducers and biorecognition elements employed in their fabrication. In general, transducer-based categorization includes electrochemical, optical, magnetic, piezoelectric, and thermoelectric biosensors, while the recognition element-based classification encompasses DNA or ribonucleic acid (RNA) biosensors, enzymatic, immunosensors, and whole-cell sensors. Biosensors first started in the1960s [29] and have since emerged as powerful modern tools for POC diagnostics, distinguished by their capacity to detect biomarkers with high sensitivity, specificity, and rapid response, as well as high throughput detection [30]. This ability makes the biosensors well-suited for identifying various diseases such as infectious diseases, cancer, and metabolic disorders. Furthermore, biosensors are portable, user-friendly, and cost-effective, rendering them ideal for POC diagnostics.

Hybrid nanomaterials, on the other hand, refer to the combination of different types of nanomaterials to create composite structures or systems that offer enhanced properties and functionalities. Nanomaterials consist of a range of elements, such as various metal nanoparticles (including copper, iron, zinc, silver, gold, and nickel), carbon-based nanomaterials (like carbon nanotubes, carbon nanofibers, graphene, graphene oxide, and fullerenes), as well as polymers (such as phenol–formaldehyde resin, resorcinol, polymethyl methacrylate, chitosan, polyvinyl alcohol, and poly(acrylic acid). These hybrid nanomaterials constitute unique chemical combinations of inorganic and organic elements, showcasing varied properties and holding significant potential for applications in nanotechnology [31]. The synergistic effects of integrating multiple nanomaterials allow for improved performance, which can enhance their sensitivity, selectivity, and stability in biosensors, enabling simultaneous detection of various analytes, offering vast advantages over traditional biosensors. Hybrid nanomaterials also lend themselves well to functionalization with biomolecules like antibodies, enzymes, or DNA, facilitating precise and targeted analyte detection.

3 Diverse Applications and Innovations in Hybrid Nanomaterials for Biomedical Technologies

Despite broad utilization of hybrid nanomaterials in environmental treatment methods like adsorption, photocatalysis, catalytic reduction, membrane filtration, and advanced oxidation processes for the elimination of inorganic and organic pollutants from wastewater [32, 33], hybrid nanomaterials also have promising potential as platforms for imaging and therapeutic purposes in biomedical applications. Inorganic nanoparticles like metal-based and metal oxide nanoparticles, semiconductor nanomaterials, ceramic nanomaterials, and lipid-based nanomaterials that are devoid of carbon atoms and enclosed within organic nano compartments, such as liposomes, micelles, poly (lactic-co-glycolic acid) (PLGA) nanoparticles, dendrimers, or chitosan nanoparticles, are classified as hybrid nanomaterials. These hybrid nanomaterials have demonstrated potential for in vivo utilization in the delivery of drugs and medical imaging [34–36].

Current studies emphasize hybrid nanomaterials like metal–organic frameworks (MOFs), graphene-based materials, and polymer-based nanoparticles, with MOFs standing out for their substantial surface area, tunable porosity, and capacity for incorporating various functional groups. Metallic gold and silver nanoparticles have found widespread application in biosensing due to their distinctive optical, electrochemical, and plasmonic properties. The high electron density of these noble metals gives rise to localized surface plasmon resonance, a phenomenon that enhances the local electromagnetic field and thus improves the signal-to-noise ratio in detection. Gold nanoparticles, in particular, are also excellent at facilitating electron transfer, making them valuable in electrochemical biosensors. For example, multiwalled carbon nanotube-gold nanoparticle (MWCNT-AuNP) composites are used for glucose detection and DNA sensing [37–40]. Carbon nanotubes (CNTs) can enhance the performance of biosensors owing to their exceptional conductivity, large surface area, and capability to promote electron transfer. Their tubular structure can be exploited to trap target molecules, enhancing specificity. The combination of gold nanoparticles with multiwalled CNTs provides a substantial surface area for biomolecule immobilization, improved electron transfer properties, and enhanced sensitivity.

Graphene-based nanomaterials, on the other hand, possess excellent electrical and mechanical characteristics, rendering them ideal for application in electrochemical biosensors. For example, magnetic nanoparticles-graphene oxide (MNPs-GO) hybrid nanomaterial has been used for DNA detection and enzyme immobilization [41, 42]. The combination of magnetic nanoparticles and graphene oxide enhances the stability and dispersibility of magnetic nanoparticles while providing an excellent platform for biorecognition elements immobilization. Previous studies have also reported using graphene hybrid nanomaterials for live cell biosensing [43].

Polymer-based nanoparticles with nanostructure characteristics often exhibit improved functionalities associated with increased surface-to-volume ratio. Their flexibility, biocompatibility, and ease of fabrication render them attractive for biomedical sensing applications [44]. An instance of such a hybrid system is the polymeric lipid hybrid nanoparticles (PLNs), which harness the distinctive properties of both polymeric nanoparticles and liposomes, capitalizing on their early therapeutic efficacy while addressing the limitations of each nanoparticle type. This hybrid approach offers a robust drug delivery platform with excellent drug release kinetics, high encapsulation efficiency, and targeted drug uptakes. PLNs have demonstrated the capability to enhance drug biocompatibility and stability, indicative of potential in advancing resilient drug delivery technologies [45]. Another instance is found in the PLGA-based hybrid nanoparticles, exemplifying their potential in drug delivery and theranostic applications [46] that can be implemented with sensing technology.

4 Role of Hybrid Nanomaterials in Advanced Point-of-Care Diagnostics

Hybrid nanomaterials have brought forth a significant impact on the field of advanced point-of-care (POC) diagnostics, particularly within the realm of healthcare. Their integration has instigated a paradigm shift in the diagnostic process, endowing it with augmented levels of precision and specificity in the detection of diverse biomarkers and analytes. This heightened accuracy plays a pivotal role in furnishing precise and prompt assessments of diseases and health conditions, consequently facilitating the implementation of targeted and efficacious treatment modalities. Hybrid nanomaterial-based biosensors also offer heightened sensitivity and selectivity, thereby exceeding the performance of conventional diagnostic methodologies. Their superior capabilities in speed, sensitivity, and portability elevate the accuracy of disease and health condition diagnosis, even at trace levels, surpassing the boundaries of traditional diagnostic practices. These devices can be easily deployed in various healthcare settings, including resource-limited environments, expanding access to advanced diagnostics where traditional methods may be impractical. The versatility in design and function of hybrid nanomaterials allows for customizing biosensors to target specific biomarkers or analytes, thereby enhancing diagnostic capabilities. This multifunctionality enhances the diagnostic capabilities, potentially addressing the limitations of traditional diagnostic approaches.

Notably, advanced biosensors employing hybrid nanomaterials enable rapid testing and real-time monitoring at the POC, thereby facilitating prompt decision-making and interventions. These capabilities hold significant value, particularly in critical healthcare settings where timely diagnoses are paramount for effective treatment and patient care. Hybrid nanomaterials, by virtue of their high surface area to volume ratio, adjustable optical and electrical properties, and bioconjugation capabilities, confer benefits such as signal amplification, enhanced detection limits, and heightened specificity and sensitivity in biosensing applications. We further explore the distinct types of hybrid nanomaterial biosensors implored widely today in healthcare settings, namely, DNA-, enzyme-, and immunosensor-based biosensors.

5 DNA-Based Hybrid Nanomaterial Biosensors

A DNA-based biosensor is composed of three key elements: a recognition unit for interaction with analytes, an amplification unit, and a transducer [47]. It includes an immobilized DNA strand for detecting the complementary sequence through DNA-DNA hybridization, with potential for DNA-RNA and RNA-RNA hybridizations [48]. To date, various DNA-based biosensors have gained traction in bioanalytical applications due to their programmability, stability, and biocompatibility [49–51]. However, in comparison to enzyme- and immunosensor-based biosensors, the availability of DNA biosensors in the market or under research and development is limited, largely due to the cost constraints. Nevertheless, unlike enzymes or antibodies, DNA offers easily synthesizable, highly stable biological recognition layers that can be reused after thermal melting of the DNA duplex [52]. These attributes, among other reasons, have led to increased attention towards DNA-based hybrid nanomaterial biosensors as crucial tools for disease detection, primarily due to their exceptional sensitivity, selectivity, and specificity.

DNA-based hybrid nanomaterial biosensors represent an innovative approach that harnesses the distinctive properties of DNA and nanomaterials for the identification and analysis of biological molecules and disease-related biomarkers, including DNA, RNA, proteins, and other small molecules. These biosensors function by immobilizing the biomarker on their surface, initiating a chemical reaction that generates a detectable signal, thereby facilitating disease detection. Their design involves integrating DNA with various nanomaterials, such as graphene, nanoparticles, and nanotubes, to create hybrid structures with enhanced sensing capabilities. The DNA component selectively binds to the target analyte through complementary base pairing or other specific recognition mechanisms, while the nanomaterial component enhances the biosensor's sensitivity, stability, and electronic properties, enabling precise and efficient detection of the target molecule. Another way to look at it is that DNA-based biosensors operate based on the specific DNA hybridization principle directly on the surface of a physical transducer, leveraging immobilized DNA strands as recognition elements to identify the presence of different target molecules and facilitate their quantification by detecting measurable signal changes.

DNA-based hybrid nanomaterial biosensors encompass diverse functional forms in the field of bioanalytical technology with various transduction mechanisms. They can exist as electrochemical, optical, magnetic, and piezoelectric (mass-sensitive) [53], each capitalizing on distinct operational mechanisms. Electrochemical biosensors detect changes in electrical signals due to the interaction of immobilized DNA with the target molecule during the hybridization reaction, whereas optical biosensors utilize changes in light absorption, fluorescence, or scattering, with both mechanisms being extensively employed. Electrochemical biosensors include DNA bound-enzyme detection, and label-based and label-free detections, while optical biosensors include but not limited to technologies like optical fibers, surface plasmon resonance and evanescent waves [53]. On the other hand, magnetic particles facilitate the measurement of biological interactions through magnetically labeled targets, contributing to the advancements in diagnostic techniques. They provide a significant advantage of minimal background interference in contrast to optical and electrical transducing methods [54]. Piezoelectric biosensors, also known as mass-sensitive biosensors, detect mass changes that occur as a result of DNA interaction with the target molecule. When the target DNA binds to the biosensor's surface, it causes a measurable change in the device's resonance frequency, allowing for accurate quantification of the target molecule.

The applications of DNA-based hybrid nanomaterial biosensors in disease detection are diverse and far-reaching. These biosensors demonstrate considerable efficacy in detecting a spectrum of diseases, spanning from infectious diseases such as human immunodeficiency virus (HIV), hepatitis, and tuberculosis, to non-infectious conditions such as cancer and diabetes. In the case of HIV, biosensors can be tailored to identify specific HIV biomarkers, allowing for early diagnosis and monitoring of HIV progression. Similarly, in hepatitis, biosensors can target hepatitis virus DNA or RNA sequences, enabling sensitive and specific detection of hepatitis infection. For instance, Diao et al. [55] devised a highly sensitive surface plasmon resonance DNA biosensor for enzyme-free and label-free detection of HIV-related DNA through entropy-driven strand displacement reactions and double-layer DNA tetrahedrons. Meanwhile, El-Said et al. [56] developed a rapid and highly selective Hepatitis C virus (HCV) biosensor using gold nanodots/indium tin oxide and peptide nucleic acid to detect HCV-RNA from clinical samples based on selective hybridization between the peptide nucleic acid and HCV-RNA at the 5′-UTR, monitored by Raman spectroscopy and square wave voltammetry. In tuberculosis, biosensors can be tailored to identify genetic markers associated with tuberculosis, facilitating rapid and accurate disease diagnosis. For instance, Taufiq et al. designed an electrochemical DNA biosensor for sensitive and specific detection of *M. tuberculosis* in clinical samples [57]. The method demonstrates robustness, sensitivity, specificity, and demands minimal sample volume, making it suitable for potential extension to portable POC tuberculosis diagnosis.

In the context of non-infectious diseases such as cancer and diabetes, DNA-based hybrid nanomaterial biosensors have the capability to detect cancer-specific mutations or biomarkers, thus serving as valuable tools for early cancer screening and treatment monitoring. For example, the combination of gold nanoparticles and graphene oxide has shown improved DNA sensing capabilities for the detection of breast cancer biomarkers for early diagnosis [58]. Yanik et al. [59] developed a creative electrochemical biosensor using a reduced graphene oxide-polyaminophenol matrix decorated with palladium nanoparticles for discriminating interactions between mitomycin C-DNA and acyclovir-DNA. The nanobiosensor demonstrated high specificity in distinguishing between the anticancer drug mitomycin C and acyclovir. Similarly, Chen et al. [60] fabricated a three-dimensional electrochemical DNA biosensor based on 3D graphene-silver nanoparticles for sensitive detection of CYFRA21-1 gene of non-small cell lung cancer patients. In diabetes, DNA-based hybrid nanomaterial biosensors can be engineered to identify specific genetic sequences or biomarkers associated with diabetes, providing an effective tool for precise disease monitoring and management [61]. For example, Shajaripour Jaberi et al. [62] devised a highly sensitive and label-free electrochemical DNA nanobiosensor using a nanocomposite of reduced graphene oxide and gold on a graphite sheet electrode for the specific determination of the glycated hemoglobin (HbA1c) biomarker in blood, demonstrating high sensitivity and a low detection limit (1 nM), suggesting its potential for non-invasive screening and diabetes management. Additionally, Abazar et al. [63] developed a cost-effective fabrication of chitosan-carbon quantum dots nanocomposite impedimetric aptasensor with good reproducibility and accuracy for the determination of insulin in human blood serum, for clinical applications.

In addition, DNA-based hybrid nanomaterial biosensors could also potentially be used as a platform to detect biomarkers associated with neurological disorders and immunodeficiency, enabling early diagnosis and intervention. In Alzheimer's disease, for instance, these biosensors can be tailored to pinpoint specific genetic markers or protein aggregates linked to the condition, thereby enabling early detection and potential intervention [64]. Moreover, they can be engineered to identify immunodeficiency-related biomarkers, aiding in the early identification of immune system disorders. An example is the detection of 8-hydroxy-2′-deoxyguanosine (8-OHdG), a biomarker for oxidative DNA damage and cellular stress, which is relevant to conditions such as diabetic nephropathy, cancer, and Parkinson's disease [65–68].

DNA-based hybrid nanomaterial biosensors also possess the capacity to identify biomarkers related to DNA mutations and gene expression that can be used for genetic testing and identifying genetic diseases in clinics and hospitals. These biosensors typically use DNA strands as recognition elements and can be coupled with nanomaterials like silver nanoparticles or carbon nanotubes to enhance the signal and boost sensitivity, enabling the precise detection of specific DNA sequences. For example, Xu et al. [69] introduced an economical, label-free ratiometric fluorescent biosensor that utilizes peptide nucleic acid clamp-enhanced catalyzed hairpin assembly and a hairpin DNA-templated dual-emissive silver nanocluster, facilitating the sensitive and accurate detection of single-nucleotide polymorphisms like the oncogenic *KRAS* gene with a low detection limit (9.6 pM), showcasing versatility, precision, and effectiveness in practical bioanalysis. In a similar vein, Maxwell et al. [70] devised a unique strategy for investigating single base mutations and particular DNA sequences, leveraging thiol-labeled oligonucleotide molecules attached to a 2.5 nm gold nanoparticle. This approach allows the restoration of fluorescence upon target binding, resulting in a specific and overly sensitive fluorescent signal.

Apart from all that, DNA-based hybrid nanomaterial biosensors utility extends to environmental monitoring, food safety, and forensic analysis, attesting to their versatility and widespread applicability in diverse domains. DNA-based hybrid nanomaterial biosensors can be deployed to detect harmful pathogens or contaminants in agriculture and water sources [30, 71–73], enabling timely intervention to prevent waterborne diseases as well as identify specific DNA or RNA sequences associated with foodborne pathogens, enhancing food safety measures through swift and precise [74]. For example, Xu et al. [75] engineered an electrochemical DNA biosensor using graphene oxide chitosan hybrid nanocomposites to detect *E. Coli* O157:H7 and address infection outbreaks while Chen et al. [76] devised a practical detection system for genetically modified rice via surface-enhanced Raman scattering-barcoded nanosensors for recognizing *B. thuringiensis* gene-transformed rice expressing insecticidal proteins. In DNA profiling, DNA-based hybrid nanomaterial biosensors have been utilized in sensitive detection of target DNA, playing a significant role in forensic DNA analysis through the detection and profiling of specific genetic markers, thus aiding in the identification of individuals and contributing to the resolution of criminal cases [77]. There are also reports on utilizing DNA-based hybrid nanomaterial biosensors in the identification of at-risk species, such as the Malayan box turtle [78, 79], showcasing the potential applications of these biosensors in biodiversity conservation and protection efforts.

The versatility of DNA-based hybrid nanomaterial biosensors expands their application beyond conventional disease detection, encompassing various fields such as healthcare, environmental monitoring, food safety, and forensic science. Their ability to enable early detection, intervention, and monitoring across diverse sectors highlights their pivotal role in advancing disease detection technologies.

6 Enzyme-Based Hybrid Nanomaterial Biosensors

The enzyme-based biosensor along with immunosensor-based biosensor represent the earliest iteration among the three classes of biosensors. Even though the initial commercial biosensor for glucose detection was reported in 1975 [80], the glucose-enzyme electrode concept was first introduced at a 1962 symposium in the New York Academy of Sciences by Clark and Lyons. They proposed enhancing electrochemical sensors using enzyme transducers in membrane-enclosed sandwiches to entrap glucose oxidase in a semi-permeable dialysis membrane over an oxygen electrode, allowing for the indirect determination of glucose concentration through monitoring oxygen consumption [29]. Updike and Hicks later (1967) developed an enzyme electrode for rapid and quantitative glucose determination by immobilizing glucose oxidase in a polyacrylamide gel covering an oxygen electrode [81]. Similarly, immunosensor-based biosensors also emerged in the 1970s [82]. In contrast, DNA-based biosensors gained popularity in the late 1990s, owing to advancements in modern technology [83, 84].

Compared to DNA-based biosensors that rely on molecular recognition events involving DNA strands, enzyme-based biosensors primarily rely on the catalytic properties of enzymes for detecting various biomarkers, including proteins and small molecules, resulting in a detectable signal, such as a color change, fluorescence, or electrical current when the target molecule is present [85]. They do this by detecting specific analytes via measuring changes like proton concentration (H^+), gas (such as CO_2, NH_3, O_2) release or uptake, light emissions, absorption or reflectance, and heat emissions during enzymatic reactions. These changes are then converted by a transducer into measurable signals (electrical, optical, or thermal) for identifying the target analytes [86]. Additionally, enzyme-based biosensors incorporate enzymes like glucose oxidase, laccase, horseradish peroxidase, or other biorecognition elements, immobilized on nanomaterial platforms such as carbon nanotubes, graphene, or metal nanoparticles [87, 88] to harness the catalytic properties of enzymes and the unique characteristics of nanomaterials for detecting target analytes. The immobilization of enzymes on nanomaterials not only enhances the stability and activity of the enzymes but also facilitates the efficient transfer of electrons, thereby, improves the sensitivity and selectivity of the biosensors [89]. Moreover, the large surface area and high electrical conductivity of nanomaterials support rapid electron transfer kinetics, making them ideal for biosensing applications. By integrating hybrid nanomaterials, stability and sensitivity are dramatically improved, making these biosensors highly versatile and adaptable across different applications, due to the specificity and efficiency of enzyme–substrate reactions. Similar to DNA-based sensors, enzyme-based hybrid nanomaterial biosensors can also be fabricated as electrochemical, optical, magnetic, and piezoelectric biosensors although they are more commonly reported in the literature using electrochemical and optical transducing mechanisms [86].

Enzyme-based hybrid nanomaterial biosensors have been extensively applied in diverse diagnostic processes, notably in healthcare, food safety, and environmental monitoring. Within the healthcare sector, these biosensors have played a crucial role in rapidly and accurately identifying disease biomarkers related to conditions such as diabetes, cancer, and infections. For instance, enzyme-nanomaterial hybrid glucose biosensors have revolutionized glucose monitoring for diabetic patients by providing real-time and precise measurements [80, 90–92]. Ki et al. [93] utilized a localized surface plasmon resonance biosensor to detect miR-10b, a biomarker for gastric cancer, utilizing a duplex-specific nuclease and a hybridization chain reaction. Likewise, Mansor et al. [94] engineered a tri-enzyme amperometric biosensor for rapid diagnosis of bacterial sepsis in human samples by detecting secretory phospholipase Group 2-IIA (sPLA2-IIA), an indicator for sepsis and bacterial infection. This biosensor detects sPLA2-IIA using immobilized choline kinase, choline oxidase, and horseradish peroxidase enzymes, providing reproducible POC detections in human samples. Additionally, Semwal and Gupta [95] detailed the development of a sensitive fiber-optic enzyme-based sensor utilizing the immobilization of cholesterol oxidase over silver nanoparticles coated graphene oxide nanosheets for cholesterol detection, enabling remote monitoring and assessments for patients at risk of cardiovascular diseases.

In food safety, enzymatic hybrid nanomaterial biosensors facilitate rapid and accurate detection of various biomolecules and contaminants, contributing to food safety and quality control. These biosensors enable the sensitive detection of specific enzymes, toxins, pathogens, and other relevant substances in food products, aiding in the monitoring of food processing, storage, and distribution processes. For example, Dervisevic et al. [96] introduced an innovative electrochemical xanthine biosensor. This device, which utilizes hybrid nanomaterials, including chitosan, polypyrrole, and gold nanoparticles, can effectively measure the xanthine levels in samples of fish, chicken, and beef. The results have been satisfactory, making it a reliable analytical tool for assessing meat quality. Moreover, Yang et al. [97] developed a novel strategy for detecting pyruvate in fermentation processes, leveraging the consumption of H_2O_2 through non-enzymatic decarboxylation. This innovative approach involved the design of an enhanced amperometric H_2O_2 biosensor employing HRP enzyme-inorganic hybrid nanoflowers and a thiol graphene-modified Au disk electrode. Notably, this design facilitates the direct determination of high-concentration pyruvate in real microbial fermentation processes without necessitating pretreatment. Furthermore, the versatility of nanomaterials used in these biosensors offers innovative design opportunities, making them valuable tools for addressing food industry challenges [98].

In environmental monitoring, enzyme-based hybrid nanomaterial biosensors have been leveraged to identify pollutants, toxins, and pathogens across diverse settings, significantly enhancing the early assessment of environmental health hazards [99]. For example, Bao et al. [100] pioneered the development of a highly responsive electrochemical acetylcholinesterase enzymatic biosensor for detecting organophosphorus pesticides by employing a three-dimensional graphene-copper oxide nanoflowers nanocomposite. Additionally, Fernandes et al. [101] formulated an enzymatic hybrid nanomaterial biosensor capable of electrochemically detecting bisphenol A at nanomolar levels, an endocrine disruptor associated with reproductive dysfunctions and increased cancer risks. Furthermore, a copious amount of literature exists on the utilization of acetylcholinesterase inhibition-based enzymatic biosensors for the detection of pesticides, organophosphorus compounds, mycotoxins, and other substances, owing to its cost-effectiveness, high turnover rate, and the widespread occurrence of its inhibitors in the environment [102–108]. Overall, enzyme-based hybrid nanomaterial biosensors have highlighted their remarkable adaptability and significance in healthcare, food safety, and environmental monitoring. They serve as a promising foundation for the progress of diagnostic and detection technologies, bolstering the drive towards the development of portable nanobiosensors.

7 Immunosensor-Based Hybrid Nanomaterial Biosensors

Immunosensor-based hybrid nanomaterial biosensors offer several advantages over DNA-based and enzyme-based biosensors, primarily owing to their remarkable specificity, sensitivity, and selectivity [89], which hold significant value in diagnostic applications. These biosensors capitalize on the precise binding capabilities inherent in immuno-recognition elements, specifically the complex interactions between antibodies and antigens (Ab–Ag), facilitating the accurate identification of target molecules and subsequent measurable signal variations [109]. With antibodies acting as pivotal affinity biorecognition elements, their widespread adoption over two decades stems from their versatile application and robust antigen–antibody interactions. Structurally, antibodies are characterized by the "Y" shape of immunoglobulins (Ig), constituting two heavy and two light polypeptidic chains interconnected by disulfide bonds [110], exhibiting distinct classes such as IgG, IgM, IgA, IgD, and IgE based on variations in their heavy chains. The strategic integration of nanomaterials with immunosensors serves to enhance electrochemical signals and bioactivity retention, thereby elevating the overall performance and sensitivity of these biosensors [111]. Moreover, immobilizing antibodies through covalent interactions with functional groups like carboxyl, amino, aldehyde, or sulfhydryl leads to stable and efficient binding, ultimately contributing to the superior functionality of immuno-based hybrid nanomaterial biosensors.

Similarly to both DNA- and enzyme-based biosensors, immunosensor-based hybrid nanomaterial biosensors share a diverse array of applications across different fields, including medical diagnostics, environmental monitoring, food safety, and biodefense [112, 113]. However, these biosensors are commonly reported in literature within the biomedical and clinical research fields due to their significant potential for disease detection and biomarker analysis, especially for POC testing. They are also widely developed as electrochemical sensors compared to other transducing mechanisms. Their proficiency in the quick and sensitive detection of specific biological molecules, such as proteins and enzymes, also makes them well-suited for healthcare applications. Further details about some remarkable inventions stemming from these biosensors are provided in the subsequent paragraphs.

Advancements in immunosensor technology have led to the development of groundbreaking detection methods for viruses and bacterial diseases. For instance, an innovative immunosensor utilizing a graphene-titanium dioxide hybrid nanocomposite, integrated with plant-derived antigenic glycoprotein, effectively showcases the potential of immunosensor-based hybrid nanomaterial biosensors in detecting dengue virus IgG antibodies [114]. Tung et al. [115] developed an effective method for investigating an ultra-weak binding between dengue virus and its CLEC5A receptor through a hybrid nanostructured electrochemical biosensor, which can be applicable in other infections as well. Haji-Hashemi et al. [116] introduced a label-free immunosensor featuring electrodeposited Prussian blue and gold nanoparticles on a surface resembling bamboo-like multiwall carbon nanotubes-ionic liquid nanocomposite modified glassy carbon electrode. This approach enables sensitive detection of citrus bacterial canker disease while offering high selectivity, long-term stability, and good reproducibility. Additionally, Fang et al. [117] have devised a highly sensitive immunosensing strategy for detecting procalcitonin (PCT) using zinc nanoparticles-functionalized ordered mesoporous carbon–silica nanocomposites, recognizing PCTs as the most promising inflammatory marker for diagnosing bacterial infections.

Besides that, a label-free biosensor utilizing aminated reduced graphene oxide functionalized with antiapolipoprotein B 100 was fabricated to detect low-density lipoprotein cholesterol [118]. The immunosensor demonstrates remarkable sensitivity and a swift response time and was confirmed of the long-term stability and excellent reproducibility, making it ideal for analyzing lipid molecules' antibody and antigen interactions. Arkan et al. [119] devised an ionic liquid-multiwall carbon nanotube impedimetric immunosensor enhanced with gold nanoparticles for recognizing human serum albumin (HSA), pivotal in early nephritis and chronic kidney disease diagnosis and treatment [120]. Zong et al. [121] introduced an ultra-sensitive metal-enhanced chemiluminescence immunoassay using silver nanoparticle hybrid probes for C-reactive protein (CRP) detection down to sub-pM levels, crucial in cerebrovascular disease diagnosis. Additionally, Wang et al. [122] developed a highly sensitive surface-enhanced Raman spectroscopy method based on graphene oxide/gold nanoparticles to identify Aβ 1–42 protein, a crucial marker of Alzheimer's disease neuropathology, with detection limits of 0.0232 ng mL^{-1} for monomers and 0.0192 ng mL^{-1} for fibrils. This approach, complemented by machine learning algorithms, shows promise for label-free Alzheimer's early disease diagnosis.

Immunosensor-based hybrid nanomaterial biosensors also find many uses in cancer diagnostics. For example, Yu et al. [123] innovated an electrochemical immunosensors utilizing single-wall carbon nanotube forest platforms and multi-label secondary antibody-nanotube bioconjugates for highly sensitive cancer biomarker detection, achieving a remarkable 4 pg mL^{-1} detection limit for prostate-specific antigen in undiluted calf serum. Wang et al. [124] detailed a label-free electrochemical immunosensing approach for carbohydrate antigen 199 (CA 199) detection utilizing a polyaniline derivative-gold/platinum nanocomposite. CA199 holds significant potential for detecting malignant tumors including pancreatic cancer, colorectal cancer, liver cancer, gastric cancer, and ovarian cancer. Ge et al. [125] developed a disposable electrochemical immunosensor for the simultaneous assay of breast cancer tumor markers—cancer antigen 153 (CA 153), cancer antigen 125 (CA 125), and carcinoembryonic antigen (CEA). Furthermore, Fan et al. [126] developed an electrochemical biosensor employing reduced graphene oxide/thionine/gold nanoparticles nanocomposites for CA 125 determination, offering an extensive linear range, low detection limit, and high sensitivity. Gasparotto et al. [127] developed an electrochemical immunosensor based on zinc oxide nanorods-gold nanoparticles nanohybrids for ovarian CA 125 detection, with a detection limit hundred times lower than immunoblotting test. Kovarova et al. [128] devised an electrochemical magneto-immunosensor for simultaneous detection and quantification of ovarian cancer biomarkers—human epididymis protein 4 (HE4), alpha-fetoprotein (AFP), and CA 125. The approaches taken by these sensors exhibit considerable promise in the precise detection of cancer biomarkers during clinical screening and POC diagnostics, particularly aiming at early clinical diagnoses. Apart from all of that, immunosensor-based hybrid nanomaterial biosensors have also been reported for the detections of toxins [129–131], immunoglobins [132], bacteria [133], viruses [134], and autoimmune disorders [135, 136], illustrating its substantial diagnostic capabilities in various clinical and research settings. Despite its paramount applicability, it is crucial to consider the limitations associated with these biosensors, which can primarily include complexities in nanomaterial functionalization, production costs, cross-reactivity, and the potential for false-positive results, requiring stringent quality control and careful optimization to ensure reliability.

8 The Differences and Similarities Between DNA-, Enzyme-, and Immunosensor-Based Hybrid Nanomaterials Biosensor

DNA-based, enzyme-based, and immunosensor-based hybrid nanomaterial biosensors are prominent in advanced biosensing technologies, each with their distinct features and applications. While DNA-based biosensors leverage DNA's recognition properties, enzyme-based biosensors utilize enzyme catalysis, and immunosensor-based biosensors rely on antibody-antigen interactions. Despite the individual strengths of each type of biosensor, there is not a one-size-fits-all in terms of application. Each type's benefits and optimal utilities largely hinge on the specific medical scenario in which it is applied. The specific differences between these types of biosensors are detailed in Table 1, while Table 2 showcases some of the recent advancements of hybrid nanomaterial-based biosensors across various settings.

Despite their differences, these biosensors also share several similarities: all three types can be highly integrated with hybrid nanomaterials that improve their sensitivity and stability due to a larger surface area and better electrical properties. In addition to that, they all can benefit from miniaturization and can be incorporated into portable and POC devices. Each type relies on specific biomolecule interactions—whether it is enzyme–substrate, nucleic acid complementarity, or antigen–antibody binding—providing a selective approach to sensing. These biosensors are under continuous development to enhance their performance, reduce costs, and enable more widespread applications. These comprehensive insights into the differences and similarities between DNA-based, enzyme-based, and immunosensor-based hybrid nanomaterial biosensors provide a detailed understanding of these advanced biosensing technologies.

9 Challenges and Future Perspectives

Hybrid nanomaterials play a pivotal role in advancing POC diagnostics, yet they are not without challenges and limitations that necessitate thorough scrutiny. The translation of these biosensors from laboratory-based research to real-world POC applications represents a significant challenge. Factors such as cost-effectiveness, scalability, standardization, and regulatory compliance become paramount and pose significant hurdles to the widespread adoption of hybrid nanomaterials in advanced diagnostic settings. The financial implications associated with the production and implementation of these cutting-edge materials must be carefully considered, as their affordability will directly impact their accessibility within healthcare systems. Moreover, ensuring the scalability of hybrid nanomaterial-based diagnostic technologies is imperative for their effective integration into clinical practice, as the ability to produce these materials at a large scale is essential for their efficient deployment across varied medical facilities. Standardization is another critical aspect that requires attention, as the establishment of uniform protocols and benchmarks is vital to ensure the consistency and reliability of diagnostic outcomes. Regulatory compliance poses another significant challenge, as strict adherence to healthcare regulations and quality standards is vital for the ethical and safe utilization of hybrid nanomaterials in diagnostics.

Furthermore, ensuring the safety and biocompatibility of these materials is paramount for their successful translation into clinical applications. Rigorous assessments and validations are imperative to discern and mitigate any potential adverse effects or unintended consequences associated with the use of hybrid nanomaterials in diagnostic contexts. Additionally, there is a need to develop reliable methods for characterizing hybrid nanomaterials and understanding their interactions with biological systems.

Table 1 General differences between DNA-, enzyme-, and immuno-based hybrid nanomaterial biosensors

Biosensor type	DNA-based	Enzyme-based	Immunosensor-based
Recognition element	Rely on the specific binding of single-stranded DNA probes that can hybridize with their complementary sequences	Use enzymes as the biorecognition element to catalyze specific reactions with the target analyte	Utilize antibodies or antigens to specifically bind to the respective antigen or antibody
Mechanism of detection	Work mainly through hybridization signals when the probes recognize and bind to specific nucleic acid sequences	Often generate signals through a catalytic reaction that produces or consumes a substance that can be detected	Typically rely on the strength of the antigen–antibody interaction, which can be quantified by various methods
Sensitivity	Sensitive to the hybridization efficiency and the sequence-specific interactions	Sensitive to both the presence and concentration of the substrate	Highly sensitive to the affinity between the antibody and antigen
Stability	DNA biosensors can be prone to denaturation but offer stable recognition elements when protected	Enzyme biosensors may lose activity over time due to enzyme denaturation	Immunosensors might face challenges in stability due to the fragile nature of the antigen–antibody bond but typically offer robust performance with proper stabilization
Specificity	Highly specific to the sequence of interest	Specific to the substrate for which the enzyme is specific	Offer high specificity due to the unique antigen–antibody interactions
Applications	Primarily used in genomics and molecular diagnostics, infectious disease detection, and forensic analysis due to their ability to specifically recognize genetic sequences	Find applications in clinical diagnostics, environmental monitoring, and food industry for detecting contaminants	Have applications in disease diagnostics, therapeutic monitoring, and allergen detection
Transduction mechanism	May employ optical, electrochemical, or mass-sensitive transduction methods	Often use electrochemical or optical transducers due to the enzymatic reactions	Often applied in electrochemical, optical, and piezoelectric transduction mechanisms

(continued)

Table 1 (continued)

Biosensor type	DNA-based	Enzyme-based	Immunosensor-based
Dynamic range	Typically have a narrower dynamic range, due to the binary nature of hybridization events	Have a dynamic range tied to the kinetics of the enzyme–substrate interaction	Might offer a broader dynamic range given the variable affinity between antibodies and antigens
Sample preparation	May require extensive sample preparation to isolate nucleic acids	May use crude samples but may require optimal pH and temperature conditions	Often require careful sample preparation to maintain antibody/antigen integrity
Reusability	Can be reused by denaturing and removing the bound strands, although this might affect stability	Are typically less reusable due to enzyme deactivation over time	Can be reused if the antigen–antibody dissociation can be achieved without damaging the antibody

Addressing these multifaceted challenges and limitations is crucial for harnessing the full potential of hybrid nanomaterials in advanced POC diagnostics, thereby unlocking their capacity to revolutionize medical diagnostics and positively impact patient care outcomes. Continuous research endeavors in this domain exhibit significant potential for the establishment of advanced biosensors capable of delivering precise and real-time detection of diverse analytes. It is important to note that the choice of hybrid nanomaterials depends on specific applications, and further research is needed to determine their full potential in healthcare diagnostics. Additionally, there may be other types of hybrid nanomaterials with unique advantages that are not extensively studied yet. Through the proper development and implementation of hybrid nanomaterials in biosensors, there is potential to enhance human health and safety, as well as advance scientific research in various fields.

The amalgamation of hybrid nanomaterials and biosensor technologies presents a compelling avenue for enhancing advanced POC diagnostics in healthcare. The diversified use of DNA, enzyme, and immunosensor-based designs underscores the potential of these biosensors in achieving accurate and rapid diagnostic outcomes. Despite the challenges and limitations, the research and development in this field continue to show promising results, offering new possibilities for addressing healthcare diagnostic needs. As advancements in nanotechnology and biosensing methodologies progress, these hybrid nanomaterial-based biosensors are poised to play an increasingly significant role in shaping the future of healthcare diagnostics.

Table 2 Recent advances and innovations in DNA-, enzyme-, and immuno-based hybrid nanomaterial biosensors

Biosensor type	Transduction mechanism	Hybrid nanomaterial	Analyte	Application	Limit of detection	Refs.
DNA	Optical	ESDRs–DDTs	tDNA	Enzyme-free and label-free HIV-related DNA detection	48 fM	[55]
DNA	Optical	Gly–QDs/Ab–Au NPs	HAs	Detection of influenza virus A hemagglutinin antigen	38 pM to 12 nM	[137]
DNA	Electrochemical	Au NPs–GO	ERBB2, CD24	Detection of breast cancer biomarkers for early diagnosis	0.16 nM and 0.23 nM	[58]
DNA	Electrochemical	rGO–Pd@PACP	MC–DNA, ACY–DNA	Detection and discrimination of mitomycin C-DNA and acyclovir-DNA interaction	51.3 nM	[59]
DNA	Electrochemical	3DGF–Ag NPs	CYFRA21-1 DNA	Detection of CYFRA21-1 gene in non-small cell lung cancer	1.0×10^{-14} M	[60]
DNA/PNA	Spectro electrochemical	Au NDs–ITO	HCV–RNA	Detection of Hepatitis C virus	101.5–264.5 IU/mL	[56]
DNA	Electrochemical	GS/rGO–Au	HbA1c	Determination of glycated hemoglobin in blood for low-invasive screening and diabetes management	1 nM	[62]
DNA	Electrochemical	CQD–Chit	BSA	Detection of insulin in human blood serum	106.8 pM	[63]
DNA	Electrochemical	ssDNA–GN	8-OHdG	Detection of 8-Hydroxy-2′-deoxyguanosine known to cause cancer, diabetes, and neurodegenerative diseases	0.875 nM	[67]

(continued)

Table 2 (continued)

Biosensor type	Transduction mechanism	Hybrid nanomaterial	Analyte	Application	Limit of detection	Refs.
DNA	Electrochemical	Au NPs–ZnO, Au NPs–TiO	tDNA	Detection of pathogenic fungus infection oil palms, *G. boninense*	–	[30, 138]
DNA/PNA	Optical	DNA–Ag NCs	$KRAS_{mA}$	Detection of single-nucleotide polymorphisms associated with risks surrounding genetic variation	9.6 pM	[69]
DNA	Optical	DNA–Ag NCs–Au@Pd NPs	Streptomycin	Detection of streptomycin in milk, indicating its suitability for detection in animal-derived foods	18.7 nM	[139]
DNA	Electrochemical	GO–Chit	tDNA	Detection of *E. Coli* O157:H7 causing hemorrhagic colitis and hemolytic uremic syndrome	3.58×10^{-15} M	[75]
DNA	Optical	GO–Au NPs	tDNA	Simultaneous and quantitative detection of a meat adulterant pork and an endangered species -Malayan Box Turtle	1×10^{-14} M	[79]
Enzyme	Optical	DNA–tannic acid capped Au NPs	miR-10b	Detection of gastric cancer relevant miRNAs, miR-10b	2.45 pM	[93]
Enzyme	Optical	GO–Ag NPs	Cholesterol	Assessment of the risk of developing cardiovascular disease	5.14 nm/mM, 1.131 mM	[95]

(continued)

Table 2 (continued)

Biosensor type	Transduction mechanism	Hybrid nanomaterial	Analyte	Application	Limit of detection	Refs.
Enzyme	Optical	AuVCs nanozymes	Glutathione	On-site intracellular glutathione detection to evaluate drug-resistance level	9.80 μM	[140]
Enzyme	Electrochemical	GRA–PANI	Glucose	Detection of glucose in plasma samples	2.769 μM	[90]
Enzyme	Electrochemical	Chit–PPy–Au NPs	Xanthine	Evaluation of meat freshness	0.25 μM	[96]
Enzyme	Electrochemical	HRP–CuP NFs/ tG/ Au	Pyruvate	Determination of high concentration pyruvate in real microbial fermentation processes without pretreatment	0.06 mM	[97]
Enzyme	Electrochemical	3DG–CuO NFs	Malathion	Detection of organophosphate pesticides in water samples	0.31 ppt	[100]
Enzyme	Electrochemical	rGO–Fe_3O_4 NPs	Bisphenol A	Detection of the endocrine disruptor bisphenol A	18 nM	[101]
Immunosensor	Electrochemical	Bi NPs/ Nafion–MWCNTs	Anthrax PA toxin	Detection of *B. anthracis* PA toxin in human serum for prompt diagnosis and targeted treatment of infected persons	50 pg/ml	[131]
Immunosensor	Electrochemical	Au NPs/AAO	CLEC5A–DV	Useful for investigating weak molecular interactions, such as ligand–receptor interaction in dengue fever and enterovirus, or cancer surface glycoproteins and their receptors	Viral titer of 9.5×10^5	[115]

(continued)

Table 2 (continued)

Biosensor type	Transduction mechanism	Hybrid nanomaterial	Analyte	Application	Limit of detection	Refs.
Immunosensor	Electrochemical	G/TiO_2	DENV IgG	Practical serological diagnostic tool for dengue detection in realistic applications	2.81 ng/mL	[114]
Immunosensor	Electrochemical	Au NPs/PB/ BCNT-IL	PthA	The initial electrochemical immunosensor for citrus canker disease detection, modifiable for detecting other targets	0.028 nM	[116]
Immunosensor	Electrochemical	OMCSi– Zn NPs	PCT	Potential use in clinical diagnostics for detecting procalcitonin to diagnose bacterial infections	0.013 pg/mL	[117]
Immunosensor	Electrochemical	Au NPs@MW-CILE	HSA	Beneficial screening tool for tracing HSA in biological fluids of proteinuria and diabetic patients in clinical labs	15.4 ng mL^{-1}	[119]
Immunosensor	Optical	DNA modified Ag NPs hybrid probes	CRP	Detection of C-reactive protein in human serum, a vital marker for cerebrovascular disease	0.05 ng mL^{-1}	[121]
Immunosensor	Optical	GO/Au NPs	Aβ 1–42	Label-free SERS detection of Aβ 1–42 for early Alzheimer's disease detection	0.0232 ng mL^{-1} 0.0192 ng mL^{-1}	[122]
Immunosensor	Electrochemical magneto	Si NPs–CdTe QDs, Si NPs–PbS QDs, Si NPs–Au NPs	HE4, CA-125, AFP	Simultaneous detection and quantification of ovarian cancer biomarkers	5.07 pM, 3.10 IU L^{-1}, 2.44 pg L^{-1}	[128]
Immunosensor	Optical	$g\text{-}C_3N_4$@PANI-Au	Anti-CCP	Early diagnosis of autoimmune disease, rheumatoid arthritis	0.2 pg mL^{-1}	[136]

10 Conclusion

Recent advancements and innovations in hybrid nanomaterial biosensors have significantly propelled the field forward, with materials such as graphene, carbon nanotubes, and metal–organic frameworks notably bolstering sensitivity and selectivity, enhancing diagnostic potential. Furthermore, the integration of biosensors with microfluidic systems has heralded the advent of POC devices that enable rapid and accurate diagnoses, thus holding immense promise for advancing healthcare and diagnostic capabilities. These developments underscore the profound impact of DNA-, enzyme-, and immunosensor-based hybrid nanomaterial biosensors on the landscape of disease detection and personalized medicine. To put it into perspective, four decades ago, the frequency of biosensor-related paper publications stood at around one every two years, and the global market valuation was less than US$5 million annually. By 2003, the number of papers on biosensors published each year had risen to approximately 4500, with global biosensor sales surging to roughly around US$13 billion [83]. That was eleven years ago. This exponential growth in both research and market valuation over the past decades is a testament to the transformative power of hybrid nanomaterial biosensors in healthcare diagnostics. And if the past is any indication, the future holds even more exciting advancements in this field.

So, would it be DNA-, enzyme-, or immunosensor-based hybrid nanomaterials biosensors, in advanced POC diagnostics, for healthcare? The integration of hybrid nanomaterials into biosensor platforms in advanced POC diagnostics has already demonstrated significant promise for enhancing sensitivity, selectivity, and multiplexing capabilities—all of which are crucial for disease detection and personalized medicine. The profound influence of nanotechnology advancements on the roles of these biosensors cannot be overstated. Each offers tailored advantages: DNA-based biosensors are excellent for genetic testing, enzyme-based biosensors provide versatility in detecting various analytes, and immunosensor-based biosensors proved effective in detecting infectious diseases. The future of personalized medicine is reliant on the progression of these technologies and their integration into accessible, efficient, and user-friendly tools for clinicians. It is essential to customize the selection of hybrid nanomaterials for biosensors in healthcare applications, considering the context-dependent effectiveness of each system. The ongoing advancement of scientific research continuously enhances the capabilities and accuracy of these biosensors, underscoring the significance of informed selection to optimize healthcare diagnostic outcomes. Moreover, the amalgamation of diverse nanomaterials in biosensors confers an array of benefits, surpassing the performance of standalone types and ushering in a new era of heightened efficiency in healthcare applications.

Acknowledgements Author Adilah Ayoib gratefully acknowledges the support from the Faculty of Chemical Engineering & Technology and the Institute of Nano Electronic Engineering, Universiti Malaysia Perlis (UniMAP), as well as the staff members and colleagues for valuable discussion and helpful comments. The views expressed in this publication are the authors' own interpretation based on the data collected and do not reflect in any way the views of the institutions.

Disclosure of Interest The author declares no conflicts of interest

References

1. Kool, L., et al.: Gold and silver dichroic nanocomposite in the quest for 3D printing the Lycurgus cup. Beilstein J. Nanotechnol. **11**, 16–23 (2020). https://doi.org/10.3762/bjnano.11.2
2. Barber, D.J., Freestone, I.C.: An investigation of the origin of the colour of the lycurgus cup by analytical transmission electron microscopy. Archaeometry **32**(1), 33–45 (1990). https://doi.org/10.1111/j.1475-4754.1990.tb01079.x
3. The British Museum. https://www.britishmuseum.org/collection/image/36153001
4. Chartres Day 1. https://www.flickr.com/photos/bramhall/albums/72157684065025142/
5. Mosquée El Aqsa. https://www.flickr.com/photos/lazhar/albums/72157625709964312/
6. Jin, R., Cao, Y., Mirkin, C.A., Kelly, K.L., Schatz, G.C., Zheng, J.G.: Photoinduced conversion of silver nanospheres to nanoprisms. Science (1979) **294**(5548), 1901–1903 (2001). https://doi.org/10.1126/science.1066541
7. Chang, K.: Tiny Is Beautiful: Translating 'Nano' Into Practical. The New York Times (2005). https://archive.nytimes.com/www.nytimes.com/imagepages/2005/02/21/science/20050222_NANO1_GRAPHIC.html
8. The Corning Museum of Glass. https://www.cmog.org/sites/default/files/images/orange.jpg
9. Teapot. https://www.artic.edu/artworks/117267/teapot
10. Musée du Louvre. https://collections.louvre.fr/en/recherche?q=qing&limit=100&collection%5B0%5D=7
11. The MET Museum. https://www.metmuseum.org/art/collection/search/32739
12. Bayda, S., Adeel, M., Tuccinardi, T., Cordani, M., Rizzolio, F.: The history of nanoscience and nanotechnology: from chemical-physical applications to nanomedicine. Molecules **25**(1), 112 (2019). https://doi.org/10.3390/molecules25010112
13. Cionti, C., Stucchi, M., Meroni, D.: Mimicking stained glass: a hands-on activity for the preparation and characterization of silica films colored with noble metal ions and nanoparticles. J. Chem. Educ. **99**(3), 1516–1522 (2022). https://doi.org/10.1021/acs.jchemed.1c01141
14. Molina, G., Murcia, S., Molera, J., Roldan, C., Crespo, D., Pradell, T.: Color and dichroism of silver-stained glasses. J. Nanopart. Res. **15**(9), 1932 (2013). https://doi.org/10.1007/s11051-013-1932-7
15. Bandiera, M., Lehuédé, P., Verità, M., Alves, L., Biron, I., Vilarigues, M.: Nanotechnology in Roman Opaque Red glass from the 2nd century AD. archaeometric investigation in red sectilia from the decoration of the Lucius Verus Villa in Rome. Heritage **2**(3), 2597–2611 (2019). https://doi.org/10.3390/heritage2030159
16. Pradell, T., Molera, J., Smith, A.D., Tite, M.S.: Early Islamic lustre from Egypt, Syria and Iran (10th to 13th century AD). J. Archaeol. Sci. **35**(9), 2649–2662 (2008). https://doi.org/10.1016/j.jas.2008.05.011
17. Borgia, I., et al.: Characterisation of decorations on Iranian (10th–13th century) lustreware. Appl. Phys. A **79**(2), 257–261 (2004). https://doi.org/10.1007/s00339-004-2519-z
18. Pradell, T., Molera, J.: Ceramic technology. How to characterise ceramic glazes. Archaeol. Anthropol. Sci. **12**(8), 189 (2020). https://doi.org/10.1007/s12520-020-01136-9
19. Colomban, P., Kırmızı, B., Zhao, B., Clais, J.-B., Yang, Y., Droguet, V.: Non-invasive on-site raman study of pigments and glassy matrix of 17th–18th century painted enamelled Chinese metal wares: comparison with French enamelling technology. Coatings **10**(5), 471 (2020). https://doi.org/10.3390/coatings10050471
20. Reibold, M., Paufler, P., Levin, A.A., Kochmann, W., Pätzke, N., Meyer, D.C.: Carbon nanotubes in an ancient Damascus sabre. Nature **444**(7117), 286–286 (2006). https://doi.org/10.1038/444286a

21. Sanderson, K.: Sharpest cut from nanotube sword. Nature (2006). https://doi.org/10.1038/news061113-11
22. Feynman, R.P.: There's plenty of room at the bottom. Resonance **16**(9), 890–905 (2011). https://doi.org/10.1007/s12045-011-0109-x
23. Taniguchi, N., Arakawa, C., Kobayashi, T.: On the basic concept of nano-technology. In: Proceedings of the International Conference on Production Engineering. Tokyo, Japan (1974)
24. Fotiadis, D.: Atomic force microscopy for the study of membrane proteins. Curr. Opin. Biotechnol. **23**(4), 510–515 (2012). https://doi.org/10.1016/j.copbio.2011.11.032
25. Seeman, N.C., Sleiman, H.F.: DNA nanotechnology. Nat. Rev. Mater. **3**(1), 17068 (2017). https://doi.org/10.1038/natrevmats.2017.68
26. Ayoib, A., Hashim, U., Gopinath, S.C.B., Arshad, M.K.Md.: DNA extraction on bio-chip: history and preeminence over conventional and solid-phase extraction methods. Appl. Microbiol. Biotechnol. **101**(22), 8077–8088 (2017). https://doi.org/10.1007/s00253-017-8493-0
27. United States government. National Nanotechnology Initiative. https://www.nano.gov/timeline
28. Thuecks, D.J., Demas, H.A.: Modeling the effect of air-intake aperture size in the ping-pong ball cannon. Am. J. Phys. **87**(2), 136–140 (2019). https://doi.org/10.1119/1.5086614
29. Clark, L.C., Lyons, C.: Electrode systems for continuous monitoring in cardiovascular surgery. Ann. N. Y. Acad. Sci. **102**(1), 29–45 (1962). https://doi.org/10.1111/j.1749-6632.1962.tb13623.x
30. Ayoib, A., Hashim, U., Gopinath, S.C.B.S.C.B.: Automated, high-throughput DNA extraction protocol for disposable label free, microfluidics integrating DNA biosensor for oil palm pathogen, Ganoderma boninense. Process Biochem. **92**(January), 447–456 (2020). https://doi.org/10.1016/j.procbio.2020.02.003
31. Vargas-Bernal, R.: Introductory Chapter: hybrid nanomaterials. In: Hybrid Nanomaterials—Flexible Electronics Materials, IntechOpen (2020). https://doi.org/10.5772/intechopen.92012
32. Sharma, A., Chauhan, A.K., Kataria, N., Garg, V.K.: Hybrid nanomaterials for the removal of organic pollutants from wastewater. In: Hybrid Nanomaterials for Sustainable Applications, pp. 71–102. Elsevier (2023). https://doi.org/10.1016/B978-0-323-98371-6.00022-7
33. Nishu, Kumar, S.: Smart and innovative nanotechnology applications for water purification. Hybrid Adv. **3**, 100044 (2023). https://doi.org/10.1016/j.hybadv.2023.100044.
34. Taylor-Pashow, K.M.L., Della Rocca, J., Huxford, R.C., Lin, W.: Hybrid nanomaterials for biomedical applications. Chem. Commun. **46**(32), 5832 (2010). https://doi.org/10.1039/c002073g
35. Khan, I., Saeed, K., Khan, I.: Nanoparticles: properties, applications and toxicities. Arab. J. Chem. **12**(7), 908–931 (2019). https://doi.org/10.1016/j.arabjc.2017.05.011
36. Gu, Q., Lv, J., Mo, X., Jiang, X.: High aspect ratio metamaterials and their applications. Sens. Actuators A Phys. **335**, 113376 (2022). https://doi.org/10.1016/j.sna.2022.113376
37. Gougis, M., Tabet-Aoul, A., Ma, D., Mohamedi, M.: Laser synthesis and tailor-design of nanosized gold onto carbon nanotubes for non-enzymatic electrochemical glucose sensor. Sens. Actuators B Chem. **193**, 363–369 (2014). https://doi.org/10.1016/j.snb.2013.12.008
38. Wu, B.-Y., et al.: Amperometric glucose biosensor based on multilayer films via layer-by-layer self-assembly of multi-wall carbon nanotubes, gold nanoparticles and glucose oxidase on the Pt electrode. Biosens. Bioelectron. **22**(12), 2854–2860 (2007). https://doi.org/10.1016/j.bios.2006.11.028
39. Dong, X., Lu, X., Zhang, K., Zhang, Y.: Chronocoulometric DNA biosensor based on a glassy carbon electrode modified with gold nanoparticles, poly(dopamine) and carbon nanotubes. Microchim. Acta **180**(1–2), 101–108 (2013). https://doi.org/10.1007/s00604-012-0900-8
40. Zhang, Y., Wang, J., Xu, M.: A sensitive DNA biosensor fabricated with gold nanoparticles/ploy (p-aminobenzoic acid)/carbon nanotubes modified electrode. Colloids Surf. B Biointerfaces **75**(1), 179–185 (2010). https://doi.org/10.1016/j.colsurfb.2009.08.030
41. Vinothini, K., Rajendran, N.K., Rajan, M., Ramu, A., Marraiki, N., Elgorban, A.M.: A magnetic nanoparticle functionalized reduced graphene oxide-based drug carrier system for

a chemo-photodynamic cancer therapy. New J. Chem. **44**(14), 5265–5277 (2020). https://doi.org/10.1039/D0NJ00049C
42. He, Y., Yi, C., Zhang, X., Zhao, W., Yu, D.: Magnetic graphene oxide: Synthesis approaches, physicochemical characteristics, and biomedical applications. TrAC, Trends Anal. Chem. **136**, 116191 (2021). https://doi.org/10.1016/j.trac.2021.116191
43. Kim, T.-H., Lee, D., Choi, J.-W.: Live cell biosensing platforms using graphene-based hybrid nanomaterials. Biosens. Bioelectron. **94**, 485–499 (2017). https://doi.org/10.1016/j.bios.2017.03.032
44. Domingues, J.M., Miranda, C.S., Homem, N.C., Felgueiras, H.P., Antunes, J.C.: Nanoparticle synthesis and their integration into polymer-based fibers for biomedical applications. Biomedicines **11**(7), 1862 (2023). https://doi.org/10.3390/biomedicines11071862
45. Sivadasan, D., Sultan, M.H., Madkhali, O., Almoshari, Y., Thangavel, N.: Polymeric Lipid Hybrid Nanoparticles (PLNs) as emerging drug delivery platform—a comprehensive review of their properties, preparation methods, and therapeutic applications. Pharmaceutics **13**(8), 1291 (2021). https://doi.org/10.3390/pharmaceutics13081291
46. Tonbul, H., Çapan, Y.: Hybrid PLGA nanoparticles as advanced drug delivery and theranostic applications. In: Poly(lactic-co-glycolic acid) (PLGA) Nanoparticles for Drug Delivery, pp. 417–431. Elsevier (2023). https://doi.org/10.1016/B978-0-323-91215-0.00017-0
47. Mo, L., et al.: Recent progress in the development of DNA-based biosensors integrated with hybridization chain reaction or catalytic hairpin assembly. Front. Chem. **11** (2023). https://doi.org/10.3389/fchem.2023.1134863
48. Zhai, J., Cui, H., Yang, R.: DNA based biosensors. Biotechnol. Adv. **15**(1), 43–58 (1997). https://doi.org/10.1016/S0734-9750(97)00003-7
49. Mo, L., Liang, D., He, W., Yang, C., Lin, W.: Ratiometric and amplified fluorescence nanosensor based on a DNA tetrahedron for miRNA imaging in living cells. J. Mater. Chem. B **9**(39), 8341–8347 (2021). https://doi.org/10.1039/D1TB01537K
50. Mo, L., Liang, D., Mo, M., Yang, C., Lin, W.: Dual-detection of miRNAs in living cells via hybridization chain reaction on DNA tetrahedron. Sens. Actuators B Chem. **375**, 132955 (2023). https://doi.org/10.1016/j.snb.2022.132955
51. Yuan, B., et al.: Enhanced imaging of specific cell-surface glycosylation based on Multi-FRET. Anal. Chem. **90**(10), 6131–6137 (2018). https://doi.org/10.1021/acs.analchem.8b00424
52. Wang, J.: From DNA biosensors to gene chips. Nucleic Acids Res. **28**(16), 3011–3016 (2000). https://doi.org/10.1093/nar/28.16.3011
53. Teles, F., Fonseca, L.: Trends in DNA biosensors. Talanta **77**(2), 606–623 (2008). https://doi.org/10.1016/j.talanta.2008.07.024
54. Chen, Y.-T., Kolhatkar, A.G., Zenasni, O., Xu, S., Lee, T.R.: Biosensing using magnetic particle detection techniques. Sensors **17**(10), 2300 (2017). https://doi.org/10.3390/s17102300
55. Diao, W., et al.: Highly sensitive surface plasmon resonance biosensor for the detection of HIV-related DNA based on dynamic and structural DNA nanodevices. Biosens. Bioelectron. **100**, 228–234 (2018). https://doi.org/10.1016/j.bios.2017.08.042
56. El-Said, W.A., Choi, J.: High selective spectroelectrochemical biosensor for HCV-RNA detection based on a specific peptide nucleic acid. Spectrochim. Acta A Mol. Biomol. Spectrosc. **217**, 288–293 (2019). https://doi.org/10.1016/j.saa.2019.03.115
57. Taufiq, S., Waqar, M., Sharif, M.N., Abbas, S.R.: Towards portable rapid TB biosensor: detecting Mycobacterium tuberculosis in raw sputum samples using functionalized screen printed electrodes. Bioelectrochemistry **150**, 108353 (2023). https://doi.org/10.1016/j.bioelechem.2022.108353
58. Saeed, A.A., Sánchez, J.L.A., O'Sullivan, C.K., Abbas, M.N.: DNA biosensors based on gold nanoparticles-modified graphene oxide for the detection of breast cancer biomarkers for early diagnosis. Bioelectrochemistry **118**, 91–99 (2017). https://doi.org/10.1016/j.bioelechem.2017.07.002
59. Yanik, S., et al.: A novel electrochemical biosensor based on palladium nanoparticles decorated on reduced graphene oxide-polyaminophenol matrix for the detection and discrimination

of mitomycin C-DNA and acyclovir-DNA interaction. J. Pharm. Biomed. Anal. **234**, 115524 (2023). https://doi.org/10.1016/j.jpba.2023.115524

60. Chen, M., et al.: Three-dimensional electrochemical DNA biosensor based on 3D graphene-Ag nanoparticles for sensitive detection of CYFRA21-1 in non-small cell lung cancer. Sens. Actuators B Chem. **255**, 2910–2918 (2018). https://doi.org/10.1016/j.snb.2017.09.111
61. Thakur, A., Kumar, A.: Exploring the potential of ionic liquid-based electrochemical biosensors for real-time biomolecule monitoring in pharmaceutical applications: from lab to life. Results Eng. **20**, 101533 (2023). https://doi.org/10.1016/j.rineng.2023.101533
62. Shajaripour Jaberi, S.Y., Ghaffarinejad, A., Omidinia, E.: An electrochemical paper based nano-genosensor modified with reduced graphene oxide-gold nanostructure for determination of glycated hemoglobin in blood. Anal. Chim. Acta **1078**, 42–52 (2019). https://doi.org/10.1016/j.aca.2019.06.018
63. Abazar, F., Noorbakhsh, A.: Chitosan-carbon quantum dots as a new platform for highly sensitive insulin impedimetric aptasensor. Sens. Actuators B Chem. **304**, 127281 (2020). https://doi.org/10.1016/j.snb.2019.127281
64. Georganopoulou, D.G., et al.: Nanoparticle-based detection in cerebral spinal fluid of a soluble pathogenic biomarker for Alzheimer's disease. Proc. Natl. Acad. Sci. **102**(7), 2273–2276 (2005). https://doi.org/10.1073/pnas.0409336102
65. Shi, S.-S., Jia, L.-P., Ma, R.-N., Jia, W.-L., Wang, H.-S.: A label-free electrochemical aptasensor for 8-hydroxy-2′-deoxyguanosine detection. J. Electroanal. Chem. **759**, 107–112 (2015). https://doi.org/10.1016/j.jelechem.2015.10.040
66. Jia, L.-P., et al.: An electrochemical aptasensor for the highly sensitive detection of 8-hydroxy-2′-deoxyguanosine based on the hybridization chain reaction. Talanta **179**, 414–419 (2018). https://doi.org/10.1016/j.talanta.2017.11.036
67. Jia, L.-P., Liu, J.-F., Wang, H.-S.: Electrochemical performance and detection of 8-Hydroxy-2′-deoxyguanosine at single-stranded DNA functionalized graphene modified glassy carbon electrode. Biosens. Bioelectron. **67**, 139–145 (2015). https://doi.org/10.1016/j.bios.2014.07.073
68. Zheng, J., Gao, T., Shi, H., Huang, Y., Xiang, Y., Li, G.: Electrochemical analysis of 8-hydroxy-2′-deoxyguanosine with enhanced sensitivity based on exonuclease-mediated functional nucleic acid. Talanta **199**, 324–328 (2019). https://doi.org/10.1016/j.talanta.2019.02.080
69. Xu, M., Wang, X., Tian, J., Chen, J., Wei, X., Li, W.: A clamp-improved universal amplified system for ratiometric fluorescent detection of single-nucleotide polymorphisms coupled with a novel dual-emissive silver nanocluster. Sens. Actuators B Chem. **367**, 132151 (2022). https://doi.org/10.1016/j.snb.2022.132151
70. Maxwell, D.J., Taylor, J.R., Nie, S.: Self-assembled nanoparticle probes for recognition and detection of biomolecules. J. Am. Chem. Soc. **124**(32), 9606–9612 (2002). https://doi.org/10.1021/ja025814p
71. Ayoib, A., et al.: Low cost design and fabrication of PDMS microfluidics micromixers for DNA extraction. In: 2017 IEEE Regional Symposium on Micro and Nanoelectronics (RSM), pp. 227–230 (2017). https://doi.org/10.1109/RSM.2017.8069158
72. Ayoib, A., Hashim, U., Gopinath, S.C.B., Thivina, V., Arshad, M.K.M.: Design and fabrication of PDMS microfluidics device for rapid and label-free DNA detection. Appl. Phys. A **126**(3), 193 (2020). https://doi.org/10.1007/s00339-020-3337-7
73. Nordin, N.K.S., Hashim, U., Ayoib, A., Thivina, V.: Surface morphological and electrical characterization of silicon nanowires (SiNWs) by conventional photolithography method. In: Proceedings of the 2017 IEEE Regional Symposium on Micro and Nanoelectronics, RSM 2017, (2017). https://doi.org/10.1109/RSM.2017.8069142
74. Zhou, J., et al.: Nanomaterial-based fluorescent biosensor for food safety analysis. Biosensors (Basel) **12**(12), 1072 (2022). https://doi.org/10.3390/bios12121072
75. Xu, S., Zhang, Y., Dong, K., Wen, J., Zheng, C., Zhao, S.: Electrochemical DNA biosensor based on graphene oxide-chitosan hybrid nanocomposites for detection of escherichia coli O157:H7. Int. J. Electrochem. Sci. **12**(4), 3443–3458 (2017). https://doi.org/10.20964/2017.04.16

76. Chen, K., Han, H., Luo, Z., Wang, Y., Wang, X.: A practicable detection system for genetically modified rice by SERS-barcoded nanosensors. Biosens. Bioelectron. **34**(1), 118–124 (2012). https://doi.org/10.1016/j.bios.2012.01.029
77. Costanzo, H., Gooch, J., Frascione, N.: Nanomaterials for optical biosensors in forensic analysis. Talanta **253**, 123945 (2023). https://doi.org/10.1016/j.talanta.2022.123945
78. Khalil, I., et al.: Graphene oxide and gold nanoparticle based dual platform with short DNA probe for the PCR free DNA biosensing using surface-enhanced Raman scattering. Biosens. Bioelectron. **131**, 214–223 (2019). https://doi.org/10.1016/j.bios.2019.02.028
79. Khalil, I., et al.: Simultaneous detection of dual food adulterants using graphene oxide and gold nanoparticle based surface enhanced Raman scattering duplex DNA biosensor. Vib. Spectrosc. **116**, 103293 (2021). https://doi.org/10.1016/j.vibspec.2021.103293
80. Yoo, E.-H., Lee, S.-Y.: Glucose biosensors: an overview of use in clinical practice. Sensors **10**(5), 4558–4576 (2010). https://doi.org/10.3390/s100504558
81. Updike, S.J., Hicks, G.P.: The enzyme electrode. Nature **214**(5092), 986–988 (1967). https://doi.org/10.1038/214986a0
82. Suzuki, S., Takahashi, F., Satoh, I., Sonobe, N.: Ethanol and lactic acid sensors using electrodes coated with dehydrogenase—collagen membranes. Bull. Chem. Soc. Jpn **48**(11), 3246–3249 (1975). https://doi.org/10.1246/bcsj.48.3246
83. Turner, A.: Biosensors: then and now. Trends Biotechnol. **31**(3), 119–120 (2013). https://doi.org/10.1016/j.tibtech.2012.10.002
84. Vo-Dinh, T.: Development of a DNA biochip: principle and applications. Sens. Actuators B Chem. **51**(1–3), 52–59 (1998). https://doi.org/10.1016/S0925-4005(98)00182-8
85. Karunakaran, C., Madasamy, T., Sethy, N.K.: Enzymatic biosensors. In: Biosensors and Bioelectronics, pp. 133–204. Elsevier (2015). https://doi.org/10.1016/B978-0-12-803100-1.00003-7
86. Kaur, J., Choudhary, S., Chaudhari, R., Jayant, R.D., Joshi, A.: Enzyme-based biosensors. In: Bioelectronics and Medical Devices, pp. 211–240. Elsevier (2019). https://doi.org/10.1016/B978-0-08-102420-1.00013-3
87. Azmi, N.S., Kader, Md.A., AKM, K.: Applications of nanomaterials for biosensor fabrication based on redox enzyme and protein: a short review. Curr. Sci. Technol. **2**(2), 20–28 (2023). https://doi.org/10.15282/cst.v2i2.9291
88. Peña-Bahamonde, J., Nguyen, H.N., Fanourakis, S.K., Rodrigues, D.F.: Recent advances in graphene-based biosensor technology with applications in life sciences. J. Nanobiotechnology **16**(1), 75 (2018). https://doi.org/10.1186/s12951-018-0400-z
89. Malhotra, B.D., Ali, Md.A.: Nanomaterials in biosensors. In: Nanomaterials for Biosensors, pp. 1–74. Elsevier (2018). https://doi.org/10.1016/B978-0-323-44923-6.00001-7
90. Feng, X., Cheng, H., Pan, Y., Zheng, H.: Development of glucose biosensors based on nanostructured graphene-conducting polyaniline composite. Biosens. Bioelectron. **70**, 411–417 (2015). https://doi.org/10.1016/j.bios.2015.03.046
91. Mansouri, N., Babadi, A.A., Bagheri, S., Hamid, S.B.A.: Immobilization of glucose oxidase on 3D graphene thin film: novel glucose bioanalytical sensing platform. Int. J. Hydrogen Energy **42**(2), 1337–1343 (2017). https://doi.org/10.1016/j.ijhydene.2016.10.002
92. Haghighi, N., Hallaj, R., Salimi, A.: Immobilization of glucose oxidase onto a novel platform based on modified TiO_2 and graphene oxide, direct electrochemistry, catalytic and photocatalytic activity. Mater. Sci. Eng. C **73**, 417–424 (2017). https://doi.org/10.1016/j.msec.2016.12.015
93. Ki, J., Young Lee, H., Son, H.Y., Huh, Y.-M., Haam, S.: Sensitive plasmonic detection of miR-10b in biological samples using enzyme-assisted target recycling and developed LSPR probe. ACS Appl. Mater. Interfaces **11**(21), 18923–18929 (2019). https://doi.org/10.1021/acsami.9b03005
94. Nik Mansor, N., et al.: An amperometric biosensor for the determination of bacterial sepsis biomarker, secretory phospholipase group 2-IIA using a Tri-Enzyme system. Sensors **18**(3), 686 (2018). https://doi.org/10.3390/s18030686

95. Semwal, V., Gupta, B.D.: LSPR- and SPR-based fiber-optic cholesterol sensor using immobilization of cholesterol oxidase over silver nanoparticles coated graphene oxide nanosheets. IEEE Sens. J. **18**(3), 1039–1046 (2018). https://doi.org/10.1109/JSEN.2017.2779519
96. Dervisevic, M., Dervisevic, E., Çevik, E., Şenel, M.: Novel electrochemical xanthine biosensor based on chitosan–polypyrrole–gold nanoparticles hybrid bio-nanocomposite platform. J. Food Drug Anal. **25**(3), 510–519 (2017). https://doi.org/10.1016/j.jfda.2016.12.005
97. Yang, L., et al.: A novel strategy for the detection of pyruvate in fermentation processes based on well-distributed enzyme-inorganic hybrid nanoflowers on thiol graphene modified gold electrodes. Electrochim. Acta **427**, 140855 (2022). https://doi.org/10.1016/j.electacta.2022.140855
98. Thakur, M., Wang, B., Verma, M.L.: Development and applications of nanobiosensors for sustainable agricultural and food industries: Recent developments, challenges and perspectives. Environ. Technol. Innov. **26**, 102371 (2022). https://doi.org/10.1016/j.eti.2022.102371
99. Zhang, J., Lei, J., Liu, Z., Chu, Z., Jin, W.: Nanomaterial-based electrochemical enzymatic biosensors for recognizing phenolic compounds in aqueous effluents. Environ. Res. **214**, 113858 (2022). https://doi.org/10.1016/j.envres.2022.113858
100. Bao, J., et al.: 3D graphene/copper oxide nano-flowers based acetylcholinesterase biosensor for sensitive detection of organophosphate pesticides. Sens. Actuators B Chem. **279**, 95–101 (2019). https://doi.org/10.1016/j.snb.2018.09.118
101. Fernandes, P.M.V., Campiña, J.M., Silva, A.F.: A layered nanocomposite of laccase, chitosan, and Fe3O4 nanoparticles-reduced graphene oxide for the nanomolar electrochemical detection of bisphenol A. Microchim. Acta **187**(5), 262 (2020). https://doi.org/10.1007/s00604-020-4223-x
102. Zhang, Q., Xu, Q., Guo, Y., Sun, X., Wang, X.: Acetylcholinesterase biosensor based on the mesoporous carbon/ferroferric oxide modified electrode for detecting organophosphorus pesticides. RSC Adv. **6**(29), 24698–24703 (2016). https://doi.org/10.1039/C5RA21799G
103. Zhao, H., et al.: An ultra-sensitive acetylcholinesterase biosensor based on reduced graphene oxide-Au nanoparticles-β-cyclodextrin/Prussian blue-chitosan nanocomposites for organophosphorus pesticides detection. Biosens. Bioelectron. **65**, 23–30 (2015). https://doi.org/10.1016/j.bios.2014.10.007
104. Cui, H.-F., Wu, W.-W., Li, M.-M., Song, X., Lv, Y., Zhang, T.-T.: A highly stable acetylcholinesterase biosensor based on chitosan-TiO_2-graphene nanocomposites for detection of organophosphate pesticides. Biosens. Bioelectron. **99**, 223–229 (2018). https://doi.org/10.1016/j.bios.2017.07.068
105. Rodrigues, N., Neto, S., Luz, R., Damos, F., Yamanaka, H.: Ultrasensitive determination of malathion using acetylcholinesterase immobilized on chitosan-functionalized magnetic iron nanoparticles. Biosensors (Basel) **8**(1), 16 (2018). https://doi.org/10.3390/bios8010016
106. Li, H., Yan, X., Lu, G., Su, X.: Carbon dot-based bioplatform for dual colorimetric and fluorometric sensing of organophosphate pesticides. Sens. Actuators B Chem. **260**, 563–570 (2018). https://doi.org/10.1016/j.snb.2017.12.170
107. Dzudzevic Cancar, H., et al.: A novel acetylcholinesterase biosensor: core–shell magnetic nanoparticles incorporating a conjugated polymer for the detection of organophosphorus pesticides. ACS Appl. Mater. Interfaces **8**(12), 8058–8067 (2016). https://doi.org/10.1021/acsami.5b12383
108. Gong, N.C., Le Li, Y., Jiang, X., Zheng, X.F., Wang, Y.Y., Huan, S.Y.: Fluorescence resonance energy transfer-based biosensor composed of nitrogen-doped carbon dots and gold nanoparticles for the highly sensitive detection of organophosphorus pesticides. Anal. Sci. **32**(9), 951–956 (2016). https://doi.org/10.2116/analsci.32.951
109. Harshavardhan, S., Rajadas, S.E., Vijayakumar, K.K., Durai, W.A., Ramu, A., Mariappan, R.: Electrochemical immunosensors. In: Bioelectrochemical Interface Engineering, pp. 343–369. Wiley, New York (2019). https://doi.org/10.1002/9781119611103.ch18
110. Schroeder, H.W., Cavacini, L.: Structure and function of immunoglobulins. J. Allergy Clin. Immunol. **125**(2), S41–S52 (2010). https://doi.org/10.1016/j.jaci.2009.09.046

111. Varnakavi, N., Lee, N.: a review on biosensors and recent development of nanostructured materials-enabled biosensors. Sensors **21**(4), 1109 (2021). https://doi.org/10.3390/s21041109
112. Savas, S., Altintas, Z.: Graphene quantum dots as nanozymes for electrochemical sensing of yersinia enterocolitica in milk and human serum. Materials **12**(13), 2189 (2019). https://doi.org/10.3390/ma12132189
113. Alves, G.F., Lisboa, T.P., Matos, R.C.: Organic-inorganic hybrid nanomaterials in biosensing applications. In: Handbook of Nanobioelectrochemistry, pp. 363–382. Springer Nature Singapore, Singapore (2023). https://doi.org/10.1007/978-981-19-9437-1_17
114. Siew, Q.Y., Pang, E.L., Loh, H.-S., Tan, M.T.T.: Highly sensitive and specific graphene/TiO_2 impedimetric immunosensor based on plant-derived tetravalent envelope glycoprotein domain III (EDIII) probe antigen for dengue diagnosis. Biosens. Bioelectron. **176**, 112895 (2021). https://doi.org/10.1016/j.bios.2020.112895
115. Tung, Y.-T., Wu, M.-F., Wang, G.-J., Hsieh, S.-L.: Nanostructured electrochemical biosensor for th0065 detection of the weak binding between the dengue virus and the CLEC5A receptor. Nanomedicine **10**(6), 1335–1341 (2014). https://doi.org/10.1016/j.nano.2014.03.009
116. Haji-Hashemi, H., Habibi, M.M., Safarnejad, M.R., Norouzi, P., Ganjali, M.R.: Label-free electrochemical immunosensor based on electrodeposited Prussian blue and gold nanoparticles for sensitive detection of citrus bacterial canker disease. Sens. Actuators B Chem. **275**, 61–68 (2018). https://doi.org/10.1016/j.snb.2018.07.148
117. Fang, Y., Hu, Q., Yu, X., Wang, L.: Ultrasensitive electrochemical immunosensor for procalcitonin with signal enhancement based on zinc nanoparticles functionalized ordered mesoporous carbon-silica nanocomposites. Sens. Actuators B Chem. **258**, 238–245 (2018). https://doi.org/10.1016/j.snb.2017.11.062
118. Ali, Md.A., Kamil Reza, K., Srivastava, S., Agrawal, V.V., John, R., Malhotra, B.D.: Lipid–lipid interactions in aminated reduced graphene oxide interface for biosensing application. Langmuir **30**(14), 4192–4201 (2014). https://doi.org/10.1021/la4049852
119. Arkan, E., Saber, R., Karimi, Z., Mostafaie, A., Shamsipur, M.: Multiwall carbon nanotube-ionic liquid electrode modified with gold nanoparticles as a base for preparation of a novel impedimetric immunosensor for low level detection of human serum albumin in biological fluids. J. Pharm. Biomed. Anal. **92**, 74–81 (2014). https://doi.org/10.1016/j.jpba.2014.01.005
120. Cheng, T., Wang, X., Han, Y., Hao, J., Hu, H., Hao, L.: The level of serum albumin is associated with renal prognosis and renal function decline in patients with chronic kidney disease. BMC Nephrol. **24**(1), 57 (2023). https://doi.org/10.1186/s12882-023-03110-8
121. Zong, C., Zhang, D., Jiang, F., Yang, H., Liu, S., Li, P.: Metal-enhanced chemiluminescence detection of C-reaction protein based on silver nanoparticle hybrid probes. Talanta **199**, 164–169 (2019). https://doi.org/10.1016/j.talanta.2019.02.060
122. Wang, L., Chen, H., Ma, S., Chang, M., Zhang, X.: Ultra-sensitive SERS detection of Aβ 1–42 for Alzheimer's disease using graphene oxide/gold nanohybrids. Vib. Spectrosc. **129**, 103614 (2023). https://doi.org/10.1016/j.vibspec.2023.103614
123. Yu, X., et al.: Carbon nanotube amplification strategies for highly sensitive immunodetection of cancer biomarkers. J. Am. Chem. Soc. **128**(34), 11199–11205 (2006). https://doi.org/10.1021/ja062117e
124. Wang, L., Shan, J., Feng, F., Ma, Z.: Novel redox species polyaniline derivative-Au/Pt as sensing platform for label-free electrochemical immunoassay of carbohydrate antigen 199. Anal. Chim. Acta **911**, 108–113 (2016). https://doi.org/10.1016/j.aca.2016.01.016
125. Ge, S., Yu, F., Ge, L., Yan, M., Yu, J., Chen, D.: Disposable electrochemical immunosensor for simultaneous assay of a panel of breast cancer tumor markers. Analyst **137**(20), 4727 (2012). https://doi.org/10.1039/c2an35967g
126. Fan, Y., Shi, S., Ma, J., Guo, Y.: A paper-based electrochemical immunosensor with reduced graphene oxide/thionine/gold nanoparticles nanocomposites modification for the detection of cancer antigen 125. Biosens. Bioelectron. **135**, 1–7 (2019). https://doi.org/10.1016/j.bios.2019.03.063
127. Gasparotto, G., Costa, J.P.C., Costa, P.I., Zaghete, M.A., Mazon, T.: Electrochemical immunosensor based on ZnO nanorods-Au nanoparticles nanohybrids for ovarian cancer

antigen CA-125 detection. Mater. Sci. Eng. C **76**, 1240–1247 (2017). https://doi.org/10.1016/j.msec.2017.02.031
128. Kovarova, A., et al.: Biosensor with electrochemically active nanocomposites for signal amplification and simultaneous detection of three ovarian cancer biomarkers. Electrochim. Acta **469**, 143213 (2023). https://doi.org/10.1016/j.electacta.2023.143213
129. Bhardwaj, H., Sumana, G., Marquette, C.A.: Gold nanobipyramids integrated ultrasensitive optical and electrochemical biosensor for Aflatoxin B1 detection. Talanta **222**, 121578 (2021). https://doi.org/10.1016/j.talanta.2020.121578
130. Riberi, W.I., Tarditto, L.V., Zon, M.A., Arévalo, F.J., Fernández, H.: Development of an electrochemical immunosensor to determine zearalenone in maize using carbon screen printed electrodes modified with multi-walled carbon nanotubes/polyethyleneimine dispersions. Sens. Actuators B Chem. **254**, 1271–1277 (2018). https://doi.org/10.1016/j.snb.2017.07.113
131. Sharma, M.K., Narayanan, J., Upadhyay, S., Goel, A.K.: Electrochemical immunosensor based on bismuth nanocomposite film and cadmium ions functionalized titanium phosphates for the detection of anthrax protective antigen toxin. Biosens. Bioelectron. **74**, 299–304 (2015). https://doi.org/10.1016/j.bios.2015.06.015
132. Zhang, H., She, Z., Su, H., Kerman, K., Kraatz, H.-B.: Effects of bipyramidal gold nanoparticles and gold nanorods on the detection of immunoglobulins. Analyst **141**(21), 6080–6086 (2016). https://doi.org/10.1039/C6AN01111J
133. Pandey, C.M., Tiwari, I., Singh, V.N., Sood, K.N., Sumana, G., Malhotra, B.D.: Highly sensitive electrochemical immunosensor based on graphene-wrapped copper oxide-cysteine hierarchical structure for detection of pathogenic bacteria. Sens. Actuators B Chem **238**, 1060–1069 (2017). https://doi.org/10.1016/j.snb.2016.07.121
134. Han, J.-H., Lee, D., Chew, C.H.C., Kim, T., Pak, J.J.: A multi-virus detectable microfluidic electrochemical immunosensor for simultaneous detection of H1N1, H5N1, and H7N9 virus using ZnO nanorods for sensitivity enhancement. Sens. Actuators B Chem **228**, 36–42 (2016). https://doi.org/10.1016/j.snb.2015.07.068
135. Karachaliou, C.-E., Livaniou, E.: Immunosensors for autoimmune-disease-related biomarkers: a literature review. Sensors **23**(15), 6770 (2023). https://doi.org/10.3390/s23156770
136. Zhao, Y., et al.: Label-free ECL immunosensor for the early diagnosis of rheumatoid arthritis based on asymmetric heterogeneous polyaniline-gold nanomaterial. Sens. Actuators B Chem. **257**, 354–361 (2018). https://doi.org/10.1016/j.snb.2017.10.184
137. Chen, L., Neethirajan, S.: A homogenous fluorescence quenching based assay for specific and sensitive detection of influenza virus A hemagglutinin antigen. Sensors **15**(4), 8852–8865 (2015). https://doi.org/10.3390/s150408852
138. Thivina, V., et al.: Distinct detection of ganoderma boninense on metal oxides-gold nanoparticle composite deposited interdigitated electrode DNA sensor. J. Phys. Conf. Ser. **2129**(1), 012050 (2021). https://doi.org/10.1088/1742-6596/2129/1/012050
139. Wang, X., et al.: A novel fluorescence sensor for streptomycin detection based on hairpin DNA–AgNCs and core-shell gold-palladium nanoparticles. Sens. Actuators B Chem. **373**, 132735 (2022). https://doi.org/10.1016/j.snb.2022.132735
140. Pang, H.-H., et al.: A new lateral flow plasmonic biosensor based on gold-viral biomineralized nanozyme for on-site intracellular glutathione detection to evaluate drug-resistance level. Biosens. Bioelectron.Bioelectron. **165**, 112325 (2020). https://doi.org/10.1016/j.bios.2020.112325

Development of a Smart Spirometry Filter Utilizing Silver Nanoparticles Coated on Pulmonary Function Filters to Prevent Bacterial Adhesion

Palaniyandi Velusamy, Sasikumar Pattabi, and Kaliyaperumal Prabhu

1 Introduction

Spirometry serves as a fundamental diagnostic tool in assessing pulmonary function by measuring the volume and flow of inhaled and exhaled air [1]. Central to the accuracy and reliability of spirometry tests are the filters employed to capture respiratory particulates and prevent contamination between patients and healthcare providers. Traditional filters, typically composed of cellulose or synthetic polymers, effectively fulfill this role but are prone to microbial adhesion and subsequent biofilm formation [2]. The presence of bacteria on spirometry filters not only compromises their efficacy but also poses significant risks, including healthcare-associated infections and erroneous test results. To mitigate these challenges, researchers have explored innovative strategies such as incorporating silver nanoparticles (AgNPs) onto the surface of pulmonary function filters. Silver nanoparticles are renowned for their potent antimicrobial properties, capable of inhibiting a broad spectrum of bacteria and fungi through mechanisms including membrane disruption, reactive oxygen species generation, and DNA binding [3].

P. Velusamy (✉)
Innovation and Incubation Centre for Health Sciences (IICHS), Sree Balaji Medical College and Hospital (SBMCH), Bharath Institute of Higher Education and Research (BIHER), Chromepet, Chennai 600 044, Tamil Nadu, India
e-mail: vels73@gmail.com

S. Pattabi
Department of General Surgery, Sree Balaji Medical College and Hospital (SBMCH), Bharath Institute of Higher Education and Research (BIHER), Chromepet, Chennai 600 044, Tamil Nadu, India

K. Prabhu
Department of Anatomy, Sree Balaji Medical College and Hospital (SBMCH), Bharath Institute of Higher Education and Research (BIHER), Chromepet, Chennai 600 044, Tamil Nadu, India

S. C. B. Gopinath and M. M. Ramli (eds.), *Hybrid-Nanomaterials*, Engineering Materials, https://doi.org/10.1007/978-981-97-9022-7_8

In pulmonary function testing (PFT) machines, filters are an essential component. These filters serve to maintain the cleanliness and sterility of the equipment by trapping bacteria, viruses, and other contaminants present in the air exhaled by patients during testing [4]. By preventing the transmission of pathogens, filters not only protect the integrity of the machine but also safeguard the health of both patients and healthcare providers. Therefore, the use of high-quality filters in PFT machines is paramount for accurate and safe respiratory testing. Further, the filter used in pulmonary function test machines plays a critical role in ensuring accurate and reliable measurements of lung function [5]. Typically, these filters are made of fine materials such as nylon, which effectively captures particles and prevents them from entering the machine during testing. However, due to their fine nature, nylon filters have a high risk of microbial contamination, posing potential health risks to individuals undergoing pulmonary function tests. To address this issue, advancements have been made in filter technology. One notable innovation involves replacing the conventional nylon filter with a metallic stainless-steel filter coated with silver nanoparticles. This modification offers several advantages over traditional filters. The metallic stainless-steel filter provides durability and longevity, ensuring repeated and long-term use without degradation. Additionally, the incorporation of silver nanoparticles imparts antimicrobial properties to the filter, effectively inhibiting the growth and proliferation of harmful microbes. As shown in Fig. 1, a schematic representation of design and fabrication of smart spirometry filter.

Silver nanoparticles are known for their broad-spectrum antimicrobial activity, making them effective against a wide range of microorganisms, including bacteria,

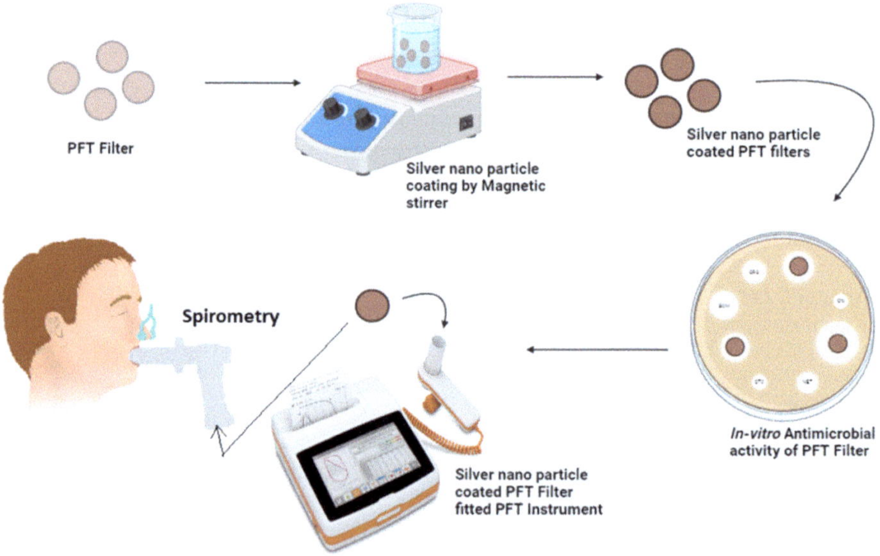

Fig. 1 Preparation of smart spirometry filter using silver nanoparticles

viruses, and fungi [6]. By incorporating silver nanoparticles into the filter, microbial contamination is significantly reduced, minimizing the risk of disease transmission during pulmonary function testing. Furthermore, the efficacy of the silver nanoparticle-coated filter filter has been validated through various studies. These studies have demonstrated a substantial reduction in microbial contamination compared to conventional filters [7–11]. Techniques such as spread plate assays have confirmed the effectiveness of the silver nanoparticle-coated filter in limiting the growth of microbial colonies, further substantiating its utility in pulmonary function testing [12].

Overall, the filter used in pulmonary function test machines, particularly that incorporating metallic stainless-steel filter coated with silver nanoparticles, represents a significant advancement in ensuring the safety and accuracy of lung function assessments. This chapter explores the fabrication of smart spirometry filters with a focus on the integration of silver nanoparticles. It discusses the rationale behind using AgNPs, details the fabrication processes involved, elucidates the mechanisms by which silver nanoparticles prevent bacterial adhesion, and examines the potential benefits of these advanced filters in clinical settings. By enhancing the antimicrobial efficacy of spirometry filters, silver nanoparticle coatings represent a promising advancement towards improving diagnostic accuracy and reducing the transmission of respiratory infections in healthcare environments.

1.1 Overview of Pulmonary Function Tests and Spirometry

Pulmonary function tests (PFTs) are critical diagnostic tools used to evaluate the respiratory system's functionality. Spirometry, one of the most common PFTs, measures the volume and flow of air that can be inhaled and exhaled, providing essential information about lung function. These tests are indispensable for diagnosing and monitoring conditions such as asthma, chronic obstructive pulmonary disease (COPD), and other respiratory disorders [13].

1.2 Importance of Preventing Bacterial Adhesion in Spirometry

In clinical settings, spirometry equipment is used by multiple patients, necessitating stringent hygiene practices to prevent cross-contamination. Traditional spirometry filters act as barriers to pathogens; however, they are often insufficient in preventing bacterial adhesion and subsequent infection [14]. This highlights the need for enhanced filter designs that not only block pathogens but also actively prevent their adhesion and growth.

1.3 Role of Silver Nanoparticles in Antibacterial Applications

Silver nanoparticles (AgNPs) have gained significant attention due to their potent antibacterial properties. Their high surface area to volume ratio allows for effective interaction with microbial cells, leading to enhanced antimicrobial activity [15]. AgNPs can disrupt bacterial cell membranes, interfere with cellular metabolism, and inhibit biofilm formation, making them ideal candidates for enhancing the antibacterial efficacy of spirometry filters.

1.4 Objectives of the Chapter

This chapter aims to explore the fabrication of a smart spirometry filter utilizing silver nanoparticles coated on a pulmonary function filter. The primary objective is to evaluate the effectiveness of AgNPs in preventing bacterial adhesion and enhancing the overall safety of spirometry tests. Through detailed analysis of the coating process, characterization techniques, and antibacterial testing, this chapter will provide comprehensive insights into the development and potential clinical application of AgNP-coated spirometry filters.

2 Background and Literature Review

2.1 Conventional Spirometry Filters

Conventional spirometry filters are designed to prevent the passage of pathogens between patients and the spirometer. These filters typically consist of layers of nonwoven materials that capture airborne particles [16–19]. However, their effectiveness in preventing bacterial adhesion and proliferation is limited, necessitating the development of more advanced filtration technologies.

2.2 Advances in Nanotechnology for Medical Devices

Nanotechnology has revolutionized the medical field by enabling the development of materials with enhanced properties. Nanomaterials, such as silver nanoparticles, offer unique advantages, including high surface reactivity and the ability to be functionalized for specific applications [20]. These properties make them ideal for improving the performance of medical devices, including spirometry filters.

2.3 Antibacterial Properties of Silver Nanoparticles

Silver nanoparticles exhibit broad-spectrum antibacterial activity. They can interact with bacterial cell walls and membranes, leading to structural damage and cell death. Additionally, silver ions released from AgNPs can penetrate bacterial cells and disrupt metabolic processes. These mechanisms contribute to the strong antibacterial efficacy of AgNPs, making them suitable for use in preventing bacterial adhesion on spirometry filters [21].

2.4 Previous Research on Silver Nanoparticles in Medical Filters

Previous studies have demonstrated the effectiveness of silver nanoparticles in enhancing the antibacterial properties of various medical filters. Research has shown that AgNP-coated filters can significantly reduce bacterial colonization and biofilm formation. These findings support the potential of AgNPs to improve the performance of spirometry filters in clinical settings.

3 Implementation of Novel Filters

3.1 Materials Used

3.1.1 Pulmonary Function Filter Specifications

The pulmonary function filters used in this study are commercially available, medical-grade filters designed for use in spirometry. These filters are composed of nonwoven fabric layers with a pore size adequate for trapping airborne particles and pathogens.

3.1.2 Silver Nanoparticles: Properties and Preparation

Silver nanoparticles used in this study are synthesized using chemical reduction methods, resulting in particles with an average size of 10–20 nm. The nanoparticles are characterized by their high surface area and potent antibacterial activity.

3.2 Coating Procedure

3.2.1 Surface Preparation of Pulmonary Function Filter

Prior to coating, the pulmonary function filters are cleaned and dried to remove any contaminants. This step ensures optimal adhesion of the silver nanoparticles to the filter surface.

3.2.2 Silver Nanoparticle Coating Techniques

The coating process involves immersing the filters in a silver nanoparticle suspension, followed by drying under controlled conditions. Various techniques, such as dip coating, spray coating, and electrostatic deposition, are explored to determine the most effective method for uniform nanoparticle distribution.

3.2.3 Optimization of Coating Parameters

The coating parameters, including nanoparticle concentration, immersion time, and drying conditions, are optimized to achieve maximum antibacterial efficacy without compromising the filter's airflow resistance.

3.3 Analysis of Coated Materials

3.3.1 Scanning Electron Microscopy (SEM)

SEM is used to examine the surface morphology of the coated filters, providing detailed images of the nanoparticle distribution and coverage.

3.3.2 Energy Dispersive X-ray Spectroscopy (EDX)

EDX analysis is performed to confirm the presence and elemental composition of silver on the filter surface.

3.3.3 Antibacterial Efficacy Testing

The antibacterial activity of the coated filters is evaluated using standard microbiological assays. Filters are tested against common respiratory pathogens, including Staphylococcus aureus and Escherichia coli.

3.3.4 Airflow Resistance and Durability Testing

The impact of the silver nanoparticle coating on the filter's airflow resistance is measured to ensure that the coating does not impede respiratory function. Additionally, the durability of the coating is assessed by subjecting the filters to repeated airflow cycles.

4 Evaluation of Coated Filters

4.1 Design and Fabrication of Smart Spirometry Filters

4.1.1 Basics of Spirometry Filters

Spirometry filters play a critical role in ensuring accurate pulmonary function tests by preventing contaminants and ensuring clean air passages [22–25]. Traditional spirometry filters are typically made from materials such as polypropylene, cellulose acetate, or fiberglass. They work by physically trapping particles and pathogens from the patient's exhaled breath, thus protecting both the patient and the device.

4.1.2 Types and Materials Used

- **Polypropylene**: Widely used due to its good filtration efficiency and low cost.
- **Cellulose Acetate**: Known for its biodegradability and decent filtration properties.
- **Fiberglass**: Provides high filtration efficiency but can be more expensive and less environmentally friendly.

4.1.3 Performance Criteria

Key performance criteria for spirometry filters include:

- **Filtration Efficiency**: The ability to trap airborne particles and pathogens.
- **Breathability**: Minimal resistance to airflow, ensuring patient comfort.
- **Biocompatibility**: Non-toxic and safe for use in medical devices.
- **Durability**: Capable of withstanding repeated use without degradation.

4.1.4 Incorporating Silver Nanoparticles

Silver nanoparticles (AgNPs) are renowned for their antimicrobial properties, making them an ideal addition to spirometry filters to enhance infection control.

4.1.5 Coating Techniques and Methodologies

Several techniques can be employed to coat spirometry filters with silver nanoparticles:

- Physical Vapor Deposition (PVD): A process where silver is vaporized and then condensed onto the filter surface, creating a uniform nanoparticle layer.
- Chemical Vapor Deposition (CVD): Similar to PVD but involves chemical reactions to deposit silver nanoparticles on the filter.
- Electrospinning: This technique produces nanofibers embedded with silver nanoparticles, which can be integrated into the filter material.
- Dip Coating: Filters are dipped into a silver nanoparticle solution and then dried, resulting in a coating of nanoparticles.

4.1.6 Optimization of Nanoparticle Size and Distribution

- Size Matters: Smaller nanoparticles have a larger surface area-to-volume ratio, enhancing antimicrobial effectiveness. Typically, sizes range from 10 to 100 nm.
- Uniform Distribution: Ensuring even distribution of nanoparticles on the filter surface is crucial for consistent antimicrobial activity. Techniques such as ultrasonic spraying can help achieve this.

4.1.7 Fabrication Process

Creating a smart spirometry filter involves several steps, from material selection to the final quality control checks.

4.1.8 Step-By-Step Fabrication

- Material Preparation: Selecting and preparing the base filter material, ensuring it meets the required performance criteria.
- Nanoparticle Synthesis: Producing silver nanoparticles using methods like chemical reduction or physical synthesis. Ensuring nanoparticles are of the desired size and purity.
- Coating Application: Applying the silver nanoparticles to the filter using one of the techniques mentioned above. Ensuring a uniform and adequate coating.
- Curing and Stabilization: Allowing the coating to stabilize and bond with the filter material, often through a curing process involving heat or UV light.
- Assembly and Integration: Integrating the coated filter into the spirometry device, ensuring compatibility and proper function.

4.1.9 Quality Control and Testing

- Ensuring the filters meet stringent quality standards is essential for both efficacy and safety.
- Filtration Efficiency Tests: Measuring the filter's ability to trap particles and pathogens.
- Antimicrobial Efficacy Tests: Assessing the filter's ability to inhibit microbial growth, typically using standardized tests such as ASTM E2149.
- Breathability Tests: Ensuring the filter does not impede airflow, maintaining patient comfort.
- Durability Tests: Testing the filter's ability to withstand repeated use without degradation or loss of performance.

4.2 Morphological Analysis of Coated Filters

4.2.1 Presence of Silver Nanoparticles Using UV Spectrophotometry Analysis

Confirmation of the presence of silver nanoparticles using UV spectrophotometry is a widely employed technique in nanoscience and materials research. UV spectrophotometry measures the absorption of light by a sample at different wavelengths. Silver nanoparticles exhibit.

unique optical properties due to surface plasmon resonance (SPR), resulting in a characteristic absorption peak in the UV–visible spectrum. To confirm the presence of silver nanoparticles, a sample containing the nanoparticles is analysed using a UV spectrophotometer [26]. The UV–Vis spectrum of the sample is examined for the presence of a strong absorption peak typically observed in the range of 420 nm, indicative of silver nanoparticles (Fig. 2). The intensity and position of this peak depend on factors such as nanoparticle size, shape, concentration, and surrounding environment [27]. Comparing the UV–Vis spectrum of the sample with that of a reference solution or pure solvent helps differentiate the absorption peak attributed to silver nanoparticles from other absorbance features. Additionally, the stability of the absorption peak upon dilution or changes in experimental conditions can provide further confirmation of the presence of silver nanoparticles [28–30]. Overall, UV spectrophotometry is a valuable tool for confirming the synthesis and characterization of silver nanoparticles, offering insights into their optical properties, and enabling researchers to tailor their applications in various fields such as catalysis, sensing, and biomedicine.

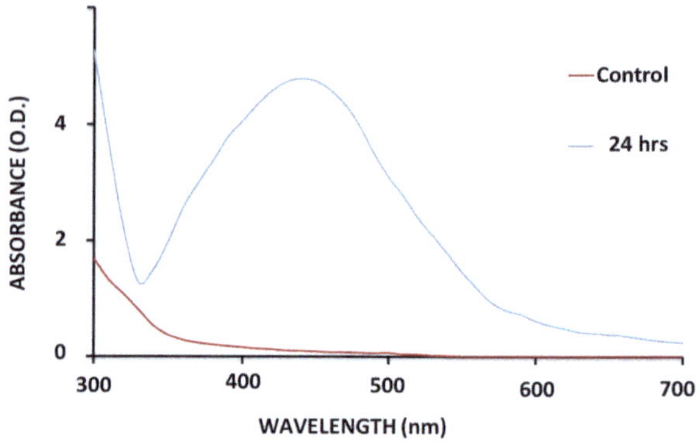

Fig. 2 UV spectrophotometry analysis for detection of silver nanoparticles at 420 nm using 24 hrs

4.2.2 SEM Image Confirmation for Nanoparticle Formation

Confirmation of the presence and characterization of silver nanoparticles is crucial in various scientific and industrial applications. Scanning Electron Microscopy (SEM) is a powerful technique used for visualizing the morphology and size distribution of nanoparticles [31]. In the case of silver nanoparticles with sizes ranging from 17.2 nm to 64.7 nm, SEM images provide valuable insights into their structure and uniformity (Fig. 3).

4.3 Elemental Composition and Distribution of Silver

EDX analysis confirms the presence of silver on the filter surface, with a homogeneous distribution of nanoparticles [32]. The elemental mapping shows a high concentration of silver in the coated areas.

Fig. 3 Scanning Electron Microscopy (SEM) analysis for confirmation of silver nanoparticles

4.4 *Antibacterial Efficacy*

4.4.1 In Vitro Testing Against Common Respiratory Pathogens

The coated filters demonstrate significant antibacterial activity, with a reduction in bacterial colonies by over 99% compared to uncoated filters. The silver nanoparticles effectively inhibit the growth of both *Pseudomonas aeruginosa* and *Staphylococcus aureus*. These measurements were compared with a standard antibacterial agent to assess the effectiveness of the silver-coated filter in inhibiting bacterial growth [33–36]. The results demonstrate that increasing the concentration of silver nitrate led to a corresponding increase in the diameter of the zone of inhibition, indicating enhanced antibacterial activity [37]. This suggests that the silver-coated filter have potential applications in inhibiting bacterial growth and could be further optimized for various biomedical and environmental applications.

The results of the antibacterial activity assay revealed promising inhibitory effects against both *P. aeruginosa* and *S. aureus* with 1 mM silver nanoparticles. Chloramphenicol demonstrated notable inhibitory activity, with mean inhibition zones of 8.4 $\pm$ 1.0 and 9.0 $\pm$ 0.6 mm against *P. aeruginosa* and *S. aureus*, respectively. Interestingly, the silver nanoparticle-coated filter exhibited enhanced antimicrobial efficacy, demonstrating mean inhibition zones of 10.4 $\pm$ 0.5 and 11.2 $\pm$ 1.0 mm against *P. aeruginosa* and *S. aureus*, while the non-coated filter has no zone of inhibition, respectively (Table 1). These findings underscore the potential of silver nanoparticles as coating to the filter in the PFT machine as effective antimicrobial agents against these pathogenic bacteria [38–40].

As showed in Fig. 4, the un-coated pulmonary function testing (PFT) filter has no inhibition of bacterial growth, whereas silver-coated filter in inhibiting bacterial growth [41, 42]. This result suggests that the confirmation of antibacterial transmission when we using silver-coated pulmonary function testing (PFT) filter. This study is the first reported for the coating PFT for the uses of pulmonary test in clinical patients.

Table 1. Efficacy of AgNPs filter for antibacterial activity against bacterial pathogens by disc diffusion assays

Bacterial pathogens	Chloramphenicol	Non-coated filter	AgNPs coated filter
The mean diameter of inhibitory zone (mm)			
P. aeruginosa	8.4 $\pm$ 1.0	0	10.4 $\pm$ 0.5
S. aureus	9.0 $\pm$ 0.6	0	11.2 $\pm$ 1.0

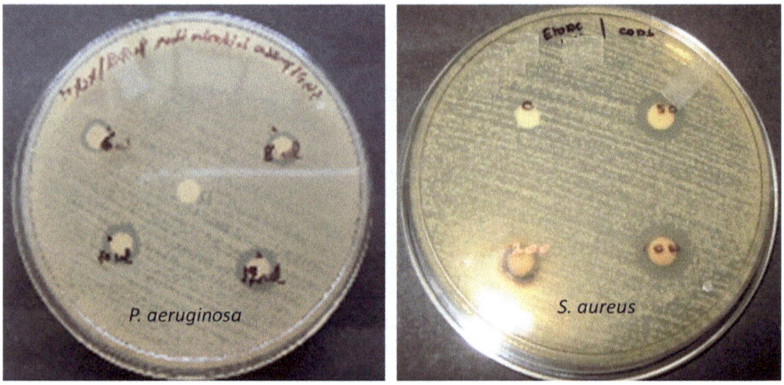

Fig. 4 Antibacterial activity assay was tested against *Pseudomonas aeruginosa* and *Staphylococcus aureus* by pulmonary function testing (PFT) filter

4.4.2 Comparison with Uncoated Filters

Uncoated filters show minimal antibacterial activity, highlighting the enhanced performance of the silver nanoparticle-coated filters in preventing bacterial adhesion and growth.

4.5 Impact on Airflow Resistance

Airflow resistance tests indicate that the silver nanoparticle coating does not significantly affect the filter's performance. The coated filters maintain acceptable resistance levels, ensuring they do not hinder respiratory function during spirometry tests.

4.6 Durability and Longevity of the Coating

Durability tests show that the silver nanoparticle coating remains intact after multiple airflow cycles, indicating its potential for long-term use in clinical settings.

4.7 Discussion on Practical Implications and Potential Challenges

The results suggest that silver nanoparticle-coated filters offer a promising solution for preventing bacterial adhesion in spirometry. However, potential challenges such as cost, scalability, and regulatory approval need to be addressed before widespread clinical implementation.

5 Conclusion

5.1 Summary of Findings

This study demonstrates the successful fabrication of a smart spirometry filter using silver nanoparticles. The integration of silver nanoparticles into spirometry filters represents a significant advancement in pulmonary testing technology. The coated filters exhibit significant antibacterial activity, effectively preventing bacterial adhesion without compromising airflow resistance. By enhancing antimicrobial properties and maintaining high filtration efficiency, these smart filters offer improved safety and reliability in clinical settings. The careful design and meticulous fabrication process ensure that these filters meet the high standards required for medical devices, ultimately contributing to better patient outcomes and more accurate diagnostic results.

5.2 Advantages of Using Silver Nanoparticle-Coated Filters

The use of silver nanoparticles enhances the antibacterial properties of spirometry filters, reducing the risk of cross-contamination and infection in clinical settings. The coated filters demonstrate significantly reduced the bacterial cross contamination and it can be showed potent antibacterial activity, with a reduction in bacterial colonies by over 99% compared to uncoated filters. The silver nanoparticles effectively inhibit the growth of both *Staphylococcus aureus* and *Escherichia coli.*

5.3 Future Research Directions

Future research should focus on large-scale production techniques, long-term efficacy studies, and comprehensive safety evaluations to support the clinical adoption of silver nanoparticle-coated spirometry filters. Further, it can be assessed the airflow resistance and filtration efficiency of the smart spirometry filter post-coating to ensure it meets spirometry standards without compromising accuracy or patient comfort. The assembled a prototype smart spirometry filter could be incorporating optimized materials and processes. The prototype can be validated the performance of the prototype filter through in vitro and potentially in vivo studies, ensuring it meets regulatory standards and clinical requirements for spirometry applications.

5.4 Potential for Clinical Implementation

The development of a smart spirometry filter coated with silver nanoparticles (AgNPs) holds significant potential for clinical implementation, primarily due to silver's well-documented antibacterial properties. The implementation of silver nanoparticle-coated spirometry filters in clinical settings offers promising benefits in preventing bacterial adhesion and enhancing patient safety. By addressing key factors such as biocompatibility, regulatory approval, cost, and performance, these advanced filters can potentially revolutionize pulmonary function testing by reducing the risk of infection transmission and improving overall hygiene standards in healthcare facilities.

References

1. Leung, J.M., Sin, D.D.: Quantification of airflow obstruction in chronic obstructive pulmonary disease. Chest **143**(4), 1065–1073 (2013)
2. Miller, M.R., Hankinson, J., Brusasco, V., Burgos, F., Casaburi, R., Coates, A., Viegi, G.: Standardisation of spirometry. Eur. Respir. J. **26**(2), 319–338 (2005)
3. Velusamy, P., Su, C.-H., Kannan, K., Kumar, G.V., Anbu, P., Gopinath, S.C.B.: Surface engineered iron oxide nanoparticles as efficient materials for antibiofilm application. Biotechnol. Appl. Biochem. **69**(2), 714–725 (2022)
4. Hoppin, F.G.: The promise and reality of nanotechnology in medicine. J. Clin. Investig. **112**(8), 1197–1203 (2002)
5. Kim, J.S., Kuk, E., Yu, K.N., Kim, J.H., Park, S.J., Lee, H.J., Cho, M.H.: Antimicrobial effects of silver nanoparticles. Nanomed. Nanotechnol. Biol. Med. **3**(1), 95–101 (2007)
6. Velusamy, P., Kiruba, K., Su, C.-H., Around, V., Anbu, P., Gopinath, S.C.B., Vaseeharan, B.: SARS-CoV-2 spike protein: site-specific breakpoints for the development of COVID-19 vaccines. J. King Saud Univ. Sci. **33**(8), 101648 (2021)
7. Rai, M., Yadav, A., Gade, A.: Silver nanoparticles as a new generation of antimicrobials. Biotechnol. Adv. **27**(1), 76–83 (2009)
8. Alexander, J.W.: History of the medical use of silver. Surg. Infect. **10**(3), 289–292 (2009)
9. Velusamy, P., Su, C.-H., Ramasamy, P., Arun, V., Rajnish, N., Senthil Kumar, S.M., Pachaiappan, R., Gopinath, S.C.B., Vaseeharan, B.: Volatile organic compounds as potential biomarkers for noninvasive disease detection by nanosensors: a comprehensive review. Crit. Rev. Anal. Chem. (2022). https://doi.org/10.1080/10408347.2022.2043145
10. Samuel, U., Guggenbichler, J.P.: Prevention of catheter-related infections: the potential of a new nano-silver impregnated catheter. Int. J. Antimicrob. Agents **23**(Suppl 1), S75–S78 (2004)
11. Lansdown, A.B.: Silver in healthcare: its antimicrobial efficacy and safety in use. Biofunctional Textiles Skin **33**, 17–34 (2006)
12. Morones, J.R., Elechiguerra, J.L., Camacho, A., Holt, K., Kouri, J.B., Ramírez, J.T., Yacaman, M.J.: The bactericidal effect of silver nanoparticles. Nanotechnology **16**(10), 2346–2353 (2005)
13. Velusamy, P., Srinivasa, C.M., Kumar, G.V., Qurishi, Y., Su, C.-H., Gopinath, S.C.B.: A pH stimuli thiol modified mesoporous silica nanoparticles: doxorubicin carrier for cancer therapy. J. Taiwan Inst. Chem. Eng. **87**, 264–271 (2018)
14. Pal, S., Tak, Y.K., Song, J.M.: Does the antibacterial activity of silver nanoparticles depend on the shape of the nanoparticle? a study of the Gram-negative bacterium Escherichia coli. Appl. Environ. Microbiol. **73**(6), 1712–1720 (2007)

15. Shrivastava, S., Bera, T., Roy, A., Singh, G., Ramachandrarao, P., Dash, D.: Characterization of enhanced antibacterial effects of novel silver nanoparticles. Nanotechnology **18**(22), 225103 (2007)
16. Retchkiman-Schabes, P.S., Canizal, G., Becerra-Herrera, R., Zorrilla, C., Liu, H.B., Ascencio, J.A.: Biosynthesis and characterization of Ti/Ni bimetallic nanoparticles. Micro Nano Lett. **1**(2), 80–83 (2006)
17. Velusamy, P., Su, C.-H., Shritama, A., Kumar, G.V., Pandian, K.: Biopolymers regulate silver nanoparticle under microwave irradiation for effective antibacterial and antibiofilm activities. PLoS ONE **11**(6), e0157612 (2016)
18. Sondi, I., Salopek-Sondi, B.: Silver nanoparticles as antimicrobial agent: a case study on E. coli as a model for Gram-negative bacteria. J. Colloid Interface Sci. **275**(1), 177–182 (2004)
19. Panáček, A., Kvítek, L., Prucek, R., Kolář, M., Večeřová, R., Pizúrová, N., Zbořil, R.: Silver colloid nanoparticles: synthesis, characterization, and their antibacterial activity. J. Phys. Chem. B **110**(33), 16248–16253 (2006)
20. Velusamy, P., Kumar, G.V., Jeyanthi, V., Das, J., Pachaiappan, R.: Bio-inspired green nanoparticles: synthesis, mechanism, and antibacterial application. Toxicol. Res. **32**, 95–102 (2016)
21. Franci, G., Falanga, A., Galdiero, S., Palomba, L., Rai, M., Morelli, G., Galdiero, M.: Silver nanoparticles as potential antibacterial agents. Molecules **20**(5), 8856–8874 (2015)
22. Velusamy, P., Jeyanthi, V., Pachaiappan, R., Anbu, P., Gopinath, S.C.B.: Secretion of 2,4 di-tertbutylphenol by a new Pseudomonas strain SBMCH11: a tert-butyl substituted phenolic compound displayed antibacterial efficacy. Results Chem. **8**, 101593 (2024)
23. Li, X.Z., Nikaido, H., Williams, K.E.: Silver-resistant mutants of Escherichia coli display active efflux of Ag+ and are deficient in porins. J. Bacteriol. **179**(19), 6127–6132 (1997)
24. Zhang, H., Chen, G.: Potent antibacterial activities of Ag/TiO_2 nanocomposite powders synthesized by a one-pot sol-gel method. Environ. Sci. Technol. **43**(8), 2905–2910 (2009)
25. Velusamy, P., Su, C.-H., Shritama, A., Kumar, G.V., Jeyanthi, V., Pandian, K.: Synthesis of oleic acid coated iron oxide nanoparticles and its role in anti-biofilm activity against clinical isolates of bacterial pathogens. J. Taiwan Inst. Chem. Eng. **59**, 450–456 (2016)
26. Guzman, M., Dille, J., Godet, S.: Synthesis and antibacterial activity of silver nanoparticles against gram-positive and gram-negative bacteria. Nanomed. Nanotechnol. Biol. Med. **8**(1), 37–45 (2012)
27. Velusamy, P., Das, J., Pachaiappan, R., Vaseeharan, B., Pandian, K.: Greener approach for synthesis of antibacterial silver nanoparticles using aqueous solution of neem gum (Azadirachta indica L.). Ind. Crops Prod. **66**, 103–109 (2015)
28. Choi, O., Deng, K.K., Kim, N.J., Ross, L., Surampalli, R.Y., Hu, Z.: The inhibitory effects of silver nanoparticles, silver ions, and silver chloride colloids on microbial growth. Water Res. **42**(12), 3066–3074 (2008)
29. Chou, W.L., Yu, D.G., Yang, M.C.: The preparation and characterization of silver-loading cellulose acetate hollow fiber membrane for water treatment. Polym. Adv. Technol. **16**(8), 600–607 (2005)
30. Gopinath, V., Velusamy, P.: Extracellular biosynthesis of silver nanoparticles using Bacillus sp. GP-23 and evaluation of their antifungal activity towards Fusarium oxysporum. Spectrochim. Acta Part A Mol. Biomol. Spectrosc. **106**, 170–174 (2013)
31. Gaikwad, S.C., Birajdar, M.S., Shinde, R.B.: Synthesis of silver nanoparticles by using tea leaf extract. Asian J. Chem. **24**(12), 5775–5778 (2012)
32. Ip, M., Lui, S.L., Poon, V.K., Lung, I., Burd, A.: Antimicrobial activities of silver dressings: an in vitro comparison. J. Med. Microbiol. **55**(1), 59–63 (2006)
33. Jeong, S.H., Yeo, S.Y., Yi, S.C.: The effect of filler particle size on the antibacterial properties of compounded polymer/silver fibers. J. Mater. Sci. **40**(20), 5407–5411 (2005)
34. Lara, H.H., Ayala-Nuñez, N.V., Ixtepan-Turrent, L., Rodriguez-Padilla, C.: Mode of antiviral action of silver nanoparticles against HIV-1. J. Nanobiotechnol. **8**(1), 1 (2010)
35. Priyadarshini, S., Gopinath, V., Meera Priyadharsshini, N., MubarakAli, D., Velusamy, P.: Synthesis of anisotropic silver nanoparticles using novel strain, Bacillus flexus and its biomedical application. Colloids Surf. B **102**, 232–237 (2013)

36. Lu, L., Sun, R.W., Chen, R., Hui, C.K., Ho, C.M., Luk, J.M., Wong, K.Y.: Silver nanoparticles inhibit hepatitis B virus replication. Antivir. Ther. **13**(2), 253–262 (2008)
37. Martin, P., Nunan, R.: Cellular and molecular mechanisms of repair in acute and chronic wound healing. Br. J. Dermatol. **173**(2), 370–378 (2015)
38. McShan, D., Ray, P.C., Yu, H.: Molecular toxicity mechanism of nanosilver. J. Food Drug Anal. **22**(1), 116–127 (2014)
39. Panáček, A., Smékalová, M., Večeřová, R., Bogdanová, K., Röderová, M., Kolář, M., Zbořil, R.: Silver nanoparticles strongly enhance and restore bactericidal activity of inactive antibiotics against multiresistant Enterobacteriaceae. Colloids Surf. B **142**, 392–399 (2016)
40. Anbu, P., Jayanthi, S., Velusamy, P.: Characterization of nanoparticles using nano-analytical techniques. In: Gopinath, S., Gang, F. (eds.) Nanoparticles in Analytical Medical Devices, 1st edn., vol. 1, pp. 56–74. Elsevier (2021)
41. Saha, S., Tomaro-Duchesneau, C.: Probiotics as oral health biotherapeutics. Expert Opin. Biol. Ther.Ther. **13**(11), 1493–1509 (2013)
42. Thomas, V., Nair, S.: A brief review on the antimicrobial properties of silver nanoparticles. Int. Nano Lett. **3**(1), 1–10 (2013)

Preparation and Applications of Silicon and Silica Hybrid Materials

Chun Hong Voon, Tijjani Adam, Shahidah Arina Shamsuddin, and Bee Ying Lim

1 Introduction

Hybrid materials are intensively studied and reported recently because hybrid materials offer endless possibilities for novel properties and structures. Hybrid materials are nanocomposites at the nanoscale or molecular scale with a combination of organic and/or inorganic components. According to the International Union of Pure and Applied Chemistry, a hybrid material is composed of an intimate mixture of inorganic components, organic components, or both interpenetrate on scales of less than 1 μm. This makes hybrid materials different from traditional composites in which the fillers are at the macroscopic level. Therefore, the properties of the hybrid materials are not just the result of the individual contributions of their components, but also from the strong synergy produced by a hybrid interface, which is distinctively different from their individual components or the traditional composites of these components. The diversity of the materials and their properties used in the production of hybrid materials results in various multifunctional materials.

The history of hybrid materials can be dated back to Stone Age, in which ancient people combined organic and inorganic natural resources from plants, animals and rocks to make various tools, weapons and shelter to help them, for protection and for hunting in their daily life. Out of these hybrid materials, the Maya Blue, an

C. H. Voon (✉) · T. Adam · S. A. Shamsuddin
Institute of Nano Electronic Engineering (INEE), Universiti Malaysia Perlis (UniMAP), 02600 Arau, Perlis, Malaysia
e-mail: chvoon@unimap.edu.my

T. Adam · S. A. Shamsuddin
Faculty of Electronic Engineering & Technology, Universiti Malaysia Perlis (UniMAP), 02600 Arau, Perlis, Malaysia

B. Y. Lim
Faculty of Chemical Engineering & Technology, Universiti Malaysia Perlis (UniMAP), 02600 Arau, Perlis, Malaysia

S. C. B. Gopinath and M. M. Ramli (eds.), *Hybrid-Nanomaterials*, Engineering Materials, https://doi.org/10.1007/978-981-97-9022-7_9

Fig. 1 Presence of Maya blue pigment in the wall painting at Cacaxtla, Mexico [2]

unique pigment from 8th century, has survived more than twelve centuries of harsh jungle environment. This hybrid materials was a mixture of vegetable dye "indigo", inorganic clay mineral "palygorskite" and copal resin [1]. Maya Blue can be traced back to the first Millennium AD in the Maya area. Traces of Maya blue was also be found in deposits in Mesoamerica. Figure 1 shows the presence of Maya Blue pigment in one of the wall paintings at Cacaxtla, Mexico.

Today, various hybrid materials were engineered from combination of various materials for wide range applications because the necessities of man have increased tremendously, and this has developed the technology of hybrid materials. These hybrid materials are developed according to their use. One of the most successfully commercialized hybrid materials is the glass laminate aluminum reinforced epoxy (GLARE®), which is a fiber-metal laminate that is widely employed in aircraft industry replacing conventional aluminum. This hybrid materials composed of several thin metal sheets, mainly made of aluminum, as interlayers between several glass fiber/ epoxy prepreg layers. This unique structure provides GLARE® a higher impact and fatigue resistance, being lighter and cheaper, and better corrosion and fire resistance than the conventional aluminum counterpart, leading to the utilization of GLARE® in aircraft industry since year 2001.

Previous studies show that hybrid materials can be obtained by varying the material combinations and processing techniques. The choice of component materials and processing techniques significantly affect the properties and applications of these hybrid materials because the properties of hybrid materials are not only affected by the respective properties of their components but are also affected by the interfaces between these components. The component materials in these hybrid materials are distributed homogeneously while in some hybrid materials, heterogeneous systems are preferred. According to Kickelbick, hybrid materials can be categorized according to the strength of the interactions of the organic and inorganic building blocks into two classes, namely Class I (weak interactions) and Class II (strong interactions) hybrids [3]. Weak interactions such as electrostatic forces, van der Waals and hydrogen bonding exist between phases in Class I hybrid materials. Only physical

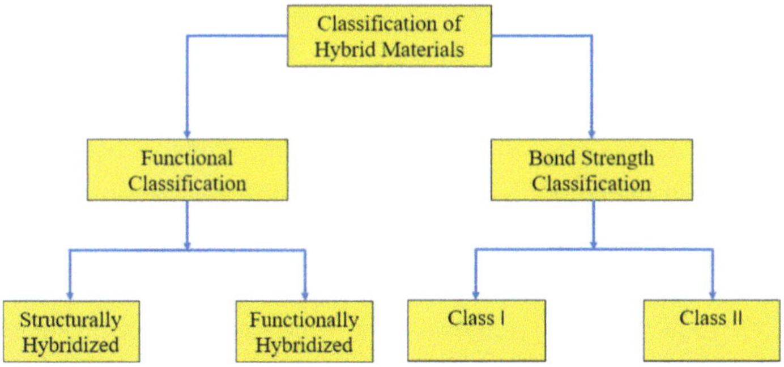

Fig. 2 Classification of hybrid materials

interactions are available between phases in Class I hybrid materials. Strong chemical bonds hold two phases together in Class II hybrid materials. Hybrid materials can also be classified structurally and functionally hybridized materials. Composites are structurally hybridized materials since the properties of these materials can be understood from the combination of properties of the respective component materials. Hybridization based on the macroscopic structure is the purpose of the combining these component materials. However, for functionally hybridized materials, new functions or super functions are created when combining the hybridization of the functions of each component materials. The classification of the hybrid materials is summarised in Fig. 2.

Recently, gradual transition from hybrid materials to nanohybrid materials is observed. Structurally, nanohybrid materials are similar to the hybrid materials, except that one or more of the component materials are in nanoscale. Nanohybrid materials show characteristics of the component materials and in same case, novel properties are observed. Nanocomposite is one of the most common nanohybrid materials. Functional nanomaterials such as nanoparticles and nanowires are dispersed in various matrices such as polymers, ceramics or metals for the fabrication of nanocomposites. Nanohybrid materials in which two or more component materials are in the nanoscale such as core–shell nanoparticles [4] or nanoparticles decorated nanowires [5] were also reported previously.

Particularly, in this chapter, recent progress on the synthesis and preparation of silicon and silica hybrid materials are highlighted. This chapter presents most widely reported processes of silicon and silica hybrid materials preparation, with relatively few descriptions about less reported, new processes as and when felt necessary. This chapter is organized into three sections. After a brief introduction of silicon and silica hybrid materials in Sects. 1.1 and 1.2 describes various applications and potentials of silicon and silica hybrid materials. Following that, Sect. 1.3 focuses on the various approaches for rational synthesis and preparation of various silicon and silica hybrid materials, which can be divided into liquid phase processing and vapor phase processing. This chapter ends in Sect. 2 with a concluding remark on the recent progress and future challenges in this field.

1.1 Silicon and Silica

Silicon is the most abundant element in the earth's crust after oxygen, makes up 27.7 wt % of the Earth's crust. It has a simple cubic crystal structure. Silicon is a very useful element to mankind. In nature, silicon never occurs free, and it always combines with oxygen to form silica or silicates. Therefore, commercially available elemental silicon is obtained commercially by carbothermal reduction. Highly pure silicon for electronic industry is produced by thermal decomposition of trichlorosilane at 1,100 °C followed by recrystallisation. Most of the elemental silicon are used to produce aluminium–silicon, ferrosilicon or silicocalcium alloys. On the other hand, monocrystalline silicon of 99.9999999% purity is produced by the Czochralski process. Monocrystalline silicon is the most commonly used material in the semiconductor industry for the fabrication of electronics and photovoltaic devices. Silicon is also used for in the polymer industry for the production of silicone. Silicone are polymers of siloxanes in which silicon-oxygen chains with two methyl groups are attached to each silicon atom. Silicone is used mostly in manufacturing industry, automotive industry and kitchenware industry and as waterproofing sealants.

As silicon does not occur naturally, it normally combines with oxygen to form silica or silicon dioxide. Silica is one of the most common minerals in the earth's crust. Silica can be crystalline, naturally in the form of quartz in sand and granite and it can be non-crystalline in the form of glass. Figure 3 shows the structure of crystalline and amorphous silica. Crystalline silica is composed of SiO_4 tetrahedra, in which four O atoms are covalently bonded to a single Si atom, and these tetrahedra form three-dimensional framework. In contrast, amorphous silica, as in Fig. 3, shows no orderly lattice structure. Due to the difference of the structure, the properties of amorphous and crystalline silica are significantly different. For example, crystalline silica is toxic to human body while amorphous silica is less hazardous. Silica is widely used in industry to produce ceramics, catalysts, adhesives, biosensors support, composites, detergents, enzyme/drug carriers and pharmaceutical products.

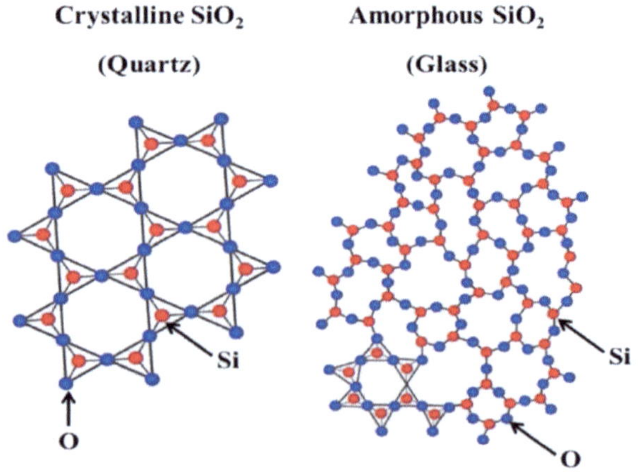

Fig. 3 Structure of crystalline and amorphous silica [6]

1.2 Applications of Silicon and Silica Hybrid Materials

Both silicon and silica are generally reported for the fabrication of hybrid materials for similar applications. For example, both silicon and silica have been popular materials for the fabrication of various hybrid materials in various domains due to their excellent properties. For example, mesoporous silicon was used as mono-component additive and synergistic agents in traditional intumescent and phosphinate systems to form hybrid material as highly efficient fire retardant due to its remarkable thermal stability and unique surface diffusion characteristic [7]. The addition of silicon into polycarbonate (PC) [8] and poly (butylene adipate-co-terephthalate) (PBAT) [9] for example can promote the self-charring effect of these matrices for fire retardancy. Silica was also used in the form of inorganic–organic hybrid system for fire retardancy. For instance, silica were combined with phytic acid to improve the fire resistance of wood [10, 11] and bamboo [12]. In another study, silica were also reported to combine with coupling agent zinc borate for the preparation of hybrid sol as coating to improve the flame retardancy of the cotton fabric [13].

Silicon can also be used in the fabrication of hybrid materials for dye removal through adsorption. Meng et al. reported the fabrication of micro-porous silicon–carbon–nitrogen hybrid material for selective adsorption of organic dyes [14]. In another study, cyclodextrin-functionalized silicon nanoparticles hybrid material was used for the adsorption of methylene blue (MB) [15]. Silica was also used for the fabrication of hybrid materials for the removal of textile dyes by adsorption. In the study reported by Mohamed S. Hasanin, porous silica was extracted from a black liquor by-product and was used as a precursor material to prepare the hybrid material based on polyvinyl alcohol (PVA) as a binder polymer and functionalized by nicotinic acid [16]. This silica hybrid material was used as multi adsorbent for textile wastewater treatment and shows improved adsorption of dyes comparing to conventional materials.

Besides, silicon nanomaterials are also a good photocatalyst with excellent optical absorption, large surface to volume ratio, and enhanced charge transport phenomenon, in addition to being abundant and environmentally benign. For this reason, silicon nanomaterials were used for fabrication of hybrid photocatalyst for dye degradation. The most reported silicon nanomaterials for the fabrication of hybrid materials for dye degradation are silicon nanowires. Silicon nanowires decorated with titanium oxide nanoparticles [17], graphene oxide [18], PbS nanoparticles [5] and a combination of graphene oxide and copper nanoparticles [19] were used as the hybrid materials for photocatalytic dye degradation. Other than silicon nanowires, other silicon nanomaterials such as silicon nanocrystals were also used to decorate other nanomaterials such as ZnO nanowires [20] for photocatalytic dye degradation. Silica was also used to prepare hybrid materials for the photocatalytic removal of dyes. Silica were combined with different materials, such as metal [21, 22], titania [23] and chitosan [24] for the fabrication of photocatalyst for dye degradation. Interestingly, silica were combined with Fe_3O_4 for the fabrication of a reusable magnetic hybrid photocatalyst for dye degradation [25]. Similarly, silicon and silica based hybrid materials can also be used as photocatalyst for hydrogen production [26, 27] and carbon dioxide reduction [28]. Due to the excellent properties of silicon and silica such as highly biocompatible, easily surface functionalized, degrade into nontoxic silicic acid and can be processed into various forms, silicon and silica were widely studied for fabrication of hybrid materials for biomedical applications such as drug delivery, regenerative medicine, therapeutics, disease diagnosis, and medical implantation. For example, by combining silicon and silica building blocks with various materials with different characteristics, various silicon and silica-based hybrid vehicles [6, 29, 30] were developed.

1.3 Synthesis and Preparation of Silicon and Silica Hybrid Materials

Significant progress of synthesis and preparation of various silicon and silica hybrid materials has been achieved. Silicon and silica hybrid materials for various applications can be synthesized and fabricated in many ways. The methods for the preparation and synthesis of silicon and silica hybrid materials can be broadly categorized into liquid-phase processing [31] and vapor phase processing. The liquid-phase processing methods are mostly based on coprecipitation, microemulsion, thermal decomposition, solvothermal routes and chemical reduction, while vapor phase processing is based on the physical vapour deposition, molecular beam epitaxy, chemical vapour deposition, and atomic layer deposition. For these methods, two approaches are employed for the synthesis and preparation of silicon and silica hybrid materials, namely, bottom-up approach and top-down approach. A bottom-up approach is also known as the self-assembly process, which involves the assembly of smaller building blocks to form the desired structure, while the top-down approach involves the removal or etching of bulk matter for the formation of the smaller desired structure. In this section, recent progress in the synthesis and preparation of silicon and silica hybrid materials using liquid-phase processing and vapor-phase processing are addressed.

1.3.1 Liquid-Phase Processing

When handling, treating, or manipulating materials in a liquid state, a variety of procedures and methodologies are referred to as liquid-phase processing. These technologies are employed in a wide range of sectors. Liquids including water, oils, solvents, and other liquid substances can be included in this. Many industries, such as chemical engineering, materials engineering, pharmaceuticals, food and beverage, environmental engineering, and more, use liquid-phase processing [32]. To develop materials with certain qualities, silicon-based compounds or silica (silicon dioxide) are manipulated in a liquid media during the liquid-phase processing of silicon and silica hybrid composites. In the fields of electronics, optics, and energy storage, this procedure is frequently used in materials science. As a result, this section addresses several liquid-phase processing synthesis techniques, their benefits and drawbacks, and the characteristics of the resulting hybrid materials [33]. A wide variety of methods are used in liquid-phase synthetic procedures to create materials. The following are some distinct liquid-phase synthesis approaches: chemical reduction, solvothermal pathways, thermal breakdown, coprecipitation, and microemulsion [34]. Liquid-phase synthetic methodologies provide several methods for customizing the structures and properties of materials according to application needs. Based on the required properties of the finished product and the limitations of the synthesis circumstances, researchers select the best approach.

Coprecipitation Method

The common method for creating hybrid materials, such as silicon and silica hybrids, is coprecipitation. Coprecipitation is the process by which two or more compounds precipitate out of a solution at the same time, creating a composite material. A flexible technique for creating silicon and silica hybrid materials, coprecipitation gives you control over the end product's structure and content [35]. In order to obtain particular material qualities appropriate for a variety of applications, researchers frequently modify the process parameters [36].

The initial identification of appropriate precursors for silicon and silica is the first step in the general coprecipitation process for silicon and silica hybrid products. Sodium silicate or silicic acid solutions are frequently employed for silica, while silicon alkoxides such as tetraethyl orthosilicate (TEOS) are common precursors for silicon. A solvent is used to dissolve the precursors of silicon and silica [37]. The solvent of choice is determined by the precursors' solubility and reactivity. Water or ethanol are typically utilized. The pH level of the solution is changed to regulate the precipitation process. The solubility of the precursors and the subsequent synthesis of the hybrid material are influenced by changes in pH. Precursor solutions are combined in regulated environments, particularly with regard to the temperature. A chemical reaction, typically involving the hydrolysis and condensation of the silicon precursor, causes the coprecipitation to occur. Temperature is a key factor in the coprecipitation process, which occurs when the silicon and silica species combine and precipitate as particles or gels. Depending on the desired characteristics of the hybrid material, either room temperature or higher temperatures can be used for the reaction. After that, the coprecipitated substance might age or mature thanks to the aging process. Aging contributes to the final structure and characteristics of the hybrid material by assisting in the growth and consolidation of particles. Moreover, Typically, ethanol or water is used. To control the precipitation process, the solution pH is adjusted. pH manipulation influences the solubility of the precursors and the subsequent formation of the hybrid material. The precursor solutions are mixed under controlled conditions especially the temperature. The coprecipitation occurs because of a chemical reaction, usually involving the hydrolysis and condensation

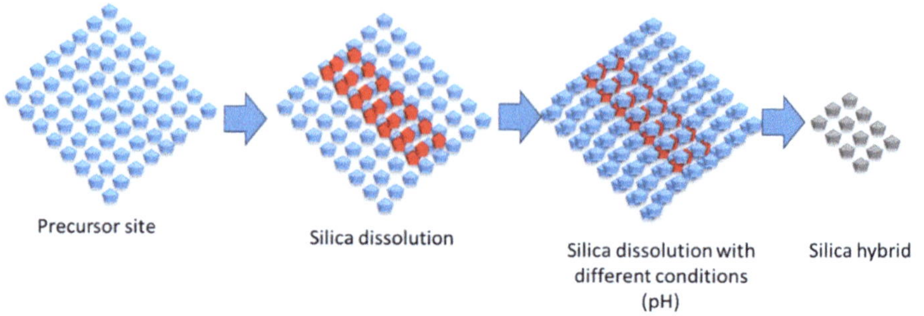

Fig. 4 Steps in coprecipitation method for the preparation of silicon and silica hybrid materials

of the silicon precursor. The silicon and silica species come together and precipitate as particles or gels and temperature plays a crucial role in the coprecipitation process. The reaction may be conducted at room temperature or elevated temperatures, depending on the desired properties of the hybrid material. This followed by aging process, which allows the coprecipitated material to age or mature. Aging helps in the consolidation and growth of particles, contributing to the final structure and properties of the hybrid material [38]. The coprecipitated material is separated from the liquid phase through filtration. Then following by washing to obtain pure solid by removing any unreacted precursors or by-products. The coprecipitated material is separated from the liquid phase through filtration. Then following by washing to obtain pure solid by removing any unreacted precursors or by-products. To obtain the final hybrid material, the coprecipitated material is dried to remove residual solvent and obtain the final product in a solid form as in Fig. 4.

Several researchers have presented silica/silicon hybrid materials were produced in recent time. Among a study on silica/silicon hybrid which was synthesised based on Fe_3O_4/a–SiO_2, the hybrid nanoparticles were produced by co-precipitation with natural materials. The findings demonstrated that the diffraction pattern changed when silica was added to Fe_3O_4, resulting in nanometric-sized nanocomposites with a core–shell structure. The Si–O–Fe bound was found around 555–568 cm^{-1} in the FTIR spectra [39]. Another study was presented, exploring surface modification of silica using silane-coupling agents, potentially enabling nanostructured silica to be grafted or conjugated with polymers or proteins to form silica/silicon hybrid for future applications in biotechnology and medicine [40]. In another study, a low-density, renewable, and high reinforcing filler of silica decorated cellulose nanocrystal for natural rubber compounds [41]. This cellulose nanocrystal-based hybrid filler was synthesized by decorating the surface of cellulose nanocrystal with silica coprecipitation method. This rod-like hybrid material has high aspect ratio and lower density than silica and was used as filler to replace silica in rubber compounds. Ali et al. also reported that this hybrid material has better compatibility with polymer matrix due to well-developed silica–silane technology to form strong percolation network.

The precursors utilized, the circumstances of the reaction and the post-treatment procedures affect the characteristics of the hybrid materials created by the coprecipitation process. Desired composition, ideal structure and surface area, desirable size and shape are some broad attributes that are frequently linked to hybrid materials created through coprecipitation [42]. One of the many benefits of the coprecipitation process is that the size, shape, and content of the coprecipitated particles may be controlled by varying the reaction conditions and the choice of precursors and solvent. The surface characteristics of the resultant hybrid material can be altered by adding surfactants or additives during coprecipitation. For the synthesis of various silicon and silica hybrids, including bulk solids, films, and nanoparticles, the coprecipitation process can be modified [43]. Coprecipitation procedures are a popular and flexible way to create silicon and silica hybrid materials, but they also have significant drawbacks. Some of the downsides of the coprecipitation methods include difficulties in controlling the size of the particles and the shape of the produced materials, which can lead to a lack of exact control. Variations in the final product can be resulted from difficult to control particle shape and size distribution. Furthermore, in coprecipitated materials, homogeneity can be difficult to achieve. Heterogeneity in the result can arise from variations in composition and structure within the majority of the material in variety of chemical reactions, hybrid materials with clearly specified structures and compositions are used as catalysts. These materials are appropriate for catalytic applications due to their high surface area and regulated reactivity. As a result, catalytic applications use a considerable portion of the hybrid materials created by the coprecipitation process [42].

Microemulsion Method

Microemulsion is another method of synthesising silica/silicon hybrid and using microemulsions as reaction media, the microemulsion process can be used to create hybrid materials comprising silicon and silica. Surfactants, and occasionally co-surfactants, sustain isotropic dispersions of water and oil that are thermodynamically stable. The synthesis of materials at the nanoscale is made possible by this technology in a regulated environment. Selecting appropriate silicon precursors and silica sources for the microemulsion system is the first step in the microemulsion process of creating silicon and silica hybrid materials. Silicon alkoxides like tetraethyl orthosilicate (TEOS) or organ silanes are frequently employed as precursors in the microemulsion technique [44]. After that, suitable surfactants and co-surfactants are chosen to create a stable microemulsion system. The amphiphilic molecules are essential for stabilizing the oil-in-water or water-in-oil microemulsion and for preparing it by the precise mixing of surfactants, co-surfactants, water, and oil (organic phase) in predetermined ratios [45]. As the silicon precursors are introduced into the microemulsion system, the system should be clear and stable, signifying the creation of a microemulsion. The incorporation can be done by adding the precursors directly or by using a preformed precursor-containing solution. The reaction between the silicon precursors within the microemulsion initiated and gelation or precipitation of silicon and

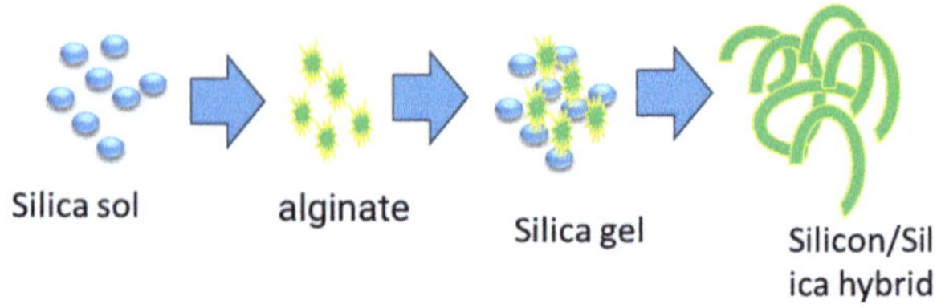

Fig. 5 Microemulsion method for producing alginate within a silica sol–gel to form silicon and silica hybrid

silica hybrid materials occurs within the microemulsion, leading to the formation of nanoscale structures hybrid material, as in Fig. 5 [46].

The next step involves allowing the system to age to ensure the complete formation and stabilization of the silicon and silica hybrid material. Aging can take place within the microemulsion or after its extraction. After this, the silicon and silica hybrid material can be extracted from the microemulsion system. Solvent extraction and centrifugation are two popular techniques for removing silicon and silica hybrid material. After that, any remaining solvent is eliminated by drying the resultant substance. A study was conducted on producing silica hybrid using microemulsion method, the method was used to synthesized hybrid silica-alginate nanoparticles. The nanoparticles synthesized in response to the molar ratio of water to surfactant and alginate concentration. Studies on the release of silica-alginate nanoparticles in vitro suggested that they have the potential to be used as a pH-responsive drug carrier for controlled drug release [47]. Another study explored the interaction between silica and proteins, which are crucial for the formation of biosilica and functional hybrid materials. The researchers found that lysozyme, a non-incorporated component of silica, enhanced the aggregation of silicon particles due to electrostatic and hydrophobic interactions. This led to composites with distinct characteristics from pure silica. The study also revealed that lysozyme affected particle growth and polymerization during co-precipitation. These findings enhanced understanding of organic macromolecule interactions with silica [48]. In another study, monodisperse organic–inorganic hybrid hollow mesoporous silica spheres were synthesized by the co-condensation of TEOS and 1,2-bis(triethoxysilyl)ethane (BTEE) using simple microemulsion method [49]. This process is relatively simple, involving acid-promoted pre-hydrolysis of TEOS and BTEE to obtain amphiphilic organosilanes to emulsify the oil droplets followed by using composite micelles of CTAB and pore expanders such as the anion surfactant sodium salicylate (NaSaL), non-ionic surfactants such as P123 and F127, hydrophobic oils such as cyclohexane (CyH) and 1,3,5-trimethyl benzene (TMB) to jointly construct a stable O/W microemulsion to obtain a hollow structure. The diameters, shell thicknesses, and pore sizes of this hybrid material could be controlled by adjusting the amounts of expanders during the microemulsion process.

Nanoscale-sized silicon and silica hybrid materials can be created using the microemulsion technique. Because the particles in the resulting materials are nanoscale in nature, they frequently have high surface area. The regulated reaction that occurs within the microemulsion system produces nanoparticles or nanocomposite structures. Applications including sensing, adsorption, and catalysis benefit from this feature. The synthesis is conducted in a controlled environment with a rather homogeneous particle size distribution thanks to the microemulsion process. For situations where constant qualities are needed, this uniformity may be crucial [50].

The microemulsion method offers advantages because it allows for precise control over particle size, shape, and content, the microemulsion handle is a flexible way to create silicon and silica hybrid materials with specific characteristics [51]. The microemulsion process has disadvantages when it comes to creating silicon and silica hybrid products, though. Previous research has shown that the size and stability of the resulting hybrid material can be affected by the composition of the microemulsion, including the water content and surfactant-to-oil ratio. Nevertheless, achieving the ideal microemulsion composition might be challenging. Furthermore, maintaining the appropriate characteristics of the silicon and silica hybrid material during microemulsion formation and the reaction depends on managing the pH and temperature. However, control is challenging while the process is underway. It is also known that the reaction kinetics and the ultimate structure of the hybrid material can be affected by the solvents chosen for the oil phase and for dissolving the precursors. However, it is difficult to identify the appropriate solvent for this purpose [6]. Hybrid materials comprising silicon and silica are appropriate for catalytic applications due to their high surface area and carefully regulated composition. They can be utilized directly as catalytic materials or as supports for catalysts in processes like organic synthesis and environmental cleanup [52].

Thermal Decomposition Method

The thermal decomposition method is a technique for producing silicon and silica hybrid materials through the controlled heating and decomposition of precursor compounds. In this context, silicon compounds, often organosilanes, are subjected to thermal treatment to yield silicon and silica hybrid materials [53]. Thermal decomposition method starts with choosing suitable silicon precursors. Organosilanes, such as tetramethylsilane (TMS), tetraethylsilane (TES), or other silicon-containing organic compounds, are commonly used. The selected silicon precursor is dissolved in a suitable solvent. The choice of solvent depends on the precursor's solubility and the desired properties of the final hybrid material. The precursor solution is then applied onto a substrate to form a thin film or coating. Techniques such as spin coating, dip coating, or spray coating may be used for uniform deposition [53]. The coated substrate is then dried to the solvent from the precursor solution. This step is crucial for the subsequent thermal decomposition process. The heating of coated substrate is controlled to initiate the thermal decomposition of the silicon precursor. This process results in the release of volatile organic components and the formation of silicon and silica hybrid structures. The oxidizing atmosphere is reduced, such as air or oxygen, during the thermal treatment to facilitate the oxidation of silicon species and the formation of silica.

Silicon and silica hybrid materials produced through thermal decomposition can be employed in the fabrication of electronic devices and semiconductors. Their controlled structure and composition contribute to their electrical properties [54]. Thermal decomposition method offers a versatile approach for producing silicon and silica hybrid materials with tailored properties, making them suitable for various applications, including electronics, sensors, and coatings. Optimization of process parameters is essential to achieve the desired characteristics of the final material [55]. However, obtaining desired crystal shape and size is difficult by controlling the heating rate and temperature during the thermal decomposition process to achieve the desired structure and properties because higher temperatures may lead to increased crystallinity [56]. The thermal decomposition method for producing silicon and silica hybrid materials has found application in various fields due to the unique properties and structures it can yield [57]. Silicon and silica hybrid materials synthesized using the thermal decomposition method are commonly employed for electronics and semiconductors, sensors and sensing devices, photovoltaics and solar cells [53].

1.3.2 Vapor-Phase Processing

Due to the unique properties, hybrid materials have generated considerable interest in materials research including silicon and silica. Silica, have remarkable properties to be offered. Optical stability properties, heat resistance, and chemical reactivity properties have made them very useful in electronics and photonics applications. On the other hand, silicon possesses excellent semiconducting properties. By combining silicon and silica with other materials through vapor phase techniques, advanced hybrid materials with synergistic properties which enhance their capability of multiple usages across a variety of applications can be produced. Unlike liquid phase processing technique, in vapor phase techniques (or also known as vapour phase deposition), target material is vaporised into an atomic or molecule gaseous and then reaching out to the substrate surfaces to be adhered implant or deposited. Dense and compact layers of thin film can be formed at high level of purity and crystallinity. Due the adhesion strength of the thin film formed vapor phase technique is commonly used in producing thin films, coatings, and nanostructures with precise characteristics. There are several types of vapour phase techniques, including physical vapour deposition, molecular beam epitaxy, chemical vapour deposition, and atomic layer deposition. These methods operate on distinct principles with precise control of deposition parameters which are temperature, pressure, precursor gases, and reaction kinetics. Their main speciality is the ability to tailor the properties of materials at the atomic and nanoscale levels [58, 59]. Parameters such as film thickness, composition, crystallinity, and structure can be adjusted accordingly to the applications. Vapour phase techniques enable the fabrication of intricate nanostructures, uniform thin films, and conformal coatings, hence facilitating the advancement of state-of-the-art electronics with enhanced reliability and efficiency. Understanding and employing vapour phase techniques in producing silicon and silica hybrid materials provide an opportunity to develop new innovative materials that meet the needs of advanced technology and industries.

The properties of silicon and silica hybrid materials synthesized using vapour phase processes are influenced by their composition, structure, and fabrication method. Different characteristic can be engineered and obtained suitable with their application needs. For the structural and mechanical properties of hybrid materials, homogenous, defect and pit free, hardness, elasticity strength and durability play a critical role in defining their structural integrity and long-lasting use. Selection of precursor is also crucial in determining the quality of final product [60–62]. Meantime, utilisation of vapour phase techniques allows for the durable coatings and structures that have protection against the corrosion, oxidation, and high-temperature conditions. Consequently, hybrid materials that are fabricated using these technologies are well-suited for applications that require protective coatings, catalysis, and operations at elevated temperatures. The electrical conductivity or resistivity properties of silicon and silica hybrids materials are crucial in electronics, sensors, photovoltaics, and semiconductor devices applications. Doping levels, bandgap engineering, and interface properties of the hybrid materials, are all affected by the vapour phase deposition parameters which in turn affecting their electrical characteristics [63]. Other important properties of silicon and silica hybrids that can be adjusted depending on the vapour phase process is optical features, such as transparency, refractive index, and light emission characteristics [64]. These qualities are crucial in optical coatings, photonic devices, and optoelectronics. By utilising vapour phase processes parameters such as deposition temperature [65], one may engineer the composition and thickness of the thin film layers, hence adjusting the optical properties that suited for their applications such as optical coatings, waveguides, and sensors.

Due to the semiconductor properties of silicon and stability of silica, their hybrid materials can bring benefit to the microelectronic industries. They have potential to be functioning as dielectric materials, insulators, protective coatings, optical coatings, and waveguides in integrated circuits, sensors, solar, optical fibres, semiconductor and photovoltaic devices applications [66–69]. In the fields of photonics and optoelectronics, optical characteristics of silicon hybrids such as refractive indices, transparency, and light-emission properties are adjustable and can be utilised in optical coatings, waveguides, photonic devices, and optical fibres [64, 70]. In biomedical applications, silicon and silica hybrids material offers biocompatibility and stability characteristic. Their surface ability to be engineered and customised to specific needs is a bonus. These materials are utilised in drug delivery systems, biosensing platforms, tissue engineering scaffolds, chemical sensors, biosensors, gas sensors and bioimaging applications [71].

With good mechanical and chemical characteristics, silicon and silica hybrid material are commonly chosen to be used as protective coatings to combat corrosion, wear, and environmental deterioration [72]. Vapour phase techniques allow for coatings these hybrid materials onto different types of substrates, resulting in improved surface characteristics and prolonged durability of components in industries such as aerospace, automotive, nuclear reactors and manufacturing. Silicon and silica hybrid materials are recently being utilised in high end latest technologies industries, including flexible electronics, quantum computing, energy storage, and nanotechnology which making them possibly seen as highly promising advanced materials for new technologies that can solve any technical obstacles with their customised qualities abilities.

Chemical Vapor Deposition (CVD)

Chemical Vapour Deposition (CVD) is well known as a powerful technology for producing high-quality solid thin films and coatings. In general, generation of thin film by CVD method occurs by a chemical reaction involving gaseous source components that are combined together and deposited onto the surface of a substrate. During the process, a gas mixture containing selected precursors and reactants is delivered into a reaction chamber as the initial substances. There are one or two substrates located inside the chamber where the thin film coating is going to be deposited. The precursor gases then undergo chemical processes that are activated by the heat from the substrate, resulting in the nucleation and growth mechanism that leads to the formation of new compounds thin film on the surfaces of substrate. Deposited thin film can be in the form of crystalline, amorphous or a combination of two. The whole process of CVD is generally carried out at elevated temperature ranging from 900 to 1400 °C.

Hybrid materials produced using CVD display a wide range of properties such as great adherence to substrates, high purity, controlled porosity, and tuneable mechanical, electrical, and optical properties. All these properties are varying depending on the specific process parameters used including the composition, selection of precursors, deposition temperature, pressure, and time. Silicon and silica hybrid materials produced via CVD are utilised mostly in microelectronics, optoelectronics, MEMS (Micro-Electro-Mechanical Systems), photovoltaics, and sensors fields and functioning as the protective coatings, antireflective coatings, functional layers, or fundamental components for sophisticated devices [65, 73]. To produce silicon or silica-based hybrid material using CVD technique, precursors such as silane (SiH_4), tetraethyl orthosilicate (TEOS), or dichlorosilane (SiH_2Cl_2), has commonly been used [74, 75].

The two main categories of CVD are low-pressure chemical vapour deposition (LPCVD) and plasma-enhanced chemical vapour deposition (PECVD). LPCVD is conducted under low atmospheric pressures. LPCVD allows for the deposition of high-quality films with fine control over thickness by manipulating factors such as temperature and precursor flow rates. Meanwhile, PECVD incorporates plasma as the heat supply into the deposition process. Plasma activation increases the responsiveness of precursors, allowing for deposition to be happened at lower temperatures [76]. The other categories of CVD techniques are includes horizontal chemical vapour deposition (CVD) and vertical CVD, atmospheric pressure CVD, Hot-wall CVD and cold-wall CVD, photo-assisted CVD and laser-assisted CVD, metal–organic CVD

(MOCVD), hot filament/wire CVD, Initiated CVD and oxidative CVD [60]. In one study, a hybrid silica films was synthesized on porous polyamideimide substrates from an organically bridged monomer, 1,2-bis(triethoxysilyl)ethane using PECVD [77]. This film structure and composition can be altered by changing the flow rate of monomer. PECVD was also used for the fabrication of hybrid silica membrane using hexamethyldisiloxane as a silicon precursor and annealed at 400 °C in N_2 [78]. Hybrid silica membrane was obtained by plasma-deposition of hexamethyldisiloxane mixed with a second precursor (1,5-cyclooctadiene) or a carbon-rich silicon precursor (triethylsilane). Nagasawa et al. also investigated the effects of membrane preparation parameters such as thermal annealing temperature and precursor on their gas permeation properties. This hybrid material is potentially useful for the separation of molecular mixtures.

Highlighting on the advantages, CVD is capable to deposit compounds onto a wide range of substrates, such as silicon wafers, glass, ceramics, and polymers. By using CVD, thickness of the thin film, composition, and structure of the films can precisely be control, thus enabling the user to tailor the material properties according to the application requirements. Despite the benefits, there are still drawbacks and limitations. The cost to purchase and maintaining CVD systems can be very expensive due to the specialized requirements and the complexity of the equipment. In addition, the requirement of CVD processes to work at elevated temperatures has limiting their suitability for substrates that are sensitive to the high heat. The toxicity of the precursors and by products gaseous has resulted in safety concerns issue and possibility to cause hazardous environments [73, 76, 79].

Atomic Layer Deposition (ALD)

Atomic Layer Deposition (ALD) is a precise and regulated method used for depositing thin films for a wide range of applications. One of the unique selling points of ALD is its capability to atomic-scale control over film composition, thickness, and homogeneity. Generally, ALD working principle is based on deposition of thin films through a series of self-limiting chemical reactions on the surface of the substrate. In an ALD setting, two or more precursor gases are added (one at a time) into a reaction chamber where a substrate is in it. The first precursor gas flows into the chamber and once the whole substrate surface is saturated and covered with the initial gas monolayer, the excess gas is pumped out from the chamber. Second precursor gas is then introduced into the chamber, and they chemisorb with the initial monolayer, resulting in the formation of the second additional monolayer. The excess gas is purged away, and this process sequence can be repeated for several times until the desired thickness is reached. This self-limiting, alternating cycle process allows for conformality and thickness uniformity of the film at the atomic-level even on powder samples, huge surfaces or complicated 3D structures [80]. Conformal is referring to the coatings ability to follow the surface's topography curve while keeping the uniform thickness over the surface regardless of roughness. The growth rate of ALD is directly influenced by the flux of the precursor at the substrate. Determining the growth rate can be a complex task, since it is influenced by various factors such as temperature, precursor concentration, and purge periods. Materials produced using ALD provide outstanding properties such as consistency, stronger adherence, precise thickness, adjustable characteristics, encompass regulated surface shape, accurate stoichiometry, and customised functions. ALD-produced silicon and silica hybrid materials have diverse applications in advanced semiconductor devices, nanotechnology, biotechnology, catalysis, energy storage (photovoltaics, fuel cells, batteries) and protective coatings [81, 82]. Their finely regulated features make them highly desirable for new technologies that require high-performance materials with specialised properties.

Instead of ALD, CVD and PVD have been used widely in the industries. However, ALD has gained recognition as the main developing technology due to its ability to produce ultra-thickness thin film at nanometre-scale, homogenously with pinholes free. Unlike CVD which requires high temperatures to deposit the precursor on the substrate, ALD operations are conducted at lower temperatures, even at room temperature [80]. Saito et al. have deposited multiple layers of aluminium silicate and silicon dioxide using ALD at room temperature to produce a new material with enhanced cation sorption. Tris(dimethylamino)silane and trimethyl aluminum (TMA) have been used as the precursors. Plasma excited humidifier with Argon gas is used for the oxidizing. The deposited aluminium silicate thin film was tested with several types of cations in aqueous solution of 10 mM and shows an enhanced in Na, Cs and K sorption [83]. Kim et al. demonstrated the successful deposition of a SiO_2 thin film employing aminodisilane precursor, 1,2-bis(diisopropylamino)disilane (BDIPADS), along with ozone using ALD at temperatures below 100 °C. Kim et al. have claimed that by using this type of precursor, high-quality SiO_2 films with relatively high growth rates, high film densities, and low impurity contents still can be synthesized even without the plasma enhanced or catalyst used. FTIR and XPS measurements have confirmed the production of stoichiometric SiO_2 films at temperatures below 100 °C, with no detectable impurities of nitrogen and carbon. However, at temperature less than 80 °C, Si–H moieties are slightly formed in the SiO_2 films and unfortunately, this have affect the dielectric constant, leakage currents and breakdown fields to slightly degrade [84]. In another study, using carbon nanotube (CNT) as the template, Karg et al. have successfully synthesized silicon dioxide nanotube using Low temperature ozine assisted ALD technique. Four types of silane derivatives have been used as the precursors which are (3-aminopropyl) triethoxysilane (APTES), bis(diethylamino)silane (BDEAS), diphenylaminosilane (DPAS), and triethylsilane. As the number of ALD cycles increase, carbon from the CNT structure were found to be removed, leads to the continuous growth of SiO_2 nanotube. Several morphological kinds of one-dimensional SiO_2 nanotube are obtained at the end of process including single wall, hollow wall, tube in tube and nanowires, which depended on the morphology of CNT template used [85]. Pavlenko et al. also reported the preparation of porous silicon/zinc oxide nanocomposite by a combination of metal-assisted chemical etching and atomic layer deposition methods [86]. They demonstrated that this hybrid material can be used for optical biosensor applications toward mycotoxins detection.

Similar to CVD, ALD technique allows for the precise adjustment on the properties of synthesized material such as their refractive index, electrical characteristics, porosity, and surface roughness by adjusting deposition settings and selecting appropriate precursors. Beside of having benefits that have mentioned, there are still has limitations on ALD technique. The main drawback of ALD technique is its sluggish deposition rate. ALD exhibits slow deposition rate in comparison to certain other deposition processes. The intrinsic features of the deposition cycle, which involve alternating injections of precursors and purges, resulted in longer duration of fabricating process and increasing the cost, hence not suitable for large-scale manufacturing, particularly for components with low added value. Even though this method could offer homogenous, and pinholes defect free which leads to the fabrication of new material with superior properties, however its applicability limitation in high-throughput manufacturing due to the slow deposition process could be very challenging. In the meantime, this technique also requires high vacuum condition, ultra clean surface, precise control, specialised equipment because ALD is very dependent on the reactor design, and specialised knowledge for operation and maintenance since ALD systems and its processes are very complex and expensive [80, 83].

Molecular Beam Epitaxy (MBE)

Molecular beam epitaxy (MBE) is an extremely precise method of evaporation to create crystalline thin films with atomic-scale precision in very tight controllable vacuum environment. MBE has been widely utilised in semiconductor industry to deposit crystalline thin film from several kind of solid source materials, mainly silicon, germanium, III–V semiconductors, II–VI semiconductors, and metal oxides. In this technique, solid source materials are placed in evaporation cell (or so called as effusion cell) inside the MBE chamber. To grow crystalline film, selected solid source material is first heated up to 1500–1600 °C. Then, a direct beam of atomic or molecular source is then direct targeting from the evaporation cell to the substrate which is also been heat up to 500–850 °C [87]. Meanwhile, the particles from the molecular beam adhere to the substrate forming a single layer of crystal atom. The whole process is held in an ultra-high vacuum system (UHV) at a pressure of 10^{-5} torr [88]. Since the UHV system did not allow any contaminants inside the chamber during the process, hence the crystallinity of the single atomic layer formed is free from any contaminants. This approach enables the sequential placement of materials through the kinetic energy of evaporating particles to generate layers with precise regulation of thickness with high purity content. These characteristics make them suitable to be used as high-performance materials in advanced semiconductor devices, quantum computing, optoelectronics, and nanotechnology [89].

MBE has the capability to deposit materials with unmatched precision and control over the surfaces in growing high purity, defect-free, sharp doping profile, high-quality crystalline films with accurate thickness control [90]. This method enables precise manipulation of interfaces between diverse materials, making it easier to create intricate heterostructures with customised features [87]. The extremely low-pressure conditions in MBE result in minimal impurity inclusion and lower defect concentrations, leading to the production of materials with superior properties. However, this has brought constraints to the MBE systems due to their requirement of ultra-high vacuum conditions and advanced equipment. Consequently, these factors render them less attainable for some research centres. The slower deposition rate of this technology also limited its usefulness for large-area deposition or industrial-high throughput scale production [58].

Physical Vapor Deposition (PVD)

Physical Vapour Deposition (PVD) refers to a vaporized coating technique where metal vapour is transferring to a substrate within a vacuum environment. PVD process is almost similar with CVD except that in PVD, the precursors are the material that about to be deposited which starting in solid form, then heated to vaporised it. Whereas in CVD, the precursors that introduced into the reaction chamber is in the gaseous form. There are two main types of PVD which are sputtering and evaporation [91]. Within a vacuum chamber, the source material either solid or liquid can be changed from condensed phase to a vapour phase through the evaporation or sputtering methods. Atoms or ions that have been vaporised move through the vacuum and then deposit on the surface of the substrate, creating a new thin film layer. PVD have four main steps as its working principles which are evaporation or sputtering the materials to be deposited to form vapour gas, transportation of the vapour gas to the substrate, reaction between the vapour gas atom with reactive gaseous during the transport stage, and deposition of thin film on the substrate as the final product.

The most common method used in PVD is evaporation. There are two kinds which are thermal and e-beam evaporation. Through thermal evaporation, heat is transferred by adjusting the electrical current through the conductor that connected to the precursor target. To start deposition, precursor material must be heated up to change them from solid to evaporating state. To control the evaporation rate, power supply can be adjusted. E-beam is the alternative method of thermal evaporation. E-beam uses electron beam to heat and evaporate the precursor target, hence the deposition rate is higher than thermal evaporation. In contrast, sputtering technique as known as magnetron sputtering is a non-thermal vaporization method. It involves the ejection or sputtering of atoms from the precursor's solid surface by accelerating collision of bombarding energetic ions that are generated from plasma [92]. The main advantage of sputtering is its capability to deposit in a large-area films for multipurpose of applications, including alloys at significantly cheaper than any other plasma method. Comparing with evaporation, sputtering technique is more versatile and the deposition of thin film can be conducted from upward, downward or even sideways, whereas evaporation could only deposit thin film by upward configuration [93].

Deposited thin film synthesized by PVD methods have potentially been used as protective coatings, functional layers, or components in a variety of devices. Liu et al. have successfully growth 145 nm diameter of SiC or SiO_2 nanowires on a carbon/carbon composite substrate using thermal evaporation method and SiO powders as the solid precursor while Ni $(NO_3)_2$ as the catalyst [94]. In another study, a hybrid dual layer surface coating consisting of a silicon (Si) underlayer and poly(ε-caprolactone) (PCL) overlayer was prepared by a combination of physical vapour deposition (PVD) and dip coating process [95]. This hybrid dual layer coating can reduce the corrosion rates of magnesium-based biomaterials and therefore a promising route to control the corrosion rate and mechanical properties of Mg-based biomaterials. Michael et al. have successfully synthesized silicon thin film for microelectromechanical systems (MEMS) application using ultra high vacuum electron beam at substrate temperature ranging from 200 to 625 °C, deposition rate of 50–400 nm/min and annealed at 600 °C with remarkably low residual stress, uniform surface morphology, and thick layers with reasonably large deposition rates. 30 μm thickness of Cantilever beams have been fabricated from the films. Rms roughness, residual stress and stress gradient were measured at c5 nm, of 20 MPa, of c1.1 MPa/μm, respectively with Young's Modulus of 169 GPa. They also claimed that the silicon thin film deposited by electron beam is better than LPCVD [66]. Instead of using evaporation, Hossain et al. have use RF magnetron sputtering to fabricate silica thin film for photovoltaic application. SiO_2 target was sputtered under different oxygen total flow ratios where this have affected the photoconversion efficiency to enhanced. Computational analysis using OptiLayer have confirmed that by coupling silica layer with a high-refractive-index oxide layer (n = 1.415), the total reflectance obtain was less than 0.4%. Hence, these silica thin films is efficient to use as antireflection and self-cleaning coatings [67]. Meantime, Sankar et al. in their work have synthesize silica thin film with the same technique but at different glancing angle deposition (45, 60, 75, 85°). The results have shown that at different angle deposition, there are transitions from featureless to columnar structure. At 85°, the reflectance is found decreases to 3.15% whereas transmittance increases to 93.43%. Meanwhile at 20°, the hydrophilicity of film surface have improved with oblique angle increase [68]. To functionalize the silica thin film, Hossain et al. have treated their e-beam evaporator deposited silica thin film with acetone in ultrasonic bath to introduced OH groups in order to improve film's to be superhydrophilic [69].

PVD technique is almost similar to the LPCVD. This method has the capability to deposit materials onto a wide range of substrates selection such as silicon wafers, glass, metals, and polymers, hence providing flexibility in various type of hybrid materials deposition. Like LPCVD, PVD allows for precise manipulation of film thickness, composition, and structure, resulting in customised material characteristics. The approach of this method guarantees on the consistency of deposition across extensive surfaces, which is crucial for production on a mass industrial production scale. But there are still certain restrictions for PVD technique. PVD technique is working by following a direct path, which restricts the evenness of the coating on intricate or complex 3D surfaces. Thin film coatings of the hybrid material synthesized by PVD technique may encounter adhesion or stress-related issues that need for extra surface treatments or optimisation such as the annealing process [96].

2 Conclusion

Hybrid material is no longer only of research interest. In fact, hybrid material has a profound impact on our daily. Hybrid materials, especially silicon and silica-based hybrid materials, have been extensively studied and reported. Significant breakthroughs were reported on the progress of research on silicon and silica-based hybrid materials. Current research aims to manipulate silicon and silica hybrid materials more precisely where the composition, structure, interface qualities, and defect manipulation are strictly adjusted at the atomic level. The implementation of modern liquid or vapour phase techniques, such as advanced reactors, precise precursor delivery systems and thorough process monitoring enables this higher degree of control to yield silicon and silica hybrid materials with synergistic properties. This chapter has reviewed the current progress on the fabrication, characterization, and potential applications of silicon and silica hybrid materials. From the published literature included in this chapter, various methods have been developed for the fabrication of silicon and silica hybrid materials, either through liquid phase or vapor phase processing. Although these methods can successfully prepare silicon and silica hybrid materials, their drawbacks are numerous. There are still processing challenges for the fabrication of silicon and silica hybrid materials with controllable parameters at high yield and low cost that can be integrated into current manufacturing processes. The upcoming research endeavours will be focused on tailoring silicon and silica hybrid materials to perfectly align with the requirements of these technologies, including sustainability, performance, and scalability. Further experimental and theoretical work is needed to achieve an adequate understanding of the fabrication of various silicon and silica hybrid materials according to the requirements of applications and to realize the tremendous potential associated with various silicon and silica hybrid materials to open new avenues for hybrid materials.

References

1. Aksit, M., Altstädt, V.: Hybrid Materials-Historical Perspective and Current Trends (2020)
2. Chiari, G., Giustetto, R., Druzik, J., Doehne, E., Ricchiardi, G.: Pre-columbian nanotechnology: reconciling the mysteries of the maya blue pigment. Appl. Phys. A **90**(1), 3–7 (2008)
3. Kickelbick, G.: introduction to hybrid materials. In: Hybrid Materials, pp. 1–48 (2006)
4. Cheng, Q., Li, Z.-H., Sun, Y.-X., Zhang, X.-Z.: Controlled synthesis of a core-shell nanohybrid for effective multimodal image-guided combined photothermal/photodynamic therapy of tumors. NPG Asia Mater. **11**(1), 63 (2019)

5. Algethami, F.K., Trabelsi, K., Hajjaji, A., Rabha, M.B., Khezami, L., Elamin, M.R., Bessais, B., El Khakani, M.A.: Photocatalytic activity of silicon nanowires decorated with pbs nanoparticles deposited by pulsed laser deposition for efficient wastewater treatment. Materials **15**(14), 4970 (2022)
6. Prabha, S., Durgalakshmi, D., Rajendran, S., Lichtfouse, E.: Plant-derived silica nanoparticles and composites for biosensors, bioimaging, drug delivery and supercapacitors: a review. Environ. Chem. Lett. **19**(2), 1667–1691 (2021)
7. Li, N., Kang, G., Liu, H., Li, M., Qiu, W., Wang, Q., Liu, L., Yu, J., Li, B., Li, F., Wu, D.: Hybrid nanoparticles based on novel Schiff Base for durable flame retardant and antibacterial properties. Compos. B Eng. **238**, 109905 (2022)
8. Mu, X., Jin, Z., Chu, F., Cai, W., Zhu, Y., Yu, B., Song, L., Hu, Y.: High-performance flame-retardant polycarbonate composites: Mechanisms investigation and fire-safety evaluation systems establishment. Compos. B Eng. **238**, 109873 (2022)
9. Zhang, L., Zhang, Q., Cao, W., Guo, Z., Fang, Z., Li, J., Chen, P.: Ultra-high flame-retardant efficiency of phosphorous-silicon hybrid microsphere in poly (butylene adipate-co-terephthalate). Compos. B Eng. **263**, 110861 (2023)
10. Lin, C.-F., Zhang, C., Karlsson, O., Martinka, J., Mantanis, G.I., Rantuch, P., Jones, D., Sandberg, D.: Phytic acid-silica system for imparting fire retardancy in wood composites. Forests **14**(5), 1021 (2023)
11. Chen, Z., Zhang, S., Ding, M., Wang, M., Xu, X.: Construction of a phytic acid-silica system in wood for highly efficient flame retardancy and smoke suppression. Materials **14**(15), 4164 (2021)
12. Peng, C., Zhong, J., Ma, X., Huang, A., Chen, G., Luo, W., Zeng, B., Yuan, C., Xu, Y., Dai, L.: Transparent, hard-wearing and bio-based organic/silica hybrid coating for bamboo with enhanced flame retardant and antifungal properties. Prog. Org. Coat. **167**, 106830 (2022)
13. Li, D., Wang, Z.H., Zhu, Y.S., You, F., Zhou, S.T., Li, G., Zhang, X.F. and Zhou, C.: Synergistically improved flame retardancy of the cotton fabric finished by silica-coupling agent-zinc borate hybrid sol. J. Ind. Textiles **51**(5_suppl), 8297S–8322S (2022)
14. Meng, L., Zhang, X., Tang, Y., Su, K., Kong, J.: Hierarchically porous silicon–carbon–nitrogen hybrid materials towards highly efficient and selective adsorption of organic dyes. Sci. Rep. **5**(1), 7910 (2015)
15. Li, J., Qiu, C., Fan, H., Bai, Y., Jin, Z., Wang, J.: A novel cyclodextrin-functionalized hybrid silicon wastewater nano-adsorbent material and its adsorption properties. Molecules **23**(6), 1485 (2018)
16. Hasanin, M.S.: Sustainable hybrid silica extracted from rice husk with polyvinyl alcohol and nicotinic acid as multi adsorbent for textile wastewater treatment. Environ. Sci. Pollut. Res. **27**(21), 26742–26749 (2020)
17. Ramachandran, K., Ziad, R., Columbus, S., Daoudi, K., Hammouche, J., El Khakani, M.A., Chidambaram, S., Gaidi, M.: TiO_2/Si nanowires hybrid system for efficient photocatalytic degradation of organic dye. J. Mater. Sci. Mater. Electron. **33**(12), 9194–9203 (2022)
18. Gaidi, M., Daoudi, K., Columbus, S., Hajjaji, A., Khakani, M.A.E., Bessais, B.: Enhanced photocatalytic activities of silicon nanowires/graphene oxide nanocomposite: Effect of etching parameters. J. Environ. Sci. **101**, 123–134 (2021)
19. Rodríguez, C.H., Bueno, J.D.J.P., Pérez, A.X.M., Flores, M.R., Oza, G.: Photoelectrocatalytic activity of silicon nanowires decorated with electroless copper nanoparticles and graphene oxide using a plasma jet for removal of methyl orange under visible light. RSC Adv. **13**(16), 10621–10635 (2023)
20. Zhang, Y., Mandal, R., Ratchford, D.C., Anthony, R., Yeom, J.: Si nanocrystals/ZnO nanowires hybrid structures as immobilized photocatalysts for photodegradation. Nanomaterials **10**(3), 491 (2020)
21. Chen, X., Zhang, L., Xu, B., Chen, T., Hu, L., Yao, W., Zhou, M., Xu, H.: Hairy silica nanosphere supported metal nanoparticles for reductive degradation of dye pollutants. Nanoscale Adv. **3**(10), 2879–2886 (2021)

22. Naz, A., Arun, S., Singh, V., Narvi, S.S., Alam, M.S., Dutta, P.K.: Efficient and reusable Cu (II)-metalated silica-based inorganic-organic hybrid catalyst for dye degradation. J. Indian Chem. Soc. **99**(1), 100296 (2022)
23. Bayal, N., Singh, R., Polshettiwar, V.: Nanostructured silica-titania hybrid using dendritic fibrous nanosilica as a photocatalyst. Chemsuschem **10**(10), 2182–2191 (2017)
24. Blachnio, M., Budnyak, T.M., Derylo-Marczewska, A., Marczewski, A.W., Tertykh, V.A.: Chitosan-silica hybrid composites for removal of sulfonated azo dyes from aqueous solutions. Langmuir **34**(6), 2258–2273 (2018)
25. Kumar, A.P., Bilehal, D., Desalegn, T., Kumar, S., Ahmed, F., Murthy, H.C.A., Kumar, D., Gupta, G., Chellappan, D.K., Singh, S.K., Dua, K., Lee, Y.-I.: Studies on synthesis and characterization of Fe₃O₄@SiO₂@Ru hybrid magnetic composites for reusable photocatalytic application. Adsorpt. Sci. Technol. **2022**, 3970287 (2022)
26. Choi, Y., Choi, S., Lee, I., Nguyen, T.V.T., Bae, S., Kim, Y.H., Ryu, J., Park, S., Ryu, J.: Solar biomass reforming and hydrogen production with earth-abundant si-based photocatalysts. Adv. Mater. **35**(30), 2301576 (2023)
27. Ryu, S.Y., Balcerski, W., Lee, T.K., Hoffmann, M.R.: Photocatalytic production of hydrogen from water with visible light using hybrid catalysts of CdS attached to microporous and mesoporous silicas. J. Phys. Chem. C **111**(49), 18195–18203 (2007)
28. Bisht, M., Singh, B., Pandey, S.: Deep eutectic solvent-modified mesoporous silica for CO_2 capture: a new generation of hybrid sorbents. J. Mol. Liquids **384**, 122203 (2023)
29. McInnes, S.J., Voelcker, N.H.: Silicon–polymer hybrid materials for drug delivery. Future Med. Chem. **1**(6), 1051–1074 (2009)
30. Bagheri, E., Ansari, L., Abnous, K., Taghdisi, S.M., Charbgoo, F., Ramezani, M., Alibolandi, M.: Silica based hybrid materials for drug delivery and bioimaging. J. Control. Release **277**, 57–76 (2018)
31. Nguyen, T.N., Huynh, T.N., Hoang, D., Nguyen, D.H., Nguyen, Q.H., Tran, T.H.: Functional nanostructured oligochitosan–silica/carboxymethyl cellulose hybrid materials: synthesis and investigation of their antifungal abilities. Polymers **11**(4), 628 (2019)
32. Soares, S.F., Daniel-da-Silva, A.L., Trindade, T.: Transition metal-modification of carrageenan-silica hybrids by a sol–gel method. J. Sol-Gel Sci. Technol. **107**(1), 201–214 (2023)
33. Xue, B., Wang, X., Yu, L., Di, B., Chen, Z., Zhu, Y., Liu, X.: Self-assembled lignin-silica hybrid material derived from rice husks as the sustainable reinforcing fillers for natural rubber. Int. J. Biol. Macromol. **145**, 410–416 (2020)
34. Handayani, M., Nafi'ah, N., Nugroho, A., Rasyida, A., Prasetyo, A.B., Febriana, E., Sulistiyono, E., Firdiyono, F.: The development of graphene/silica hybrid composites: a review for their applications and challenges. Crystals **11**(11), 1337 (2021)
35. Kamanina, O.A., Saverina, E.A., Rybochkin, P.V., Arlyapov, V.A., Vereshchagin, A.N., Ananikov, V.P.: Preparation of hybrid sol-gel materials based on living cells of microorganisms and their application in nanotechnology. Nanomaterials **12**(7), 1086 (2022)
36. Hafidh, A., Touati, F., Sediri, F.: Synthesis, charaterization and optical properties of nanostructured silica hybrid materials obtained by soft chemistry from perhydropolysilazane/1,2,4-triazole precursors. J. Mol. Struct. **1218**, 128496 (2020)
37. Mann, J., Garnweitner, G., Schilde, C.: Preparation of self-assembled nanoparticle and ndash;polymer hybrids from modified silica nanoparticles and polystyrene-block-polyacrylic acid vesicles via the co-precipitation method. Polymers **15**(2), 444 (2023)
38. Przywecka, K., Kowalczyk, K., Grzmil, B.: Sequential co-precipitation as a convenient preparation method of anticorrosive hybrid calcium phosphate/calcium silicate powder pigments. Powder Technol. **373**, 660–670 (2020)
39. Munasir, A.S., Dewanto, A., Yulianingsih, I.K.F., Saadah, Z.A.I., Supardi, Mufid, A., Taufiq, A.: Composites of Fe_3O_4/SiO_2 from natural material synthesized by co-precipitation method. IOP Conf. Ser. Mater. Sci. Eng. **202**(1), 012057 (2017)
40. Rahman, I.A., Padavettan, V.: Synthesis of silica nanoparticles by sol-gel: size-dependent properties, surface modification, and applications in silica-polymer nanocomposites—a review. J. Nanomater. **2012**, 132424 (2012)

41. Ali, S.D., Imiete, I.E., Orlandi, M.E., Castellani, L., Hanel, T., Zoia, L.: Novel CNC/silica hybrid as potential reinforcing filler for natural rubber compounds. J. Appl. Polym. Sci. **137**(5), 48332 (2020)
42. Moretto, E., Fernandes, J.P.C., Staropoli, M., Rogé, V., Steiner, P., Duez, B., Lenoble, D., Thomann, J.-S.: Dual-silane premodified silica nanoparticles–synthesis and interplay between chemical, mechanical, and curing properties of silica-rubber nanocomposites: application to tire tread compounds. ACS Omega **7**(21), 17692–17702 (2022)
43. Andolsi, A., Chaari, I., Hamzaoui, A.H.: Surface modification of magnetite using silica coating: spectroscopic, structural, morphological characterization and interaction with crystal violet dye. SILICON **15**(14), 6257–6268 (2023)
44. El Hajam, M., Kandri, N.I., Zerouale, A., Wang, X., Gustafsson, J., Wang, L., Mäkilä, E., Hupa, L., Xu, C.: Lignocellulosic nanocrystals from sawmill waste as biotemplates for free-surfactant synthesis of photocatalytically active porous silica. ACS Appl. Mater. Interfaces **14**(17), 19547–19560 (2022)
45. Peng, R., Asadullah Khan, M., Li Liu, C.: Synthesis of silica nanoparticles and their application of dielectric relaxation spectroscopy—a review. Results Chem. **6**, 101218 (2023)
46. Hao, T., Wang, Y., Liu, Z., Li, J., Shan, L., Wang, W., Liu, J., Tang, J.: Emerging applications of silica nanoparticles as multifunctional modifiers for high performance polyester composites. Nanomaterials **11**(11), 2810 (2021)
47. Fan, X., Domszy, R.C., Hu, N., Yang, A.J., Yang, J., David, A.E.: Synthesis of silica–alginate nanoparticles and their potential application as pH-responsive drug carriers. J. Sol-Gel Sci. Technol. **91**(1), 11–20 (2019)
48. van den Heuvel, D.B., Stawski, T.M., Tobler, D.J., Wirth, R., Peacock, C.L., Benning, L.G.: Formation of silica-lysozyme composites through co-precipitation and adsorption. Front. Mater. **5** (2018)
49. Sun, Y., Zhang, C., Mao, Y., Pan, D., Qi, D., Di, N.: General microemulsion synthesis of organic–inorganic hybrid hollow mesoporous silica spheres with enlarged pore size. New J. Chem. **43**(28), 11164–11170 (2019)
50. Riccò, R., Nizzero, S., Penna, E., Meneghello, A., Cretaio, E., Enrichi, F.: Ultra-small dye-doped silica nanoparticles via modified sol-gel technique. J. Nanopart. Res. **20**(5), 117 (2018)
51. Vorotnikov, Y.A., Vorotnikova, N.A., Shestopalov, M.A.: Silica-based materials containing inorganic red/NIR emitters and their application in biomedicine. Materials **16**(17), 5869 (2023)
52. Gonçalves, M.C.: Sol-Gel silica nanoparticles in medicine: a natural choice: design, synthesis and products. Molecules **23**(8), 2021 (2018)
53. Catauro, M., Šiler, P., Másilko, J., Risoluti, R., Vecchio S.: Ciprioti: synthesis, structural, morphological and thermal characterization of five different silica-polyethylene glycol-chlorogenic acid hybrid materials. Polymers **13**(10), 1586 (2021)
54. Vasquez-Zacarias, L., Ponce-Peña, P., Pérez-López, T., Franco-Urquiza, E.A., Ramirez-Galicia, G., Poisot, M.: Hybrid cellulose-silica materials from renewable secondary raw resources: an eco-friendly method. Global Chall. **2**(7), 1700119 (2018)
55. Ebisike, K., Okoronkwo, A.E., Alaneme, K.K.: Synthesis and characterization of Chitosan–silica hybrid aerogel using sol-gel method. J. King Saud Univ. Sci. **32**(1), 550–554 (2020)
56. Łyszczek, R., Vlasyuk, D., Podkościelna, B., Głuchowska, H., Piramidowicz, R., Jusza, A.: A top-down approach and thermal characterization of luminescent hybrid BPA.DA-MMA@Ln2L3 materials based on Lanthanide(III) 1H-Pyrazole-3,5-Dicarboxylates. Materials **15**(24), 8826 (2022)
57. Meftah, N., Hani, A., Merdas, A.: Extraction and physicochemical characterization of highly-pure amorphous silica nanoparticles from locally available Dunes Sand. Chem. Africa **6**(6), 3039–3048 (2023)
58. Oke, J.A., Jen, T.-C.: Atomic layer deposition and other thin film deposition techniques: from principles to film properties. J. Market. Res. **21**, 2481–2514 (2022)
59. Chaudhari, M.N., Ahirrao, R.B. and Bagul, S.D.: Thin film deposition methods: a critical review. Int. J. Res. Appl. Sci. Eng. **9**, 5215–5232 (2021)

60. Sun, L., Yuan, G., Gao, L., Yang, J., Chhowalla, M., Gharahcheshmeh, M.H., Gleason, K.K., Choi, Y.S., Hong, B.H., Liu, Z.: Chemical vapour deposition. Nat. Rev. Methods Primers **1**(1), 5 (2021)
61. Sathiyamoorthy, D., Selvakumar, J.: Prospects of chemical vapor grown silicon carbide thin films using halogen-free single sources in nuclear reactor applications: a review. J. Mater. Res. **28**(1), 136–149 (2013)
62. Kaloyeros, A.E., Goff, J., Arkles, B.: Defect- and H-free stoichiometric silicon carbide by thermal CVD from the single source precursor trisilacyclohexane. Electron. Mater. **3**(1), 27–40 (2022)
63. Oyama, S.T., Aono, H., Takagaki, A., Sugawara, T., Kikuchi, R.: Synthesis of silica membranes by chemical vapor deposition using a dimethyldimethoxysilane precursor. Membranes **10**(3), 50 (2020)
64. Muñoz-Rojas, D., Nguyen, V.H., Masse de la Huerta, C., Aghazadehchors, S., Jiménez, C., Bellet, D.: Spatial atomic layer deposition (sald), an emerging tool for energy materials: application to new-generation photovoltaic devices and transparent conductive materials. Comptes Rendus Physique **18**(7), 391–400 (2017)
65. Cai, Z., Liu, B., Zou, X., Cheng, H.-M.: Chemical vapor deposition growth and applications of two-dimensional materials and their heterostructures. Chem. Rev. **118**(13), 6091–6133 (2018)
66. Michael, A., Kwok, C.Y., Wang, P., Kazuo, O., Varlamov, S.: Investigation of E-beam evaporated silicon film properties for MEMS applications. J. Microelectromech. Syst. **24**(6), 1951–1959 (2015)
67. Hossain, M.I., Aïssa, B., Samara, A., Mansour, S.A., Broussillou, C.A., Benito, V.B.: Hydrophilic antireflection and antidust silica coatings. ACS Omega **6**(8), 5276–5286 (2021)
68. Sarkar, S., Pradhan, S.K.: Tailoring of optical and wetting properties of sputter deposited silica thin films by glancing angle deposition. Appl. Surf. Sci. **290**, 509–513 (2014)
69. Hossain, M.I., Al Kubaisi, G., Aïssa, B., Mansour, S.: Probing the hydrophilic behaviour of e-beam evaporated silica thin films for PV-soiling application. Mater. Sci. Technol. **38**(11), 753–759 (2022)
70. Hamakawa, Y.: Physics and applications of amorphous silicon carbide. In: Amorphous and Crystalline Silicon Carbide II, pp. 164–170. Berlin, Heidelberg (1989)
71. Cigané, U., Palevičius, A., Janušas, G.: Review of nanomembranes: materials, fabrications and applications in tissue engineering (bone and skin) and drug delivery systems. J. Mater. Sci. **56**(24), 13479–13498 (2021)
72. Katiyar, A.K., Hoang, A.T., Xu, D., Hong, J., Kim, B.J., Ji, S., Ahn, J.-H.: 2D Materials in flexible electronics: recent advances and future prospectives. Chem. Rev. (2023)
73. Fraga, M., Pessoa, R.: Progresses in synthesis and application of SiC films: From CVD to ALD and from MEMS to NEMS. Micromachines **11**(9), 799 (2020)
74. Issa, E.: Novel Reactor Design and Method for Atmospheric Pressure Chemical Vapor Deposition of Micro and Nano SiO_2-x Films in Photovoltaic Applications. <Ilmenau>|Universitätsverlag Ilmenau, Ilmenau (2022)
75. Gu, Y., Vaezian, B., Khatib, S.J., Oyama, S.T., Wang, Z., Achenie, L.: Hybrid H_2-selective silica membranes prepared by chemical vapor deposition. Sep. Sci. Technol. **47**(12), 1698–1708 (2012)
76. Yasaman, H., Prathyushakrishna, M., Timothy, J.B., Rajesh, R.N., Milana, C.V.: Plasma-enhanced chemical vapor deposition: where we are and the outlook for the future. In: Chemical Vapor Deposition, pp. Chap. 10. IntechOpen, Rijeka (2016)
77. Ngamou, P.H.T., Overbeek, J.P., Kreiter, R., van Veen, H.M., Vente, J.F., Wienk, I.M., Cuperus, P.F., Creatore, M.: Plasma-deposited hybrid silica membranes with a controlled retention of organic bridges. J. Mater. Chem. A **1**(18), 5567–5576 (2013)
78. Nagasawa, H., Yamamoto, Y., Kanezashi, M., Tsuru, T.: Atmospheric-pressure plasma-enhanced chemical vapor deposition of hybrid silica membranes. J. Chem. Eng. Jpn. **51**(9), 732–739 (2018)
79. Ajayakumar, A., Maruthoor, S., Fuchs, T., Rohlfing, F., Jakovlev, O., Wilde, J., Reinecke, H.: Improvement of process parameters for polycrystalline silicon carbide low pressure chemical

vapor deposition on 150mm silicon substrate using monomethylsilane as precursor. Thin Solid Films **536**, 94–98 (2013)
80. Oviroh, P.O., Akbarzadeh, R., Pan, D., Coetzee, R.A.M., Jen, T.-C.: New development of atomic layer deposition: processes, methods and applications. Sci. Technol. Adv. Mater. **20**(1), 465–496 (2019)
81. Zhang, J., Li, Y., Cao, K., Chen, R.: Advances in atomic layer deposition. Nanomanuf. Metrol. **5**(3), 191–208 (2022)
82. Pehkonen, S.O., Yuan, S.: Tailored Thin Coatings for Corrosion Inhibition Using a Molecular Approach. Academic (2018)
83. Saito, T., Yoshida, K., Saito, K., Miura, M., Kanomata, K., Ahmmad, B., Kubota, S., Hirose, F.: Multiple layers of aluminum silicate and silicon dioxide deposited by room-temperature atomic layer deposition for enhanced cation sorption. J. Vac. Sci. Technol. A **40**(4) (2022)
84. Kim, D.H., Lee, H.J., Jeong, H., Shong, B., Kim, W.-H., Park, T.J.: Thermal atomic layer deposition of device-quality SiO_2 thin films under 100 °C using an aminodisilane precursor. Chem. Mater. **31**(15), 5502–5508 (2019)
85. Karg, M., Lokare, K.S., Limberg, C., Clavel, G., Pinna, N.: Atomic layer deposition of silica on carbon nanotubes. Chem. Mater. **29**(11), 4920–4931 (2017)
86. Pavlenko, M., Myndrul, V., Gottardi, G., Coy, E., Jancelewicz, M., Iatsunskyi, I.: Porous silicon-zinc oxide nanocomposites prepared by atomic layer deposition for biophotonic applications. Materials **13**(8), 1987 (2020)
87. Edwin, A.: Thin films/properties and applications. In: Thin Films, pp. Chap. 1. IntechOpen, Rijeka (2021)
88. Ramsden, J.J.: Chapter 6—Noncarbon nanomaterials and their production. In: Nanotechnology, 2nd edn., pp. 123–154. William Andrew Publishing, Oxford (2016)
89. Nunn, W., Truttmann, T.K., Jalan, B.: A review of molecular-beam epitaxy of wide bandgap complex oxide semiconductors. J. Mater. Res. **36**(23), 4846–4864 (2021)
90. Stenin, S.I., Kanter, B.Z., Nikiforov, A.I.: Molecular-beam epitaxy of silicon. In: Growth of Crystals, pp. 69–76. Springer US, Boston, MA (1992)
91. Li, Z., Aik Khor, K.: Preparation and properties of coatings and thin films on metal implants. In: Encyclopedia of Biomedical Engineering, pp. 203–212. Elsevier, Oxford (2019)
92. Mattox, D.M.: Handbook of Physical Vapor Deposition (PVD) processing. In: Andrew, W. (ed.) (2010)
93. Sarangan, A.: 5—Nanofabrication. In: Fundamentals and Applications of Nanophotonics, pp. 149–184. Woodhead Publishing (2016)
94. Liu, B., Sun, J., Zhou, L., Zhang, P., Yan, C., Fu, Q.: Microstructure evolution and growth mechanism of core-shell silicon-based nanowires by thermal evaporation of SiO. J. Adv. Ceram. **11**(9), 1417–1430 (2022)
95. Bakhsheshi-Rad, H.R., Hamzah, E., Daroonparvar, M., Abdul Kadir, M.R., Kasiri-Asgarani, M., Staiger, M.P.: Enhancement of corrosion resistance and mechanical properties of Mg–1.2Ca–2Bi via a hybrid silicon-biopolymer coating system. Surf. Coat. Technol. **301**, 133–139 (2016)
96. Panjan, P., Drnovšek, A., Gselman, P., Čekada, M., Panjan, M.: Review of growth defects in thin films prepared by PVD techniques. Coatings **10**(5), 447 (2020)

Modified $LiNbO_3$ Films: Insights and Potential Applications

Evan T. Salim, Rawan Bashar Fadhil, Waffa K. Khalef, Makram A. Fakhri, and Subash C. B. Gopinath

1 Introduction

Lithium niobate ($LiNbO_3$) is a compound that does not naturally occur. The constituents of $LiNbO_3$ are niobium (Nb), lithium (Li), and oxygen (O). $LiNbO_3$, as a photonic crystal material, is widely studied among other materials. This material has an indirect bandgap and is extensively invested in various fields of photonics, particularly in fabricating integrated optical devices. Lithium niobate forms in two different phases with a symmetric trigonal structure, containing ten atoms in a unit cell. The ground state is classified as ferroelectric, with a space group of R3C. It has more benefits than other ferroelectric materials due to its thermal, mechanical, and chemical stability. $LiNbO_3$ also has a low dielectric constant. It has been found that reducing the grain size to the nanoscale can enhance its properties compared to its bulk form. This enhancement is because $LiNbO_3$'s properties depend on its chemical composition, particle size, shape, and crystal structure [1]. Using a laser source on single wafers of lithium niobate, common optical applications involve optical filters as well as optical modulators. $LiNbO_3$ specifically shows potential applications in optical channel waveguides and integrated optics.

E. T. Salim (✉) · R. B. Fadhil · W. K. Khalef
Applied Science Department, University of Technology-Iraq, Baghdad 10066, Iraq
e-mail: Evan.T.Salim@uotechnology.edu.iq

M. A. Fakhri (✉)
Laser and Optoelectronics Eng. Department, University of Technology-Iraq, Baghdad 10066, Iraq
e-mail: Makram.A.Fakhri@uotechnology.edu.iq

S. C. B. Gopinath (✉)
Faculty of Chemical Engineering & Technology and Institute of Nano Electronic Engineering, Universiti Malaysia Perlis (UniMAP), 02600 Arau, Perlis, Malaysia
e-mail: subash@unimap.edu.my

Center for Global Health Research, Saveetha Medical College & Hospital, Saveetha Institute of Medical and Technical Sciences (SIMATS), Thandalam, Chennai 602 105, Tamil Nadu, India

S. C. B. Gopinath and M. M. Ramli (eds.), *Hybrid-Nanomaterials*, Engineering Materials, https://doi.org/10.1007/978-981-97-9022-7_10

$LiNbO_3$ has been considered by researchers as an attractive material because of its high quality and low optical losses for waveguide applications. For instance, $LiNbO_3$ on insulator (LNOI) plays an essential role in photonic integrated circuits, controlling the behavior of solid-state devices [2], as well as adjustable filter with a broadband application [3]. There are various techniques to synthesize $LiNbO_3$ nanocrystals one of them is chemical bath deposition (CBD) (Fig. 1) [4]. The CBD technique utilizes a heterogeneous nucleus to produce films by depositing or adsorbing aqueous ions onto a solid substrate. This method is used to produce homogeneous metal chalcogenide films (commonly oxides, selenides, and sulfides) as well as a few ionic compounds. It is advantageous for industrial applications because it is simple, inexpensive, and reliable compared to other thin film deposition techniques. The CBD method requires only an aqueous solution, minimal infrastructure, and low temperatures. While it is less effective for semiconductors than methods that produce large crystals, it is beneficial for creating nanomaterials. Films synthesized via CBD have higher photovoltaic properties (band electron gap) than those formed using other techniques. $LiNbO_3$ crystals are among the most promising materials for optoelectronic applications due to their excellent optical properties [5]. The optical properties of $LiNbO_3$ crystals have not been widely invested as a result of the few disadvantages that related to the crystal growth that were not easily overcome. These disadvantages include (i) thermal shock resistance with a low value, (ii) many centers for light scattering, and (iii) localized fluctuated refractive indices in the grown crystals [6].

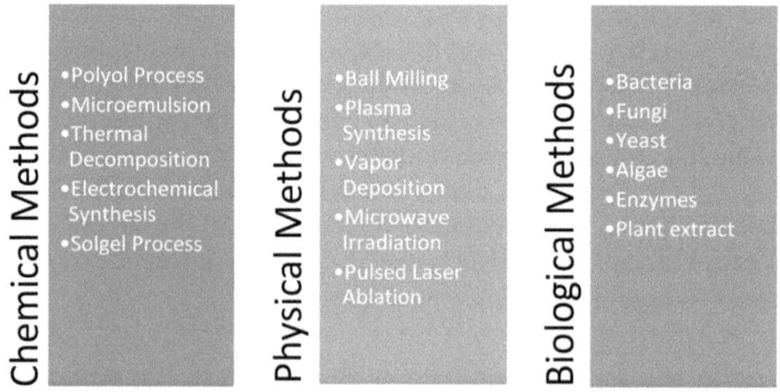

Fig. 1 Different methods involved in the depositions. These methods are directly or indirectly participated

2 Developmental Stages with $LiNbO_3$ for Different Applications

The history and impact of previous research on $LiNbO_3$ nanostructures are briefly reviewed in the following sections: In 2007, Yu et al. [7] formed $LiNbO_3$ crystals using a hydrothermal technique, employing $LiOH{\cdot}H_2O$ or $LiNbO_3$ and Nb_2O_5 as starting materials in a solution of water or ammonia. They investigated crystallization temperature, crystallization time, and the concentration of ammonia. The results confirmed that the crystallization and purity of the prepared samples increased with increasing time. In 2015, Hernández-Molina et al. [8] synthesized Lithium-Niobate structures using a wet chemical technique. This method achieved the correct stoichiometric phase, significantly increasing the surface area, controlling the phase purity, and reducing particle size. The samples were synthesized at low annealing temperatures of 500, 650, and 800 ºC. These features can be modified by controlling experimental parameters such as synthesis time and calcination. In 2017, Fakhri et al. [9] prepared Lithium-Niobate using the sol–gel technique, with quartz substrates used for the deposition process. Different molarity concentrations were used to create the solution, which was then annealed at 500 ºC. The findings showed that as the molar concentration increased, the film crystallized more uniformly, and its energy band gap increased to 4 eV. Transmission spectra between the wavelengths of 400 and 1100 nm were measured. A hexagonal form with a good atomic distribution was produced, and the crystallizations were uniform.

In 2018, Al-Douri et al. [10] prepared $LiNbO_3$ via the sol–gel technique with various stirring times of 8, 24, and 48 h. The prepared films were annealed at 500 °C. The measurements revealed that increasing the stirring time resulted in more regularly distributed structures. The results indicated that $LiNbO_3$ could be used as optical waveguides. Furthermore, the experimental measurements showed that the estimated refractive index and optical dielectric constant were in agreement. In 2018, Murillo et al. [11] formed single crystal $LiNbO_3$ nanoparticles through the aerosol-assisted chemical vapor deposition technique (AACVD). The formed nanocrystals exhibited various morphologies, such as tetrahedral, quasi-cubic, hexagonal, and polyhedral shapes, ranging in size from a few tens to hundreds of nanometers. However, rod-like nanostructures measuring 3–5 μm were also formed. In 2018, Klimin et al. [12] investigated the impact of annealing $LiNbO_3$ films in an oxygen atmosphere on their morphological and optical properties. Annealing in an oxygen atmosphere decreased the surface roughness and droplet density on the $LiNbO_3$ film surface. It was reported that annealing in an oxygen atmosphere for 1 h at 600 °C increased the oxygen content in the film. In 2019, Sosunov et al. [13] investigated the impact of pre-annealing on lithium niobate structures. The selected annealing temperatures ranged from 400 to 800 °C. Pre-annealing enhanced the lithium niobate structure of the subsurface layer. Additionally, the refractive index slightly decreased due to increased homogeneity and lower proton concentration in the crystal. The crystal structure showed more uniformity at 500 °C. The obtained results were desirable for various applications. In 2020, Liang et al. [14] synthesized $LiNbO_3$ single crystal thin films by employing

Ar + etching. $LiNbO_3$ thin film memristors were fabricated by oxygen annealing. The results demonstrated that this approach effectively decreased the breakdown possibility and increased the retention as well as the device endurance performance. In 2021, Ito et al. [15] formed nanocrystalline lithium niobate through an aqueous solution process starting from the reaction between H_2O_2 aq and NbC to form a solution of peroxo-polyniobic acid. LiOH aq was mixed with the peroxo-polyniobic acid solution and dried at 115 °C to produce amorphous $LiNbO_3$. Heating the amorphous $LiNbO_3$ to 400–500 °C for 1–24 h in air transformed it into crystalline $LiNbO_3$. In 2021, Gurpreet Singh et al. [16] synthesized a polyvinylidene difluoride (PVDF) composite film incorporating lithium niobate ceramics decorated with Ag nanoparticles using the solvent casting technique. Compared to the resulting PVDF film embedding lithium niobate and the bare PVDF film, the piezocatalytic performance of the PVDF film decorated with Ag-loaded $LiNbO_3$ was superior. In 2022, Irzaman et al. [17] produced $LiNbO_3$ thin films using the Chemical Solution Deposition (CSD) and spin coating processes on a p-type Si substrate (100). The $LiNbO_3$ thin film was reported to have energy gaps of 1.36 eV, 2.79 eV, 2.38 eV, and 1.75 eV for solubilities of 1, 2, 3, and 4 M, respectively. XRD measurements on $LiNbO_3$ thin films confirmed the rhombohedral structure with a = b lattice parameter, while the c parameter varied. In 2022, Alwazny et al. [18] used laser ablation in liquid with various laser fluences to synthesize a novel gold core $LiNbO_3$ shell (Au@$LiNbO_3$). The XRD results demonstrated that the produced Au@$LiNbO_3$ core–shell nanoparticle was polycrystalline.

3 Nanophotonic $LiNbO_3$ Synthesis and Characterization

$LiNbO_3$ photonic crystals have been discovered to be nonlinear optic materials with an electro-optic coefficient that is six times greater than that of potassium dideuterium phosphate (KDP). These ferroelectric oxide materials were produced by humans and created at Bell Labs in 1949. The attainment of tunability in their characteristics is crucial for the usage of photonic crystals in high technology applications such as optical switching and waveguiding technologies. Many other strategies have been proposed as a result. The most interesting strategy relies on utilizing the nonlinear characteristics of photonic crystals, particularly their dielectric properties, where the index of refraction varies with light intensity. As part of its tetragonal lattice, the deformed oxygen octahedral structure of $LiNbO_3$ photonic crystals is shared by other octahedral oxygen molecules. Figure 2 shows the crystalline structure of $LiNbO_3$ crystals [19].

The positions of the Li and Nb atoms are depicted along with the hexagonal unit cell in this diagram. $LiNbO_3$'s crystalline structure belongs to the 3 m space group, exhibiting mirror asymmetry across three planes separated by 60° and triple rotational asymmetry around the c-axis (also known as the Z-axis). The X and Y axes are commonly referred to in current reviews and most integrative photonic references as axes around which $LiNbO_3$ crystals exhibit mirror symmetry within

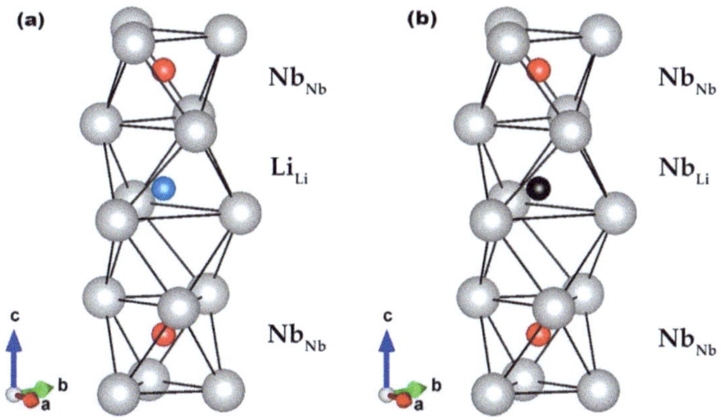

Fig. 2 $LiNbO_3$ crystal structures that are ideal and nonideal are compared. **a** Devoid of point flaws that characterise the composition in stoichiometry. **b** In the scenario when the presence of Nb antisites is most likely

the YZ plane. $LiNbO_3$ is recognized as a ferroelectric material [20]. A suitable synthesis technique must be capable of producing materials with the desired properties of $LiNbO_3$. Various deposition methods have been employed to prepare thin films of $LiNbO_3$, with chemical bath deposition (CBD) being an example of such techniques. Compared to other thin-film deposition methods, CBD is highly cost-effective, straightforward, and reliable, making it suitable for industrial applications. Figure 3 shows various XRD patterns of lithium niobate specimens produced via hydrothermal and sol–gel methods. As depicted in the image, the hexagonal $LiNbO_3$ structure is confirmed by all diffraction peaks. Furthermore, it is observed that increasing the temperature significantly enhances crystallinity (as seen in specimens produced at 600 °C compared to 500 °C). The broad diffraction peaks in the specimen synthesized at 500 °C using the sol–gel process indicate the presence of nanocrystalline powders. Specimens created using the hydrothermal and sol–gel techniques at 600 °C exhibited lattice parameters a = 5.1207 Å and c = 13.8687 Å, respectively, which closely match standard reference values (a = 5.149 Å and c = 13.86 Å). It should be noted that 240 °C is a relatively low temperature for the hydrothermal production of purified phase $LiNbO_3$ powders from basic raw materials (niobic acids and LiOH) [21].

4 Scanning Electron Microscope (SEM) Measurements

Morphological surface analysis represents a type of high-resolution imaging used in analytical microscopy, employing specialized microscopes to observe materials at high magnifications. To visualize the morphological surface, samples are exposed, and a high-energy electron beam focused on the specimen's surface by the SEM

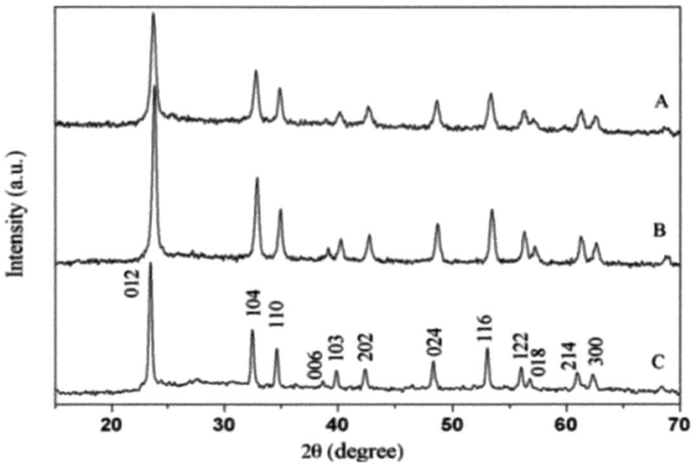

Fig. 3 Lithium niobite specimens produced using the hydro-thermic technique and the sol–gel approach at **A** 500 °C and **B** 600 °C are shown in the XRD patterning (**C**) [21]

generates signals containing data on the sample's crystal lattice, morphology, and roughness [22]. This measurement is non-destructive, allowing repeated tests on the same specimen. SEM can capture high-resolution images capable of measuring very minute objects and forms. Electrical insulating samples are typically coated with a thin layer of conductive substances such as silver, carbon, or other metals, essential for applications like elemental analysis (e.g., Ag) and high-resolution electronic imaging. However, a "low vacuum" instrument can test electrical insulating materials without the need for a conductive coat [23]. The nanostructure of $LiNbO_3$ generated by the sol–gel technique and deposited after a 48 h stirring time is depicted. Extended stirring time appears to enhance grain dispersion, resulting in a more uniform and smoother surface morphology. As mentioned earlier, longer stirring durations lead to improved grain dispersion in $LiNbO_3$ films. Examination of these samples suggests that the manufactured films resemble ice layers in form [24]. Other studies present SEM images of prepared $LiNbO_3$. Microwave radiation was utilized to optimize hydrothermal synthesis settings for producing pure nanocrystalline $LiNbO_3$. The results indicated high-quality nanoparticles with average diameters of 42 and 65 nm for synthesis durations of 2 and 3 h, respectively. Our findings highlight a time-efficient hydrothermal method supported by microwaves for synthesizing $LiNbO_3$ at the nanoscale, producing nanoparticles predominantly cubic and rhombohedral in shape [25].

5 Atomic Force Microscopy (AFM) Observations

AFM is an effective method for scanning, imaging, measuring, and analyzing surface structures of materials at the nanoscale. It can reveal single-atom arrangements and predict molecular structures due to its superior resolution compared to optical methods, making it crucial for nanoscale surface quality control [26]. AFM provides three-dimensional surface images at the nanoscale, detailing surface defects, quality, and crystal phases [27]. It operates non-destructively and does not require an expensive vacuum environment [28]. The AFM profile of spin-coated $LiNbO_3$ nanostructures shows a uniformly dense surface, indicating that grain size decreases with longer stir times. The surface topography of the $LiNbO_3$ nanostructure, as depicted in the AFM micrograph, demonstrates evenly dispersed grains across a 5×5 μm scan area, with individual columnar grains protruding. This surface characteristic is notable, as images taken 48 h later reveal it to be flatter, smoother, and more homogeneous compared to other surfaces. Larger grains appear due to the reduction in solvent fraction, correlating with an increase in mean grain size from 18.53 to 92.64 nm. The absorbance spectra of $LiNbO_3$ films created using the Czochralski method display wavelength-dependent absorption characteristics. Diffusion doping of lithium niobate crystals with metallic ions at 1073 K alters their optical properties, evidenced by distinctive ion absorption bands (e.g., Fe3+) in the spectra [29]. In addition to these findings, various studies on the UV–visible transmittance of $LiNbO_3$ films on sapphire substrates sintered at three temperatures (400, 500, and 600 °C) have been conducted. Beyond 350 nm, film transmission significantly increases with higher sintering temperatures, indicating increased transparency likely due to densification. The films exhibit optical homogeneity on sapphire (012), with no apparent pores or scattering centers that would reduce transmission [30].

6 Electrical Conductivity (Σ) and Resistivity (P)

The capacity of a substance to carry an electrical current is measured by its electric conductivity. The relationship between conductivity and temperature is one of the key instruments for examining semiconductor characteristics (T). Estimating electrical conductivity (σ) is highly helpful. The formula below can be used to calculate the conductivity [31]: $\sigma = 1/\rho$ (2–6). Where (ρ) is the resistivity (Ω cm). The physic unit for resistivity is (cm), which may be calculated by detection the material's surface-level resistivity to current flowing [32]. Surface resistivity is not influenced by the material's physic dimensions, but rather by a number of variables, including the production techniques and parameters, doping agent, as well as the annealing temperatures [31]. Employing the coming relation will allow to compute the resistivity of the thin films being deposited [32].

7 $LiNbO_3$ Nanostructure for Optoelectronic Device

The relationship between the currents flowing into an electronic instrument and the implemented voltage is referred to as the (I–V) feature. This procedure can be regarded as the ideal method for analyzing the active component positioned across two electrodes and evaluating materials characteristics including conductance, charge injecting efficiency, and charge carrier movability [33]. According to the thermionic emitting (TE) hypothesis, the (I–V) feature with forward bias is provided by the following equation [34]:

$$I = Is\ (\exp\ (qV/KBT)\ -\ 1)$$

As (V) forwards bias voltages (Is) saturation current device, and a 4 V of bias voltage, the heterojunctions' leaking current gradually rises. The diode formula was used to calculate the heterojunction's ideality coefficient. The ideality coefficients reached to 4.3, 4.1, 3, and 2.5; respectively, to the heterojunctions created with laser fluence of 1.3, 1.6, 2, and 2.2 J/cm^2. The existence of surface state and trapping center at the Au@$LiNbO_3$–Si interface, in addition to a discrepancy within lattice parameters of Au@$LiNbO_3$ or the Si substrate, are indicated by the high value of the ideality coefficient [18].

8 Spectral Responsivity (Rλ)

The measurement of an optical detector's reaction to light at a certain wavelength is known as spectral response. The results are presented according to (output currents for unit incidental optical powers) [35], and are determined using the equation [36]:

$$R\lambda = \text{ output current (I)/incident output power}$$
$$\text{(P)01} - (20\,/\text{Pph})$$
$$R\lambda = I$$

When (Iph) represents the photocurrent, (P0) stands for the source's associated emission power, and (A/W) has been the unit of measurement. The Au@LN/p–Si heterojuncting photodetector's spectrum responsiveness at a 5 V inverse bias voltage. Instruments built with low laser fluences of 1.3 and 1.6 J/cm^2 showed couple response peaks. The primary faint peak, caused by the LN film's absorbance edge, was seen at 380 nm. The secondary peak that discovered at 850 nm may have been caused by light absorbance within the silicon substrate's deplation area. In terms of maximal response, the photodetector manufactured at 1.3 J/cm^2 has values of 0.69 A/W at 380 nm and 0.65 A/W at 850 nm; respectively. The increasing thicknesses of the shells with rising laser fluence, that lowered the quantity of light can pass through the Au cores. This might be cause of the photodetector's dependency upon laser fluence for its responsiveness. The SPR of Au, on the contrary, improved the response at 500 nm [37].

At present, research on modified $LiNbO_3$ films is ongoing and has a number of uses, including: (1) Optoelectronics: Because modified $LiNbO_3$ films have electro-optic qualities that enable effective modulation of light signals, they are employed in optical waveguides and modulators. (2) Acousto-optic Devices: These films are utilised in acousto-optic modulators and deflectors, which control the propagation of light using sonic waves. (3) Piezoelectric Sensors: Modified $LiNbO_3$ films can function as sensitive piezoelectric sensors for a range of physical parameters, including acceleration and pressure. (4) Nonlinear Optics: Because of their huge nonlinear coefficients, they are essential in nonlinear optical applications including optical parametric oscillation and frequency conversion. (5) Integrated photonics: $LiNbO_3$ films are advantageous for on-chip signal processing and communication in integrated photonic circuits because of their small size and excellent efficiency. (6) Quantum information processing: Utilising their special optical and acoustic features, recent study investigates their potential in quantum information processing.

References

1. Maldonado-Orozco, M.C., Ochoa-Lara, M.T., Sosa-Márquez, J.E., Olive-Méndez, S.F., Espinosa-Magaña, F.: Synthesis and characterization of electrospun $LiNbO_3$ nanofibers. Ceram. Int. **41**(10), 14886–14889 (2015)
2. Branch, W., et al.: Investigation of a solid-state tuning behavior in lithium niobate. IEEE Trans. Ultrason. Ferroelectrics Freq. Contr. **67**(2), 365–373 (2020)
3. Dai, J., Xu, R., Lin, Y.S., Chen, C.H.: Tunable electromagnetic characteristics of suspended nanodisk metasurface. Opt. Laser Technol. **1**, 128 (2020)
4. Deon, V.G., Thesing, A., Santana, L.R., Costa, V.C., Vaz, M.O., Silva, R.M., et al.: Synthesis of $LiNbO_3$ nanocrystals by microwave-assisted hydrothermal method: formation mechanism and application to hydrogen evolution reaction. Chem. Papers **75**(8), 3807–3815 (2021)
5. Ballman, A.A.: Growth of piezoelectric and ferroelectric materials by the czochraiski technique. J. Am. Ceram. Soc. **48**(2), 112 (1965)
6. Shigemastu, K., Anzai, Y., Morita, S., Yamada, M., Yokoyama, H.: Growth conditions of subgrain free LiNbO3 single crystals by Czochralski method. Jpn. J. Appl. Phys. **26** Pt 1(12), 1988–1996 (1987)
7. Yu, J., Liu, X.: Hydrothermal synthesis and characterization of $LiNbO_3$ crystal. Mater. Lett. **61**(2), 355–358 (2007)
8. Hernández-Molina, R.O.S.A.B.R.I.L., Hernández-Márquez, J.A., Enríquez-Carrejo, J.L., Farias-Mancilla, J.R., Mani-González, P.G., Vigueras Santiago, E.N.R.I.Q.U.E., et al.: Synthesis by wet chemistry and characterization of $LiNbO_3$ nanoparticles. Superficies y vacío, **28**(4), 115–118 (2015)
9. Fakhri, M.A., Wahid, M.H.A., Kadhim, S.M., Badr, B.A., Salim, E.T., Hashim, U., Salim, Z.T.: The structure and optical properties of Lithium Niobate grown on quartz for photonics application. In: EPJ Web of Conferences, vol. 162, p. 01005. EDP Sciences (2019)
10. Al-Douri, Y., Fakhri, M.A., Bouhemadou, A., Khenata, R., Ameri, M.: Stirrer time effect on optical properties of nanophotonic $LiNbO_3$. Mater. Chem. Phys. **203**, 243–248 (2018)
11. Murillo, J.G., Ocón, J.A., Herrera, G.M., Murillo-Ochoa, J.R., Ocón, G.: Growth of single-crystal $LiNbO_3$ particles by aerosol-assisted chemical vapor deposition method. In: Crystal Growth. IntechOpen (2019)

12. Klimin, V.S., et al.: Influence of annealing on nanocrystalline LiNbO3films properties. In: International Conference on Micro-and Nano–Electronics 2018, vol. 11022, p. 123. SPIE (2019)
13. Sosunov, A., Ponomarev, R., Semenova, O., Petukhov, I., Volyntsev, A.: Effect of pre-annealing of lithium niobate on the structure and optical characteristics of proton-exchanged waveguides. Opt. Mater. **88** (2019)
14. Liang, X., Chen, X., Yang, X., Ni, J.: The fabrication of $LiNbO_3$ memristors for electronic synapses using oxygen annealing. Nanotechnology **32**(2), 025706 (2020)
15. Ito, Y., Tsuyumoto, I.: Preparation of nanocrystalline $LiNbO_3$ through aqueous solution process using peroxo-Polyniobic acid. Mater. Chem. Phys. **272**, 125035 (2021)
16. Singh, G., Sharma, M., Vaish, R.: Flexible Ag@ $LiNbO_3$/PVDF composite film for piezocatalytic dye/pharmaceutical degradation and bacterial disinfection. ACS Appl. Mater. Interfaces **13**(19), 22914–22925 (2021)
17. Irzaman, K.N., N., Safutra, H., Nur Aidi, M., Alatas, H., & Hardhienata, H.: The structure and optical properties of lithium niobate thin film ($LiNbO_3$) grown on silicon for various lanthanum concentration and molarity. Ferroelectrics **589**(1), 12–21 (2022)
18. Alwazny, M.S., Ismail, R.A., Salim, E.T.: High-quantum efficiency of Au@ $LiNbO_3$ core–shell nano composite as a photodetector by two-step laser ablation in liquid. Appl. Phys. A **128**(6), 1–11 (2022)
19. Sánchez-Dena, O., Fierro-Ruiz, C.D., Villalobos-Mendoza, S.D., Carrillo Flores, D.M., Elizalde-Galindo, J.T., Farías, R.: Lithium niobate single crystals and powders reviewed—Part I. Crystals **10**(11), 973. https://doi.org/10.3390/cryst10110973
20. Gopalan, V., Dierolf, V., Scrymgeour, D.A.: Defect–domain wall interactions in trigonal ferroelectrics. Annu. Rev. Mater. Res. **37**(1), 449–489 (2007)
21. Garibay-Alvarado, J.A., Farías, R., Reyes-López, S.Y.: Sol-gel and electrospinning synthesis of lithium niobate-silica nanofibers. Coatings **9**(3), 212 (2019)
22. Swapp, S.: Scanning Electron Microscopy. University of Wyoming, USA, Geochemical Instrumentation and Analysis (2017)
23. Laboratory Testing INC Scanning electron microscopy (2016)
24. Fakhri, M.A., Salim, E.T., Wahid, M.H.A., Hashim, U., Salim, Z.T., Ismail, R.A.: Synthesis and characterization of nanostructured $LiNbO_3$ films with variation of stirring duration. J. Mater. Sci. Mater. Electron. **28**(16), 11813–11822 (2017)
25. Deon, V.G., Thesing, A., Santana, L.R., Costa, V.C., Vaz, M.O., Silva, R.M., Carreno, N.L.: Synthesis of $LiNbO_3$ nanocrystals by microwave-assisted hydrothermal method: formation mechanism and application to hydrogen evolution reaction. Chem. Pap. **75**(8), 3807–3815 (2021)
26. Oxford Instuments Asylum Research, Measuring surface rougness with Atomic Force Microscopy (2013)
27. Oxford Instruments Asylum Research, AFM characterization of thin films: high-resolution topography and functional properties (2015)
28. Ghoshal, S., Srivastava, A.: Fundamental of Bioanalytical Techniques and Instrumentation Book. PHI learning Pvt. Ltd., pp. 1–292 (2009)
29. Yakhnevych, U.V., Suchaneck, G., Eydam, A., Sugak, D.Y., Syvorotka, I.I., Haiduchok, V.G., Buryy, O.A., Ubizskii, S.B., Gerlach, G.: Investigation of Optical and Pyroelectric Properties of Lithium Niobate Single Crystals Caused by Metal Ions Diffusion (2019)
30. Hur, N.H., Park, Y.K., Won, D.H., No, K.: Effect of substrate on the growth and properties of $LiNbO_3$ films by the sol-gel method. J. Mater. Res. **9**(04), 980–985 (1994)
31. Manickam, M., Ponnuswamy, V., Sankar, C., Suresh, R.: Cobalt oxide thin films prepared by NSP technique: impact of molar concentration on the structural, optical, morphological and electrical properties. Optik **127**(13), 5278–5284 (2016)
32. Gutiérrez, M., Haiyong, L., Patton, J.: Thin Film Surface Resistivity, In partial fulfillment of course requirements for Experimental Methods in Materials Engineering, pp. 1–24 (2002)
33. Kondo, T.: Current-voltage characteristic of organic semiconductor interfacial control between organac layers and electrods, Ph.D. Thesis, Georgia institute of technology (2007)

34. Yacobi, B.G.: Types of semiconductors. In: Semiconductor Materials: An Introduction to Basic Principles, pp. 135–170 (2003)
35. Hartman, J., Lind, M.: Spectral response measurement for solar cells. Solar Cells **7**, 147–157 (1983)
36. Al-fwade, E., Al-ani, S., Al-ani, T.: Influence of doping concentration on the main parameters of CdS: Cu photoconductor detector. Iraqi J. Phys. **9**, 61–65 (2011)
37. Richards, D.A., Maruani, A., Chudasama, V.: Antibody fragments as nanoparticle targeting ligands: a step in the right direction. Chem. Sci. **8**, 63–77 (2017)

Downstream Applications of Smart and Hybrid Nanomaterials

Abinaya Kanagaraja, Pinki Kumari Singh, and Pachaiappan Raman

1 Introduction

An analytical tool consisting of a biological sensing element in close proximity to a physical transducer, which together produce a detectable electrical signal that indicates the analyte's concentration is refered as biosensor [1, 2]. The basis of biosensor technology is comprised of nanomaterials, which offer the fundamental components required to develop extremely sensitive and effective detection systems. Precision biological signal capture, recognition, and transduction are made possible by nanomaterials due to their special qualities at the nanoscale, including customised chemistry of surface, high surface (area-to-volume ratio), and extraordinary electrical or optical properties [3]. Moreover, the adaptability of nanomaterials makes it possible to combine several detection modalities and create portable, tiny devices that are appropriate for field applications and diagnostics [4]. The smart and hybrid nanomaterials offer improved functionality and versatility for a variety of applications, marking a breakthrough in nanotechnology [5]. By responding to the external stimuli like light, pH, or magnetic fields, or temperature, smart nanomaterials allow for dynamic control over their behaviours and characteristics. These stimuli-responsive materials show potential use in medication delivery applications, sensing, actuation, and environmental remediation, where precise control and modulation of material properties are desired [6]. On the other hand, hybrid nanomaterials combine distinct nanoscale components of two or more, such as nanowires, nanosheets, or nanoparticles, to synergistically integrate their unique properties and functionalities [7, 8]. The beneficial qualities of many nanoparticles are combined to produce hybrid nanomaterials

Author has equally contributed—K. Abinaya and Pinki Kumari Singh

A. Kanagaraja · P. K. Singh · P. Raman (✉)
Department of Biotechnology, School of Bioengineering, Faculty of Engineering and Technology, SRM Institute of Science and Technology, Kattankulathur, Chengalpattu District 603 203, Tamil Nadu, India
e-mail: pachaiar@srmist.edu.in; pachaitvm@gmail.com

S. C. B. Gopinath and M. M. Ramli (eds.), *Hybrid-Nanomaterials*, Engineering Materials, https://doi.org/10.1007/978-981-97-9022-7_11

provide improved stability, performance, and multifunctionality. As a result, they are highly suitable for the diverse applications, including biomedicine, energy storage, electronics, and catalysis. Innovation is still being fuelled by the creation of smart and hybrid nanomaterials, which are opening up new avenues for solving difficult problems in science, engineering, and technology [9]. By integrating nanomaterials into biosensing platforms, researchers can develop devices capable of accurate, rapid, and portable detection of biological analytes of a wide range, with implications spanning healthcare, environmental monitoring, food safety, and beyond. This chapter discuss about the incorporation of smart and hybrid nanomaterials into biosensors which is a revolutionary development that provides unmatched sensitivity and adaptability for a range of uses. This combination has enormous potential to transform environmental monitoring, disease diagnostics, and other fields. Rapid accurate, and portable detection platforms are made possible by these biosensors improved performance and multifunctionality. Smart and hybrid nanomaterial-based biosensors, which promise a better and healthier future, essentially mark the beginning of a revolutionary age in sensing technologies.

2 Fundamental of Biosensors

2.1 Biosensors Components

A biosensor is a device for analysis that uses signals corresponding to the concentration of an analyte in a reaction to evaluate biological or chemical responses [10]. Using a transducer, such as an electrochemical sensor, the concentration of the analyte is transformed into an electrical signal to facilitate this detection [11]. Employing the technique of electrochemical oxygen or hydrogen peroxide detection, Clark and Lyons in 1962 developed first biosensor for determining glucose level in biological samples [12]. Biosensors applications include the pollutants, microbes, and disease-indicating markers in physiological fluids detection, as well as disease monitoring [13]. A biosensor is made up of a biological receptor connected to a transducer and a signal processing apparatus. It works by using the process of signal transduction to convert signals. By combining these components, the biological reaction is converted into an equivalent electrical reaction and, eventually, an outcome that can be measured [14]. At first, the molecule being studied in the test sample, forms a specific bond or interaction with biological receptor, leading to a physiological alteration. The characteristics of chemical and physical transducer that is located near biological receptor is modified. The transducer's optical or electronic properties are later altered and converted into an electrical signal that can be detected [15]. A typical biosensor comprises of various component as follow:

(a) **Analyte**: The term "analyte" refers to the specific material or component that the biosensor is specifically designed to detect and quantify. These entities encompass a diverse array of biological or chemical substances, including glucose, cholesterol, DNA, proteins, infections, poisons, environmental contaminants, and several others.

(b) **Receptor**: It serves as a crucial function in identifying and detecting the analyte. It is commonly known as the "recognition element" or "biological recognition element."

(c) **Transducer**: It converts the biochemical signal produced by the bioreceptor into a quantifiable output signal of analyte. Following this, different transducer types utilised in biosensors [16],

- Electrochemical transducer: It measures changes in electrical properties, such as current or voltage etc.
- Optical Transducers: It measures changes in light intensity, wavelength, or polarization as a result of the analyte and bioreceptor's interaction. Examples include fluorescence and absorbance etc.
- Piezoelectric Transducers: It measures changes in acoustic wave properties, such as frequency or amplitude, because of binding events occurs in bioreceptor and the analyte. Examples include quartz crystal microbalance biosensors.
- Thermal Transducers: It measures temperature variations based on the biological interactions that take place between the analyte and the bioreceptor.

(d) **Electronics**: This part of the biosensor processes the transduced signal so that it can be demonstrated.

(e) **Display**: A user interpretation system powers the display, producing clear numbers or curves for the user to view, similar to what a direct printer or liquid crystal display on a computer typically generate.

2.2 Characteristics of a Biosensor

Biosensors possess several characteristics that make them unique and valuable for various applications [17, 18].

- **Selectivity**: Biosensors possess excellent selectivity, which enables them to identify a specific analyte within a complicated mixture of chemicals. Enzymes, antibodies, and nucleic acids are examples of biological recognition elements that are frequently used to achieve this selectivity.
- **Sensitivity**: Biosensors enable accurate detection and quantification since they are sensitive to even exceptionally low quantities of the target analyte. A biosensor's sensitivity is especially important, in applications like medical diagnostics where it's necessary to detect analyte concentrations that are low.
- **Rapid Response**: Since biosensors often react quickly, it enables to monitor analytes in real-time. This speedy reaction is helpful in scenarios requiring prompt decision-making, such environmental monitoring or healthcare.

- **Accuracy and Precision**: Biosensors provide dependable findings by detecting analytes with great accuracy and precision. Applications include environmental monitoring, food safety, and healthcare, accurate and exact measurements are crucial for making well-informed judgements.
- **Miniaturization**: The miniaturization of biosensors is possible by the developments in nanotechnology and microfabrication, which have resulted in smaller devices, lower sample quantities, and better integration capabilities.

2.3 Biosensor Types

Biosensors can be categorised into several kinds according to a number of factors, such as the kind of biological recognition element, transduction mechanism, and use. Some common types of biosensors are as follows:

(a) **Optical Biosensors**:

- Surface Plasmon Resonance (SPR) Biosensors: Target molecules that attach to immobilised receptors cause refractive index changes on the surface of sensor, which SPR biosensors can detect.
- Fluorescence Biosensors: Fluorescent molecules are employed by these biosensors to identify the presence of analyte binding; variations in the fluorescence intensity or wavelength signify the analyte concentration [19].

(b) **Electrochemical Biosensors**:

- Amperometric Biosensors: Current generated by redox reactions between the analyte and electrode surface is measured by Amperometric biosensors.
- Potentiometric Biosensors: Analyte binding or reaction-related changes in electrode potential are measured by potentiometric biosensors [20].

(c) **Piezoelectric Biosensors**: These biosensors track variations in a piezoelectric crystal's frequency or amplitude brought on the binding or reaction of analytes [21]
(d) **Thermal Biosensors**: Thermal biosensors use thermistors or thermocouples to measure temperature changes brought on analyte interaction or response [22]
(e) **Enzyme-Based Biosensors**: Enzymes are used by these biosensors as the biological recognition element to catalyse particular reactions with the analytes, leading in the generation of a detectable signal [23].
(f) **Nucleic Acid-Based Biosensors**: Nucleic acid biosensors, uses in genomics, molecular diagnostics, and environmental monitoring, employs DNA or RNA as the recognition element to identify particular sequences or mutations [24].

3 Nanomaterials for Biosensing

3.1 Nanomaterials Used in Biosensors

Nanomaterials play a pivotal role in biosensors, offering unique properties that enhance sensitivity, selectivity, and versatility. The emerging interest in nanomaterials because of the growing need to control desired chemicals found in the environment and human body. Nanoparticles (NPs) less than the 100 nm in the one dimension or less are known as nanomaterials [25]. The intended nanomaterials' attributes include increased electrical conductivity, nanoscale size, signal amplification capabilities, and compatibility with biological molecules [26]. Nanomaterials primarily serve as transducer materials in biosensor development, significant role in converting the biochemical signals produced into quantifiable signals through interaction between recognition element and analytes. For instance, nanomaterials like quantum dots, carbon nanotubes, graphene, and metallic NPs are commonly employed in various biosensor architectures due to their unique properties [27]. Quantum confinement of delocalised electrons causes discontinuous behaviour, which greatly impacts nanomaterial behaviour. The shape of these nanoparticles determines the majority of their features. Because of their lower binding energy and larger surface atom count, they have lower melting temperatures compared to the bulk material [28]. Quantum dots and semiconductor nanocrystals with size-tunable fluorescence, are utilized as labels in fluorescence-based biosensors, enabling multiplexed detection of biomolecules with specificity and high sensitivity. Carbon nanotubes and graphene, renowned for their exceptional large surface area and electrical conductivity, are integrated into electrochemical biosensors, facilitating rapid and selective detection of analytes through efficient electron transfer processes [29]. Other fatal and incurable diseases have also been diagnosed using CDs. Liang and colleagues created a ratio metric fluorescence (RF) assay for HIV DNA detection employing CDs and quantum dots (QD) of cadmium telluride (CdTE) [30]. Nanodiamonds (NDs) have garnered the significant attention in biosensing applications owing its discrete properties, including their fluorescent nature and biocompatibility. Among their various applications, NDs have demonstrated remarkable capabilities in detecting metal ions, offering a promising avenue for environmental monitoring, industrial safety, and healthcare diagnostics. One notable study by Shellaiah and co-workers stands out in this realm, where they pioneered the development of photoluminescent cysteamine with nanodiamonds that is modified for detection of mercury ions (Hg2+). Even at low quantities, mercury is a very poisonous heavy metal that seriously endangers both human health and ecosystem health [31, 32]. Also, graphene derivatives include graphene oxide (GO), reduced GO, and graphene quantum dots, has significant attention in the biosensing field. GO is a strongly oxidised version of graphene that has carboxyl, hydroxyl, and epoxy functional groups, among other oxygen-containing groups. In biosensing, GO is utilized as a sensing platform or signal transducer due to its ability to immobilize biomolecules and facilitate signal amplification through fluorescence quenching or enhancement mechanisms [33–35].

3.2 Properties of Nanomaterials Relevant to Biosensing

Nanomaterials are extremely relevant to biosensing applications due to their many characteristics. These qualities make it possible to detect biological analytes in a sensitive, diverse, and selective manner. Because of nanoscale dimensions it exhibits remarkably high surface (area-to-volume ratio). By increasing the surface density of recognition elements, this characteristic makes it possible to immobilise biomolecules like enzymes, antibodies, or DNA probes efficiently and increases the sensitivity of biosensors [36]. Excellent biocompatibility of several nanomaterials enables their assimilation into biological systems without creating negative consequences. This characteristic is essential for biosensing applications, especially those involving interactions with biological molecules or cells, such as medical diagnostics and therapies [37]. Certain nanomaterials have special optical characteristics like fluorescence, plasmonic resonance, and light scattering. Examples of these include graphene quantum dots, gold nanoparticles. These optical characteristics allow the sensitive and real-time detection of biomolecular interactions in biosensing experiments by label-free detection or signal amplification [38]. Graphene, carbon nanotubes, and conductive polymers are examples of nanomaterials with high electrical conductivity. This feature enables selective detection and sensitive of analytes with low detection limits in electrochemical biosensors, which assess changes in electrical signals brought about by biomolecular interactions [39] Nanoparticles have remarkable mechanical qualities, such as great strength, flexibility, and resilience. Examples of these materials are carbon nanotubes and nanocomposites. These characteristics can be used to create strong, long-lasting biosensing systems that can tolerate challenging circumstances or frequent use [40]. Many nanomaterials are stored for long time and used in biosensing applications because they exhibit chemical stability in a variety of environments. Over time, this stability guarantees the dependability and repeatability of biosensor performance [41]. Also, due to their remarkable sensitivity to changes in size, shape, and composition, nanomaterials can have tunable features that are customised for certain biosensing applications. Biosensors may be precisely controlled in terms of sensitivity, selectivity, and dynamic range due to this size-dependent behaviour [42].

4 Smart Nanomaterials in Biosensors

Nanomaterials are developed and effectively used in a number of applications like nanotechnology. To achieve superior treatments, it remains difficult to produce nanomaterials with acceptable functional characteristics, particularly in biomedical applications. Smart materials are receiving increased attention these days as a way to get around the drawbacks of conventional nanoparticles [43]. In the meantime, the term "smart materials" was defined to encompass any material that may be stimulated by outside variables to produce new useful features. Additional categories for stimuli agents include pressure, light, pH, magnetic field, electric, stress, temperature, and so on. Because of their controllable properties, smart materials are very intriguing to use in numerous applications, including as the treatment of various diseases, controlled medication release, and biosensors [44]. When compared to traditional nanoparticles, these materials show enhanced capabilities such increased surface area, better reactivity, and tunable optical and electrical characteristics. Because of this, they can be used for a variety of purposes, such as industrial, environmental, and biomedical ones. Smart nanomaterials can be engineered to precisely interact with biological targets in biomedical applications, reducing adverse effects and enhancing therapy success. Enzyme-responsive nanoparticles, for instance, can release medications at the location of a tumour, reducing the effect on surrounding healthy tissues. By dynamically adjusting to environmental changes, these materials can imitate natural systems in an equivalent way to specific plants or animals react to their surroundings. They are extremely helpful in producing materials that are adaptable and self-healing because of this quality [45].

Recently, smart material and its use has drawn more attention from researchers as a result of the advancement of cutting-edge technologies and the growing demand for novel materials that satisfy evolving specifications. Toshinori Takagi published the first comprehensive study on "intelligent materials" in April 1990 [46]. Smart nanomaterials are that material which response to an external stimulus or their surroundings, cause a dynamic and reversible change in essential properties [47]. Stated differently, "Materials that can change all of these properties on demand in response to particular stimuli" include density, shape, texture, colour, modulus, stiffness, and the toughness. These days, "intelligent material" is also known as "smart material," or "stimuli-responsive material" and its use has drawn more attention from researchers as a result of the advancement of cutting-edge technologies and the growing demand for novel materials that satisfy evolving specifications [48].

4.1 Characteristics of Smart Nanomaterials

- Various environmental stimuli, including temperature, pH, light, chemical signals, and the magnetic fields, can cause smart nanomaterials to react. A controlled and focused course of action is made possible by this responsiveness.
- Certain intelligent nanomaterials can self-heal damage by reassembling their structural components chemically or physically.
- These materials have the ability to attach to particular molecules or ions with selectivity, which makes them valuable for uses in medicine administration, sensing, and environmental cleanup.
- Intelligent nanomaterials are able to modify their characteristics in reaction to shifting external circumstances.
- Many intelligent nanomaterials are made to be biocompatible, which means that employing them in biological systems will not harm or react negatively with them.

- Smart nanomaterials' composition, structure, and synthesis characteristics can be precisely controlled to enhance or personalise their properties.

4.2 *Types of Stimuli Responsive Nanomaterials*

The applied stimuli are used to classify smart nanomaterials into distinct categories. External stimuli can modify the properties of smart nanomaterials in a controlled manner. In general, these stimuli fall into three categories: physical, chemical, and biological. When exposed to one or more stimuli, such as those that are chemical (such as pH), biological (such as enzymes), or physical (such as magnetic, light, electric field, and temperature), stimuli-responsive materials can detect a signal and alter their characteristics. A smart biomaterial may receive any stimulus directly in order to cause a change or response, and it may also simultaneously transform the input into a functional trigger for the cells. This is useful because, by tracking the stimulus in both space and time, it enables precise control of cellular response when required. Thus, smart materials have the potential to significantly advance biomedical engineering, ideally leading to the creation of innovative, reasonably priced medications that are customised to each patient's specific need [49, 50].

4.2.1 Physical Responsive Nanomaterials

Nanomaterials that are physically sensitive are a fascinating class of materials that respond differently to environmental stimuli including light, pH, temperature, mechanical stress, electric or magnetic fields, or chemical signals. These materials also display unique features or experience specific alterations. These materials are designed to take use of surface phenomena and quantum mechanical effects at the nanoscale, 1–100 nm in size. This allows for customised and frequently highly programmable reactions to stimuli [51].

The physical material reactions in this perspective that can be used to help nanoscale carriers with diagnostic imaging or medication delivery. These reactions can include activation or direct release, gatekeeping, degradation, expansion, aggregation, autonomous motion induction and morphological switches. Many of the first instances of stimuli-responsive nanomaterials, such as gatekeeping, direct release, and degradation processes, start with the cleavage of a chemical bond. The medical industry has challenges in determining which connections can be broken by a single stimulus and optimizing release kinetics for rapid, effective, and targeted material activation. Shape, size, or phase changes are caused by variations in the material's physical characteristics, such as amphiphilicity, electrostatics, or steric, which lead to responses including expansion, assembly/aggregation, or morphological switching. The apparent difficulty with these approaches is that they must produce feasible strategies to successfully transition intricate designs that the function in test tube into in vivo application [52].

Thermosensitive Responsive Nanomaterials

Thermosensitive responsive materials, sometimes referred to as thermo-responsive materials or thermo-responsive polymers when exposed to temperature changes, can change reversibly in both their structure and behaviour. At particular temperature thresholds, these materials are intended to display discrete transitions, such as swelling/collapse, phase separation, or solubility changes. This characteristic has a wide application, including drug administration, tissue engineering, sensors, and actuators [53].

Electrochemical Stimuli-Responsive

Materials that respond to electrical stimuli by altering their properties or behaviour are known as electrochemical stimulus-responsive nanomaterials. Used as drug delivery, energy storage, sensing, and electrochemical catalysis, these materials' characteristics can be precisely controlled. They are made to react to variations in charge, current, or electrical potential [54].

Light-Responsive Nanomaterials

Nanomaterials that are designed to modify their properties or behaviour in response to light are known as light-responsive nanomaterials. By manipulating electrical or optical properties, releasing trapped molecules, or causing structural alterations, these materials use light energy to achieve specific effects. Various fields, including photonics, drug delivery, sensing, and optoelectronics, have found use for light-responsive nanomaterials [55].

Magnetic-Responsive Nanomaterials

A class of materials known as magnetic-responsive nanoparticles exhibits alterations in their characteristics or behaviour in external magnetic field when exposed. These materials are engineered to undergo reversible changes in their optical, mechanical, electrical, or the magnetic properties in response to magnetic stimuli. Applications for magnetic-responsive nanoparticles are numerous and include drug administration, magnetic separation, magnetic resonance imaging (MRI), biomedicine, and sensing [56].

4.2.2 Chemical Responsive Nanomaterials

A group of materials known as responsive nanomaterials chemically is created at the nanoscale to react differently to changes in their chemical surroundings. These materials have attracted a more attention due to their prospective uses in an industries, such as catalysis, medication delivery, environmental cleanup, and sensing. When the environment is acidic or basic, smart biomaterials can be engineered to adapt to the changes in parameters of environment. This means that when the pH of the environment changes, specific organ systems or pathological states can benefit from the release of medicines. The pH-controlled release devices have the potential to restrict systemic drug concentrations when administering highly hazardous chemotherapy. The development of drug delivery by pH-sensitive devices that experience physical or chemical changes when exposed to acidic pH is made feasible by the acidic extracellular environment seen in solid tumours [57].

PH Responsive Nanomaterials

The pH responsive nanomaterials experience alterations in their characteristics or functions in reaction to fluctuations in pH levels. These materials have garnered considerable interest because of their potential uses in medication delivery, sensing, controlled release, and tissue engineering [58].

Redox Responsive Nanomaterials

It is a class of nanomaterials that exhibit alterations in their characteristics or functions in accordance with variations in the redox (reduction–oxidation) environment. Due to their potential uses, these materials have garnered a lot of interest. In medicine delivery, sensing, energy storage, catalysis, and environmental cleanup [59].

4.2.3 Biological Responsive Nanomaterials

These nanomaterials are engineered to engage with biological systems includes tissues, cells, or biomolecules that provoke targeted reactions or alterations in their characteristics. These materials are essential in the biological applications, such as tissue engineering, drug delivery, diagnostics, and therapies [60].

Enzymatic Responsive Nanomaterials

A specific type of responsive nanoparticles known as "enzymatic-responsive nanomaterials" interacts with enzymes to produce a reaction or alteration in their characteristics. The purpose of these materials is to use the catalytic activity of enzymes for a variety of purposes, such as biocatalysis, medication delivery, and sensing [61].

Glucose Responsive Nanomaterials

Glucose-responsive nanoparticles are specifically engineered to detect and react to fluctuations in glucose concentrations. These materials have the potential to be utilised in the management of diabetes and the monitoring of glucose levels [62].

4.3 Applications of Smart Nanomaterials in Biosensors

Biosensors are now widely employed in clinical research, pharmaceutical analysis, and the identification of biowarfare chemicals since they are an immensely powerful diagnostic tool [63]. Additionally, more effective glucose sensors and enhanced biocompatible sensing devices are required. These sensing systems have been continuously improved by the addition of smart nanomaterials including copper, metal alloys/adatoms, titanium, graphene, composites, and organic matter specific to glucose because of their electrocatalytic reactions to glucose oxidation. Oxygen electrode is an external dialysis membrane which is a thin layer of GOx, and inner oxygen semipermeable membrane make up this first-generation glucose biosensor [64]. Glucose oxidase was immobilised on silica nanoparticles and platinum nanomaterials to generate an extremely flexible and stable glucose biosensor. With 0.8 mmol L^{-1} of detection limit, the proposed nanomaterial-based biosensor (GOx-MS and PtNP) can be utilised to quantitatively detect glucose in human serum [65, 66]. To identify pharmaceutical medications and steroidal hormones in human fluids, numerous sensors based on nanomaterials have been investigated. For example, CDs-GO nanocomposites were utilised to create an aptamer for the biosensing of P4 progesterone hormone. This nanomaterial-based biosensor was examined for its capacity to detect glucose and H_2O_2, and it was found to function as a non-enzyme approach for quick glucose detection using spectrophotometry [67]. The chemical and physical properties of nanomaterials have made scientists develop new technologies that interface with biological systems at the nanoscale. Despite being in its infancy, nanotechnology has already had a profound effect on biology in general and medicine in particular. Specifically, the discipline of nanobiotechnology—which deals with creating nanoscale materials using biological systems and molecules—is presently yielding exciting possibilities for medicines and diagnostics [68, 69].

The first human insulin recombinant protein treatment approved by the FDA appeared before 32 years, protein therapies were rarely used. Since then, the field of protein therapies has grown exponentially, and now, protein-based drugs are crucial for treating a variety of ailments, such as metabolic disorders, cancer, and autoimmune diseases. In the meantime, the FDA has approved the use of over 130 proteins or peptides in medical procedures. Protein treatments are preferable to small-molecule medications in a number of ways because of their superior biocompatibility and extremely specific and complicated range of functions. Protein therapies are a safer option than genetically engineered cells since they do not involve random or permanent alterations to the cell's genome [70]. Even though protein therapies have advanced significantly over past few decades, there are still number of challenges that need to be addressed. Protein therapies delivered directly to body with short half-life, are immunogenic, and are unstable both in vitro and in vivo [71]. Because of this, delivery strategies built using elements with intelligence that react to stimuli have received a lot of interest recently. When exposed to external stimuli, stimuli-responsive design can undergo conformational and chemical changes that are followed by changes in its physical characteristics. In addition to facilitating the release of a drug with the desired pharmacokinetics, this action can ensure that the drug can be released at the targeted site in a specific location and at a specific time [72]. There are still several issues that need to be addressed before their application from the bench to the bedside can be fully realised. To fully fulfil therapeutic promise, coordination of pharmacokinetics and protein release payloads at the appropriate location or time called as "responsive" site is important. An efficient intracellular protein delivery mechanism, for instance, should ensure that the nanocarriers have enough time to activate and release cargoes through cellular environmental variables like redox potential or acidity. Furthermore, a difficult problem for the delivery systems of glucose-responsive insulin is to respond quickly enough to replicate the activity of normal pancreatic cells [62, 73].

5 Hybrid Nanomaterials in Biosensors

Biomolecules or bio-related molecules can have their functionalities controlled in response to desired stimuli by introducing features that respond to such stimuli. The significance of molecular design lies in the development of novel molecules and accompanying materials capable of responding to specific stimuli, offering promising prospects for valuable bio-applications. Extensive research has been conducted on molecular designs that utilise photo-caged groups or trans–cis isomerization to introduce photo-switching capabilities. A target biomolecule's activities are momentarily suppressed with the introduction of photo-caged groups, or photoremovable chemical groups; these activities are subsequently restored after the external stimulus application, such as photoirradiation, which eliminates the introduced chemical groups [74–76]. In response to the limitations of less efficient single-component nanomaterials, there has been a widespread development of hybrid structures in recent times, particularly in the domains of energy and the environment, offering exceptional and multifunctional properties.

Metal organic frameworks, or MOFs, represent a class of hybrid porous materials combining organic and inorganic components. Because MOFs have a high porosity and low density, their pores may hold and accumulate an enormous number of molecules due to their enormous surface area. Moreover, the pore size can function as mesh by effectively segregating molecules. Because MOFs are porous, it is desirable to alter the intrinsic properties of the materials. Different hybrid MOF types have been reported over the last ten years in conjunction with carbon materials, polymers, metal oxides, metal nanoparticles, and biomolecules for various applications. In addition, MOFs is employed in the production of sensors, gas separation, energy storage, medication delivery, supercapacitors, and environmental cleanup [77]. The graphene family compounds, carbon nanotubes, and more recently, novel materials are carbon-based nanomaterials. The excellent electrical conductivity, stability of chemical and the availability of carbon-based dots have made them the perfect material to combine with NMs, transition metal oxides, transition metals, and other materials. Specifically, the integration of carbon-based dots (CDs), including carbon/graphene quantum dots (GQDs or CQDs), and carbon dots (C-dots), with NMs enables effective dispersion of NMs and yields numerous advantages. These CDs feature an abundance of super conductivity, functional groups, and good photostability [78]. In heterogeneous catalysis, metal nanoparticles have garnered a lot of interest. One challenge arises during the catalytic process, where migration-coalescence could readily deactivate nanoparticles due to their elevated surface energy. With nanoscience advancing rapidly, enclosing metal nanoparticles within nanoshells or nanopores emerges as one of the most promising strategies to address this stability concern. Moreover, the activity and selectivity may be raised at the same time by making use of the encapsulating materials' molecular sieving capability as well as the metal nanoparticles' synergistic effect [79]. Nanohybrids for cancer therapy have recently been created using a range of polymers. One of the most often utilised polymers for creating nanohybrids is Polyethylene glycol (PEG), which is followed by a variety of other polymers such casein, pluronic F127, Carboxy Methyl Cellulose (CMC), chitosan, Poly(lactic-co-glycolic) Acid (PLGA), polypyrroles, and poly(ethyleneimine). The unique characteristics of the nanohybrids, like pH-sensitive release and controlled release, are a result of these polymers. The incorporation of constituents like FA renders these nanohybrids exceedingly precise [80].

5.1 Organic and Inorganic Hybrid Nanomaterials

Biomedical applications long utilised biomaterials derived from polymers, ceramics, metals, and composites for the purpose of diagnosing and treating a wide range of illnesses. Polymers are the foundation of representative organic biomaterials. Various biocompatible polymers have been developed, as they often share certain characteristics with biodegradable polymers. Clinical approval for drug delivery systems of polymer-based has already occurred, indicating the remarkable success of biomaterials which is polymer in the drug delivery sector [81, 82]. The general description of organic/inorganic hybrid nanocomposites involves a blend of materials characterized by submicron-scale dimensions. For instance, modifying the surface of inorganic nanoparticles such as gold nanoparticles (AuNPs), iron oxide nanoparticles (Fe_3O_4), semiconductor quantum dots (QDs), and carbon dots with organic polymers can render them water-soluble. This transformation enables the creation of hybrid materials suitable for therapeutic or diagnostic purposes. Consequently, innovative functional therapeutic agents with dual capabilities of drug delivery via surface polymers and biomedical imaging utilizing the physical or optical properties of inorganic nanoparticles can be developed. Moreover, the mechanical properties of composites incorporating inorganic ceramic nanoparticles can be tailored for specific applications, such as bone transplant materials [83, 84]. Nanoparticles encompass organic or inorganic materials ranging in size from 1 to 100 nm. They enable the exploration of material properties in relation to particle size variations.

Within the Ag, Cu, and Au, nanoparticles of metal category show distinct electrical, optical, and the catalytic characteristics that change based on size. These nanoparticles' characteristics are different from molecular compounds or bulk metals, and they are mostly determined by the shape, size, and the inter-particle spacing of particles and also organic shell in that case. Dispersing inorganic nanoparticles in the aqueous phase is essential for its use in bio-applications. Addition to that nanoparticle must be chemically stable and resistant to sintering and other degradative processes such partial oxidation [6, 85, 86]. Since they increase the particles' chemical stability and enable them to diffuse in the aqueous phase, polymer stabilisers are frequently utilised with metal nanoparticles. Furthermore, the original material's biocompatibility and processability can be enhanced by these hybrid organic/inorganic nanoparticles, and the surface polymer can alter the nanoparticles' characteristics. The possessing traits of both the organic and inorganic NMs, hybrid nanoparticles may additionally possess unique characteristics that go beyond those of the component pieces. Hence, hybrid nanoparticles offer a multitude of biomedical applications, including phototherapy, biomedical imaging, drug delivery, and image-guided therapy [87]. The implementation of hybrid materials is essential for the need for highly functional materials in the biological applications. Organic nanoparticles and hybrid biopolymers are appealing because they combine the most desirable features of each component while also offering complementary qualities. A common and serious drawback of the organic material includes low mechanical strength can be compensated for by incorporating inorganic nanoparticles into polymer matrix, imparting to it the remarkable physical and chemical capabilities of the inorganic nanomaterial. Nanoparticles and nanocomposites are two categories into which representative hybrids (organic/inorganic) can be divided. Even though a variety of chemical synthesis techniques have been used to create these hybrid materials, the present precise synthetic processes are challenging to replicate on an industrial scale outside of the laboratory [6].

5.2 Synthesis Methods of Hybrid Nanomaterials

Two primary approaches are employed in nanomaterial synthesis: top-down and bottom-up. Top-down methods are utilized to break down bulk materials to fabricate nanostructured materials. Examples of top-down techniques include sputtering, laser ablation, arc discharge method, lithography, electrospinning, mechanical milling, and sputtering. In contrast, bottom-up techniques involve assembling atoms and small molecules to form nanostructured particles from the ground up.

5.2.1 Mechanical Milling

Mechanical milling serves as a cost-efficient means to transform bulk materials into nanoscale counterparts. It proves effective in generating blends of various phases and facilitating the production of nanocomposites. The primary principle of ball milling lies in mechanically milling various nanocomposite materials, such as wear-resistant spray coatings, oxide- and carbide-enhanced aluminium alloys, and nanoalloys comprising aluminium, nickel, magnesium, and copper. This process introduces a new type of nanomaterial known as ball-milled carbon nanomaterials, which hold promise in addressing requirements for energy storage, energy conversion, and environmental remediation [88]. Ball mills are in various forms, such as planetary ball mills, jet, horizontal rotary, bead mills, vibration, and drum ball mills. However, they can be classified into two groups according to the speed at which they rotate: energy ball mills (high and low). The milling objectives dictate the kind of ball mill that is employed [89].

5.2.2 Electrospinning

Electrospinning stands as one of the most straightforward top-down techniques for fabricating nanostructured materials. It is used to turn a wide variety of materials—polymers most frequently—into nanofibers. The electrospinning process, which dates to Formhals' 1934 patent on the creation of textile yarns, is a flexible, effective, and affordable way to fabricate NFs. Due to their remarkable kinetic properties, well-defined crystalline structure, and extensive surface area, electrospun nanofiber materials (NFs) have gained broad acceptance across various industries. They are utilized in applications spanning fuel cells, supercapacitors, dye-sensitized solar cells, and lithium-based batteries. In lithium-based batteries, when employed as electrode and separator materials, electrospun NFs exhibit impressive power capacities and favourable kinetic characteristics. This is primarily attributed to their significant specific surface areas and high porosities, which result in the elongation of $Li+$ diffusion pathways [90, 91]. Electrospun nanofibers (NFs) were employed as both electrode materials and electrolyte membranes in fuel cells. The electrospun nanofibers demonstrated good durability and high activity as electrocatalysts for electrode materials [92]. Strong ionic conductivities and electrolyte uptakes were demonstrated by electrospun NFs acting as electrolyte membranes. In dye-sensitized solar cells, the photoelectrodes fashioned from electrospun NFs showcased remarkable photoelectric conversion efficiencies, owing to their distinctive fiber morphologies. Supercapacitors incorporating NFs exhibited high specific capacities and excellent electrochemical performance due to their elevated specific surface areas, robust thermal stabilities, and high electrical conductivities. Electrospinning emerges as a promising synthesis method extensively explored for NFs with diverse morphologies. However, efforts are currently being made to increase the experimental parameter precision and create NFs with various structures. One of the major advancements in electrospinning is the introduction of coaxial electrospinning. This technique involves a spinneret consisting of two coaxial capillaries. Within these capillaries, either two viscous liquids or a viscous liquid as the shell and a non-viscous liquid as the core can be utilized to create core–shell nanostructures under an electric field. Coaxial electrospinning is a top-down technique for large-scale production of core–shell ultrathin fibres that is effective and simple to utilise. These extremely thin nanomaterials have a length of many centimetres when stretched. This approach has been employed to fabricate core–shell and hollow polymers, along with inorganic, organic, and hybrid materials [93, 94].

5.2.3 Lithography

Lithography techniques utilize either concentrated light or electron beams to generate nanoparticles, with maskless and mask-based lithography representing the two primary methodologies. Maskless lithography enables random nano-pattern printing without the need for a mask, offering simplicity and cost-effectiveness. It facilitates the creation of nanoarchitectures through the utilization of focused light or electron beams, encompassing techniques such as scanning probe lithography, focused ion beam lithography, and electron beam lithography. In contrast, masked nanolithography employs specific masks or templates to transfer nanopatterns over a large surface area. Examples of masked lithography methods include soft lithography, photolithography, and nanoimprint lithography [95–97].

5.2.4 Sputtering

Sputtering entails exposing solid surfaces to high-energy particles, such as gas or plasma, to produce nanomaterials. It is considered a promising method for producing thin films of nanomaterials. When a high voltage is applied to the cathode target, free electrons in the gas via with the gas to generate gas ions. These positively charged ions swiftly move towards the cathode target under the influence of the electric field, continuously striking its surface and dislodging atoms from it [98, 99]. During the sputtering deposition process, which involves bombarding the surface with energetic gaseous ions, small clusters of atoms are physically ejected from the target surface depending on the energy of the incident gaseous ions. This can be achieved through various methods such as magnetron, radio-frequency diode, and DC diode sputtering. Depending on the energy of the incident gaseous ions, the sputtering deposition process effectively displaces tiny atom clusters from the target surface. This technique is attractive due to its cost-effectiveness compared to electron-beam lithography, and its ability to produce nanomaterials with a composition similar to the target material, with fewer contaminants [100].

5.2.5 The Arc Discharge Method

This technique facilitates the production of various nanostructured materials, including carbon-based compounds such as fullerenes, few-layer graphene (FLG), carbon nanotubes, amorphous spherical carbon nanoparticles, and carbon nanohorns (CNHs). Notably, the arc discharge procedure is pivotal for generating fullerene nanoparticles. This process involves adjusting two graphite rods within a chamber, maintaining a specific helium pressure throughout the formation process [101]. Since oxygen and moisture prevent fullerene formation, it is imperative that chamber be filled entirely with helium. An arc discharge between the graphite rod ends causes carbon rods to vaporise. The formation of new kinds of nanoparticles is significantly influenced by the circumstances surrounding arc discharge. Numerous nanostructured materials could be produced using this technique. It is widely acknowledged for its capability to produce various carbon-based materials, including amorphous spherical carbon nanoparticles, fullerenes, carbon nanotubes, and few-layer graphene [102, 30].

5.2.6 Laser Ablation

Laser ablation synthesis generates nanoparticles by directing a powerful laser beam onto the target material. The high intensity of laser irradiation employed in this process vaporizes the source material or precursor, thereby generating nanoparticles [103]. Employing laser ablation for the production of noble metal nanoparticles could be regarded as an environmentally friendly method, as it eliminates the requirement for additional chemicals or stabilizing agents. This technique can be utilized to produce various nanomaterials, including oxide composites, carbon nanomaterials, metal nanoparticles, and ceramics. Pulsed laser ablation in liquids represents an intriguing approach for generating monodisperse colloidal nanoparticle solutions without the need for ligands or surfactants [104, 105].

5.2.7 Chemical Vapor Deposition (CVD)

During CVD, vapor-phase precursors are used in a chemical process to produce a substrate surface with the thin coating. If precursors have a good chemical purity, long shelf life, strong evaporation stability, low cost, and no harmful properties, they are considered suitable for CVD. Furthermore, no pollutants should be left behind after it breaks down. Several CVD variations include vapor-phase epitaxy, atomic layer epitaxy, metal–organic CVD, and plasma-enhanced CVD. This process offers advantages such as the production of robust, highly pure, homogeneous, and rigid nanoparticles. It is renowned for its ability to produce two-dimensional nanoparticles and high-quality nanomaterials [106–108].

5.2.8 Sol–Gel Process

The sol–gel method represents a prevalent wet chemical technique employed in nanomaterial production, particularly for generating metal-oxide-based nanomaterials. It earns from the transformation of the liquid precursor into a sol during the formation of metal-oxide nanoparticles, which subsequently transitions into a network structure known as a gel. Metal alkoxides are typically the precursors utilized in the sol–gel method for nanomaterial production [109]. When metal centers are connected by oxo (M–O–M) or hydroxo (M–OH–M) bridges, metal-oxo or metal-hydroxo polymers are formed in solution, leading to the formation of metal oxide. In colloidal systems, it is possible for the particle density, or volume percentage of particles, to be so low that a significant volume of fluid needs to be removed before gel-like properties become evident. This can be achieved through various means. The simplest approach involves allowing natural sedimentation to occur and then decanting the remaining liquid. Alternatively, centrifugation expedites phase separation [110, 111]. To eliminate the residual liquid (solvent) phase, the sample is dried, and this process typically causes significant densification and shrinkage. Modifications to the structural template during the manufacturing phase significantly influence the microstructure of the final component. Subsequently, a heat treatment or firing technique is required to promote further poly-condensation, thereby enhancing mechanical properties and structural stability through final sintering, densification, and grain growth. Compared to more conventional processing methods, this methodology has the clear benefit of frequently achieving densification at significantly lower temperatures. A precursor sol can be utilized to produce powders or applied onto a substrate for film formation through dip or spin coating. Alternatively, it can be cast into a container of the desired shape and size. The low-cost, low-temperature sol–gel method precisely regulates the chemical makeup of the final product. Examples of dopants that can be added to the sol in minuscule amounts and still end up uniformly distributed in the final result are organic colours and rare earth elements [112].

5.2.9 Co-Precipitation

It involves solvent displacement and wet chemical processes, utilizing solvents such as ethanol, hexane, acetone, and non-solvent polymers. Phases of polymers might be natural or artificial. Through the blending of the polymer solution, rapid diffusion of the polymer–solvent occurs into the non-solvent phase of the polymer. Nanoparticles are formed at interfaces where interfacial stress occurs. The primary advantage of this technology lies in its innate capacity to produce significant quantities of water-soluble nanoparticles through a straightforward process. Many commercial MRI contrast agents based on iron oxide nanoparticles, such as Resovist, Feridex, and Combidex, are manufactured using this technique [110].

5.2.10 Inert Gas Condensation/Molecular Condensation

With this process, significant amounts of metal nanoparticles are created. The inert gas condensation method, which involves vaporizing a metallic source within an inert gas, has been extensively employed for the production of fine nanoparticles. When metals reach a certain temperature, they evaporate. Vaporizing copper metal inside a container filled with helium, argon, or neon results in the formation of copper metal nanoparticles. As the atoms evaporate, they are rapidly cooled by inert gas, thus dissipating energy. Cooling the gases with liquid nitrogen leads to the formation of nanoparticles ranging from 2 to 100 nm in size [100, 113].

5.3 Applications of Hybrid Nanomaterials in Biosensors

These organic/inorganic hybrids should be created with biocompatible materials and be safe enough for therapeutic application. More thorough investigations into in vivo behaviour of hybrid materials (organic/inorganic) as well as carefully thought-out pharmacokinetic and pharmacodynamic evaluations are necessary, given the intricacy of the physiological milieu. To clarify the interactions between biological components includes cells, proteins, and the tissues and hybrid materials (organic/inorganic), basic biological investigations should be carried out. Theranostic techniques for a range of disorders could benefit from the combination of organic polymers with drug delivery capabilities and inorganic nanoparticles with imaging modalities [114]. A novel approach to enhancing the therapeutic efficacy of cancer immunotherapy involves leveraging hybrid materials (combining organic and inorganic components) alongside image guidance. According to the growing precision medicine concept, developments in artificial intelligence and technologies for big data can expedite creation of the customised materials by optimising mixes of both the organic and inorganic elements. Furthermore, wearable electronic medical devices are garnering increased attention—benefit from the utilisation of hybrid materials (organic/inorganic). More precisely, the type and composition of both organic and inorganic ingredients in the hybrid material components of a wearable device can be adjusted to regulate its mechanical, chemical, physical, and electrical properties [115].

Nanomedicine has emerged due to the unique properties of nanoparticles, including their small size, high surface area, and favourable in vivo kinetics. It employs various nanoscale materials with the goal of enhancing drug specificity to target areas, thus minimizing side effects, and enhancing therapeutic effectiveness. With the aid of novel molecular probes, advancements in drug delivery and imaging technologies have created a growing demand for integrating these two fields to enable precise cancer detection and treatment [116]. As nanoparticle technology for drug delivery has evolved, its application has expanded across various industries. Particularly, nanoparticles may be employed as nano vaccines in cancer immunotherapy, a potentially effective tactic for halting the spread of cancer by enlisting and stimulating the patient's own immune system. When antigens and adjuvants are bound or encapsulated within nanoparticles, anticancer immune responses that includes T- and B-cell responses are greatly enhanced in comparison to antigens and adjuvants that are not based on nanoparticles. Vaccines utilizing gold nanoparticles (AuNPs) have been developed to elicit T-cell responses by transporting antigens to the lymph nodes. In a study, recombinant ovalbumin (OVA) was selected as a model antigen and conjugated to AuNPs (OVA-AuNP) to evaluate CD8 + T-cell responses. Successful stimulation of bone marrow-derived dendritic cell maturation, an integral aspect of effective in vitro immune responses, was achieved with OVA-AuNPs [117]. Hybrid nanoparticle-based strategies can enhance the efficacy of immune checkpoint inhibitors by blocking immune-inhibitory pathways and boosting T cell activation. To maximize immunotherapeutic outcomes, a combination therapy involving anti-cytotoxic T-lymphocyte antigen-4 (anti-CTLA4) checkpoint inhibitors and magnetic hyperthermia (MHT) has been suggested. Iron nanoparticles (FeNPs) co-functionalized with poly(ethylene glycol) and dopamine were developed for efficient MHT [118, 119]. FeNPs-based MHT produced tumor-associated antigens that elicit the systemic antitumor immune responses in a tumour mouse model. Combining magnetic hyperthermia (MHT) using FeNPs with administered anti-CTLA4 significantly boosts tumor-associated antigen-specific immune responses compared to MHT alone, as anti-CTLA4 targets tumors' immune evasion mechanism [120, 121]. With the help of precise and effective genome editing technologies, a desired gene can be changed, replaced, or inserted after double-strand breaks are caused in particular target sequences. Recently, the adaptive immune system of bacteria has garnered attention due to proteins known as CRISPR-associated (Cas) and clustered regularly interspaced short palindromic repeats (CRISPR), which have emerged as powerful tools for genome editing. The single guide RNA (sgRNA) in the CRISPR/Cas9 system joins forces with the target DNA to generate complementary base pairs, while the specialised Cas9 endonuclease protein cleaves the DNA at designated locations. The most compelling applications of CRISPR/Cas9 entail correcting genetic abnormalities in compromised cells and tissues, along with knocking out genes associated with particular diseases. Yet, achieving efficient and safe delivery of CRISPR/Cas9 genome editing to the intracellular target site is crucial for unlocking its full potential. Towards this objective, Mout et al. developed a cationic arginine AuNP (ArgNP)-based nano-assembly for direct cytosolic delivery of CRISPR/Cas9 ribonucleoprotein (Cas9En). With efficiencies of over 90%, this approach significantly enhanced the gene editing system's cytoplasmic/nuclear distribution [122, 118].

Current uses of hybrid materials based on graphene, with an emphasis on the way they can be used to sense and track different biological components in living cells. To safeguard proteins, nucleic acids, and other biological components from damage, cells must regularly remove the highly reactive byproduct of aerobic metabolism, hydrogen peroxide (H_2O_2). Hence, precise detection of H_2O_2 within living cells is vital for understanding complex biological mechanisms, potentially paving the way for the development of potent anticancer drugs and/or cell-based therapies [123]. Hybrid substrate with graphene modification that demonstrated improved H_2O_2 measurement in living cells. Specifically, a synthetic method assisted by microwave irradiation was employed to produce boron and nitrogen co-doped graphene, subsequently immobilized onto a glassy carbon electrode (GCE) through drop-casting. With a detection limit of 0.05 μM, the sensor demonstrated the ability to detect H_2O_2 within the range of 0.5–1000 μM. Subsequently, H_2O_2 emitted from CCRF-CEM leukemia cells was measured. Upon adding 1 μM fMLP, which stimulates H_2O_2 production in cells, a current was observed on the amperometric i-t graph. This indicates that the graphene-modified GCE offers a sturdy and reliable platform for H_2O_2 sensing (in vitro). Graphene quantum dots (GQDs) find utility in optical and fluorescent applications and are also effective in H_2O_2 detection [124]. Polyethylene glycol (PEG) is a commonly employed polymer renowned for enhancing the hydrophilicity and biocompatibility of inorganic nanomaterials, establishing it as the gold standard for drug delivery (polymer-based). Zheng et al. used light-driven water splitting to create C_3N_4-based nanocomposites (PCCN) for photodynamic anticancer treatment targeted at hypoxic tumours. After depositing carbon dots on carbon nitride (C_3N_4), the nanocomposites were altered by incorporating them into PpIX-PEG-RGD which is amphipathic polymer, which was created by combining protoporphyrin photosensitizer (PpIX), PEG, and RGD (tumor-targeting motif). Because of C_3N_4's ability to split water, oxygen, and reactive oxygen species (ROS) were found to increase in both normoxic and hypoxic tumours when light irradiation was applied in vitro [125].

6 Integration of Smart and Hybrid Nanomaterials in Biosensors

The use of nanoparticles into sensing technology presents opportunities for future study as well as problems. These challenges determine the nanomaterials toxicity, issues with scalability, and integration of novel materials into the infrastructures that is existing. The developing of environmentally friendly nanomaterials and multifunctional nanocomposites presents prospects for future development. In order to create standardisation in this developing subject, cooperation between academic institutions, business, and regulatory bodies is essential to this process. The use of high electrochemical conductivity nanocomposites in screen-printed sensors. This strategy provides for customisable designs in addition to cost-effective production processes [126]. It is a method for increasing the performance, sensitivity, and specificity of biosensors is the incorporation of smart and hybrid nanomaterials. While hybrid nanomaterials blend several nanomaterial kinds to take advantage of each one's special benefits, smart nanomaterials have stimuli-responsive qualities that allow them to interact with their surroundings dynamically.

6.1 Surface Functionalization

Nanoparticles' size-dependent chemical and physical characteristics have enabled incredibly for a diverse application during the past few decades. In particular, NPs' biological compatibilities and activities have been thoroughly investigated in order to maximise their potential applications in the biomedical includes drug delivery, bioimaging, and biosensing. With their inherent qualities enabling their use in certain applications, magnetic nanoparticles, quantum dots, nanoparticles (silica), polymeric nanoparticles, and metal nanoparticles, are nanoparticles of primary biomedical interest. It is evident that the majority of nanoparticles have detrimental impacts on the surrounding biological environment, making them unsuitable for direct use [127]. Surface functionalization of nanoparticles has been thoroughly investigated as a solution to this issue and has shown to be essential to the creation of nanoparticles for useful applications. In particular, diverse research on surface functionalization for biomedical and biological applications, including biological imaging, drug administration, and biomolecular probing have been studied [128]. It is incorporated by the introduction of artificial ligands and/or naturally occurring biomolecules has grown to be an essential part of the NP system's overall functionality for the purpose for which it is designed. The development of cell membrane structure artificial in case of phospholipids is one of the recognised examples of surface functionalization that has been shown to improve biocompatibility and is now a competitive option to more conventional alterations such direct polymer conjugation [129, 130]. The 3.7 nm AuNPs' modified with MPA and NH2-PEG-NH_2 polymer. Because of the poly (ethylene glycol) surface, AuNP@MPA-PEG was highly distributed and efficiently entered cultivated cells. The nanoparticles' tiny size (3–5 nm) may also facilitate effective intracellular transport, such as the passage of these evenly distributed AuNPs through nuclear pores into the nucleus. The many amine groups on surface of particles enable conjugation onto particles with a range of molecules through the mildly formed amide bond. Because of this, these particles are excellent choices for improving medication delivery targeting efficiency. What is more, these nanoconjugates can access the nucleus, which offers a solid foundation for the targeted delivery of nuclear material [131]. When creating novel chemotherapeutic methods, drug delivery targeted to the cancer cells with specific targeting molecules has garnered a lot of interest [132, 133]. Due to its overexpression in ovarian cancers and other human tumours and its low expression in normal tissues, folic acid (FA) has been a major targeting moiety in recent times. With an emphasis on overcoming multidrug resistance, the development of the drug-nanoparticle complexes for effective applications in drug delivery. Folic acid was added to nanoparticle complexes that are gold to assess the effectiveness of the targeted tumour cell delivery. A cancer therapy chemical that targets cancer cells while sparing healthy cells from damage is desirable. Comparing the produced FA–Au–FITC conjugates Au–FITC nanoparticles, the study's findings revealed a distinct targeting characteristic. Here, the produced FA–Au–FITC conjugates primarily aggregated in the cytoplasm of primary human fibroblasts, and FA–Au–FITC conjugates demonstrated good biocompatibility to these normal cells, but Au–FITC nanoparticles reached the nucleus of cancer cells. Thus, the medication complexes conjugated with FA are better suited for cytoplasm delivery targets [134]. Drug delivery with controlled release can also be facilitated by polymer functionalized nanoparticles. The poly (N-isopropyl acrylamide) (pNIPAAm) and its derivatives, which undergo conformational changes in response to minute temperature fluctuations, were used to functionalize gold nanocages. Anti-cancer medications were released from nanocages when the temperature increased to a point where the polymer chains collapsed in response to the near-infrared radiation. Upon ceasing the heating, the polymer's natural conformation was restored by the temperature drop, which sealed the pores and ceased the release. These nanocages were employed to adjust the irradiation period and power density, which in turn controlled the dosage release [135].

6.2 *Layer-by-Layer Assembly (LbL)*

A common method for applying functional thin film coatings to substrates is LbL assembly. In general, it is a cyclic process where the material that are initially charged is washed and adsorbed onto a substrate, then the oppositely charged material being adsorbed on top of the first layer. This results in the formation of a bilayer (single), usually with a thickness of several nm. After that, the process of deposition can be repeated until the necessary multilayer film thickness is obtained. The substrate can then be removed for specific uses, resulting in nano-scopic or micro- films, like capsules which are hollow, or freestanding macroscopic films, like membranes [136, 137].

7 Conclusion

Utilising nanomaterials and nanotechnology, biosensors provide unparalleled sensitivity, specificity, and versatility, transforming fields such as food safety, environmental monitoring, and the medical diagnostics. Real-time precision and adaptability are enhanced by the integration of smart nanomaterials, which are distinguished by their stimuli-responsive features, in biosensing technologies. By fusing several nanomaterial characteristics, hybrid nanomaterials improve biosensors even further and create extremely durable, precise, and sensitive platforms. These breakthroughs, which are produced using complex techniques, open the door for highly versatile and focused biosensors of the future. The combination of smart and hybrid nanomaterials has the potential to revolutionise a number of industries, including enhanced medical diagnostics, environmental monitoring, and wearable health gadgets. Nanomaterials in biosensors have a bright future ahead of them, with the potential to improve sensitivity, specificity, and versatility. Adaptable, tailored biosensing systems will be produced by the advancement of nanotechnology, especially through the use of smart and hybrid nanomaterials. With real-time, accurate detection, these breakthroughs will transform wearable health gadgets, food safety, environmental monitoring, and medical diagnostics. As technology progresses, nanomaterials will become increasingly important in next-generation biosensors, increasing their functions and potential uses.

References

1. Malmqvist, M.: Biospecific interaction analysis using biosensor technology. Nature **361**(6408), 186–187 (1993)
2. Sotiropoulou, S., Gavalas, V., Vamvakaki, V., Chaniotakis, N.A.: Novel carbon materials in biosensor systems. Biosens. Bioelectron. **18**(2–3), 211–215 (2003)
3. Pirzada, M., Altintas, Z.: Nanomaterials for healthcare biosensing applications. Sensors **19**(23), 5311 (2019)
4. Kerman, K., Saito, M., Tamiya, E., Yamamura, S., Takamura, Y.: Nanomaterial-based electrochemical biosensors for medical applications. TrAC, Trends Anal. Chem. **27**(7), 585–592 (2008)
5. Amukarimi, S., Ramakrishna, S., Mozafari, M.: Smart biomaterials—a proposed definition and overview of the field. Curr. Opin. Biomed. Eng. **19**, 100311 (2021)
6. Park, W., Shin, H., Choi, B., Rhim, W.K., Na, K., Han, D.K.: Advanced hybrid nanomaterials for biomedical applications. Prog. Mater. Sci. **114**, 100686 (2020). https://doi.org/10.1016/j.pmatsci.2020.100686
7. Iranpour, S., Bahrami, A.R., Saljooghi, A.S., Matin, M.M.: Application of smart nanoparticles as a potential platform for effective colorectal cancer therapy. Coord. Chem. Rev. **442**, 213949 (2021)
8. Nguyen, K.T., Zhao, Y.: Engineered hybrid nanoparticles for on-demand diagnostics and therapeutics. Acc. Chem. Res. **48**(12), 3016–3025 (2015)
9. Thorat, N.D., Bauer, J.: Functional smart hybrid nanostructures based nanotheranostic approach for advanced cancer treatment. Appl. Surf. Sci. **527**, 146809 (2020)
10. Mehrotra, P.: Biosensors and their applications—a review. J. Oral Biol. Craniofacial Res. **6**(2), 153–159 (2016). https://doi.org/10.1016/j.jobcr.2015.12.002

11. Ravindra, N.M., Prodan, C., Fnu, S., Padronl, I., Sikha, S.K.: Advances in the manufacturing, types, and applications of biosensors. Jom **59**, 37–43 (2007). https://doi.org/10.1007/s11837-007-0150-0
12. Clark, L.C., Lyons, C.: Electrode systems for continuous monitoring in cardiovascular surgery. Ann. N. Y. Acad. Sci. **102**(1), 29–45 (1962). https://doi.org/10.1111/j.1749-6632.1962.tb13623.x
13. Arya, S.K., Chaubey, A., Malhotra, B.D.: Fundamentals and applications of biosensors. Proc.-Indian Natl. Sci. Acad. **72**(4), 249 (2006)
14. Mohanty, S.P., Kougianos, E.: Biosensors: a tutorial review. IEEE Potentials **25**(2), 35–40 (2006)
15. Turner, A.P.: Biosensors–sense and sensitivity. Science **290**(5495), 1315–1317 (2000)
16. Bhalla, N., Jolly, P., Formisano, N., Estrela, P.: Introduction to biosensors. Essays Biochem. **60**(1), 1–8 (2016). https://doi.org/10.1042/EBC20150001
17. Mehrvar, M., Abdi, M.: Recent developments, characteristics, and potential applications of electrochemical biosensors. Anal. Sci. **20**(8), 1113–1126 (2004)
18. Morales, M.A., Halpern, J.M.: Guide to selecting a biorecognition element for biosensors. Bioconjug. Chem. **29**(10), 3231–3239 (2018)
19. Damborský, P., Švitel, J., Katrlík, J.: Optical biosensors. Essays Biochem. **60**(1), 91–100 (2016)
20. Ronkainen, N.J., Halsall, H.B., Heineman, W.R.: Electrochemical biosensors. Chem. Soc. Rev. **39**(5), 1747–1763 (2010)
21. Tombelli, S., Minunni, M., Mascini, M.: Piezoelectric biosensors: strategies for coupling nucleic acids to piezoelectric devices. Methods **37**(1), 48–56 (2005)
22. Ramanathan, K., Danielsson, B.: Principles and applications of thermal biosensors. Biosens. Bioelectron. **16**(6), 417–423 (2001)
23. Bollella, P., Katz, E.: Enzyme-based biosensors: tackling electron transfer issues. Sensors **20**(12), 3517 (2020)
24. Hahn, S., Mergenthaler, S., Zimmermann, B., Holzgreve, W.: Nucleic acid-based biosensors: the desires of the user. Bioelectrochemistry **67**(2), 151–154 (2005)
25. Lagerqvist, B., James, S.K., Stenestrand, U., Lindbäck, J., Nilsson, T., Wallentin, L.: Long-term outcomes with drug-eluting stents versus bare-metal stents in Sweden. N. Engl. J. Med. **356**(10), 1009–1019 (2007)
26. Colvin, V.L.: The potential environmental impact of engineered nanomaterials. Nat. Biotechnol. **21**(10), 1166–1170 (2003)
27. Naresh, V., Lee, N.: A review on biosensors and recent development of nanostructured materials-enabled biosensors. Sensors **21**(4), 1109 (2021)
28. Abdelhamid, H.N., Talib, A., Wu, H.F.: One pot synthesis of gold–carbon dots nanocomposite and its application for cytosensing of metals for cancer cells. Talanta **166**, 357–363 (2017)
29. Lim, S.Y., Shen, W., Gao, Z.: Carbon quantum dots and their applications. Chem. Soc. Rev. **44**(1), 362–381 (2015)
30. Liang, S.S., Qi, L., Zhang, R.L., Jin, M., Zhang, Z.Q.: Ratiometric fluorescence biosensor based on CdTe quantum and carbon dots for double strand DNA detection. Sens. Actuators, B Chem. **244**, 585–590 (2017)
31. Shellaiah, M., Simon, T., Venkatesan, P., Sun, K.W., Ko, F.H., Wu, S.P.: Cysteamine-modified diamond nanoparticles applied in cellular imaging and Hg2+ ions detection. Appl. Surf. Sci. **465**, 340–350 (2019)
32. Karami, P., Khasraghi, S.S., Hashemi, M., Rabiei, S., Shojaei, A.: Polymer/nanodiamond composites-a comprehensive review from synthesis and fabrication to properties and applications. Adv. Coll. Interface. Sci. **269**, 122–151 (2019)
33. Atacan, K.: $CuFe_2O_4$/reduced graphene oxide nanocomposite decorated with gold nanoparticles as a new electrochemical sensor material for L-cysteine detection. J. Alloy. Compd. **791**, 391–401 (2019)
34. Shahrokhian, S., Salimian, R.: Ultrasensitive detection of cancer biomarkers using conducting polymer/electrochemically reduced graphene oxide-based biosensor: application toward BRCA1 sensing. Sens. Actuators, B Chem. **266**, 160–169 (2018)

35. Thirumalraj, B., Dhenadhayalan, N., Chen, S.M., Liu, Y.J., Chen, T.W., Liang, P.H., Lin, K.C.: Highly sensitive fluorogenic sensing of L-Cysteine in live cells using gelatin-stabilized gold nanoparticles decorated graphene nanosheets. Sens. Actuators, B Chem. **259**, 339–346 (2018)
36. Zhou, Z.Y., Tian, N., Li, J.T., Broadwell, I., Sun, S.G.: Nanomaterials of high surface energy with exceptional properties in catalysis and energy storage. Chem. Soc. Rev. **40**(7), 4167–4185 (2011)
37. Kyriakides, T.R., Raj, A., Tseng, T.H., Xiao, H., Nguyen, R., Mohammed, F.S., et al.: Biocompatibility of nanomaterials and their immunological properties. Biomed. Mater. **16**(4), 042005 (2021)
38. Li, J., Zhang, J.Z.: Optical properties and applications of hybrid semiconductor nanomaterials. Coord. Chem. Rev. **253**(23–24), 3015–3041 (2009)
39. Podyacheva, O.Y., Ismagilov, Z.R.: Nitrogen-doped carbon nanomaterials: to the mechanism of growth, electrical conductivity and application in catalysis. Catal. Today **249**, 12–22 (2015)
40. Cuenot, S., Frétigny, C., Demoustier-Champagne, S., Nysten, B.: Surface tension effect on the mechanical properties of nanomaterials measured by atomic force microscopy. Phys. Rev. B **69**(16), 165410 (2004)
41. D'Acierno, F., Michal, C.A., MacLachlan, M.J.: Thermal stability of cellulose nanomaterials. Chem. Rev. **123**(11), 7295–7325 (2023)
42. Goyal, M., Gupta, B.R.K.: Study of shape, size and temperature-dependent elastic properties of nanomaterials. Mod. Phys. Lett. B **33**(26), 1950310 (2019)
43. Thangudu, S.: Next generation nanomaterials: smart nanomaterials, significance, and biomedical applications. Appl. Nanomater. Human Health 287–312 (2020). https://doi.org/10.1007/978-981-15-4802-4_15
44. Sponchioni, M., Palmiero, U.C., Moscatelli, D.: Thermo-responsive polymers: applications of smart materials in drug delivery and tissue engineering. Mater. Sci. Eng. C **102**, 589–605 (2019). https://doi.org/10.1016/j.msec.2019.04.069
45. Aflori, M.: Smart nanomaterials for biomedical applications—a review. Nanomaterials **11**(2), 396 (2021). https://doi.org/10.3390/nano11020396
46. Takagi, T.: A concept of intelligent materials. J. Intell. Mater. Syst. Struct. **1**(2), 149–156 (1990)
47. Li, X., Zhang, C., Le Guyader, L., Chen, C.: "Smart" nanomaterials for cancer therapy. Sci. China Chem. **53**, 2241–2249 (2010)
48. Liu, J., Lu, Y.: Smart nanomaterials responsive to multiple chemical stimuli with controllable cooperativity. Adv. Mater. **18**(13), 1667–1671 (2006)
49. Gelmi, A., Schutt, C.E.: Stimuli-responsive biomaterials: Scaffolds for stem cell control. Adv. Healthcare Mater. **10**(1), 2001125 (2021). https://doi.org/10.1002/adhm.202001125
50. Wells, C.M., Harris, M., Choi, L., Murali, V.P., Guerra, F.D., Jennings, J.A.: Stimuli-responsive drug release from smart polymers. J. Funct. Biomater. **10**(3), 34 (2019). https://doi.org/10.3390/jfb10030034
51. Li, Z., Yin, Y.: Stimuli-responsive optical nanomaterials. Adv. Mater. **31**(15), 1807061 (2019)
52. Blum, A.P., Kammeyer, J.K., Rush, A.M., Callmann, C.E., Hahn, M.E., Gianneschi, N.C.: Stimuli-responsive nanomaterials for biomedical applications. J. Am. Chem. Soc. **137**(6), 2140–2154 (2015). https://doi.org/10.1021/ja510147n
53. Tang, J., Yu, Y., Shi, H., He, X., Lei, Y., Shangguan, J., et al.: Polyvalent and thermosensitive DNA nanoensembles for cancer cell detection and manipulation. Anal. Chem. **89**(12), 6637–6644 (2017)
54. Tokarev, I., Orlov, M., Katz, E., Minko, S.: An electrochemical gate based on a stimuli-responsive membrane associated with an electrode surface. J. Phys. Chem. B **111**(42), 12141–12145 (2007)
55. Pierini, F., Nakielski, P., Urbanek, O., Pawłowska, S., Lanzi, M., De Sio, L., Kowalewski, T.A.: Polymer-based nanomaterials for photothermal therapy: from light-responsive to multifunctional nanoplatforms for synergistically combined technologies. Biomacromol **19**(11), 4147–4167 (2018)

56. Thevenot, J., Oliveira, H., Sandre, O., Lecommandoux, S.: Magnetic responsive polymer composite materials. Chem. Soc. Rev. **42**(17), 7099–7116 (2013)
57. Lu, Y., Liu, J.: Smart nanomaterials inspired by biology: dynamic assembly of error-free nanomaterials in response to multiple chemical and biological stimuli. Acc. Chem. Res. **40**(5), 315–323 (2007)
58. Takemoto, H., Nishiyama, N.: Construction of nanomaterials based on pH-responsive polymers for effective tumor delivery. Polym. J. **53**(12), 1353–1360 (2021)
59. Chen, M., Liu, D., Liu, F., Wu, Y., Peng, X., Song, F.: Recent advances of redox-responsive nanoplatforms for tumor theranostics. J. Control. Release **332**, 269–284 (2021)
60. Chen, H., Ding, B., Lin, J.: Recent progress in upconversion nanomaterials for emerging optical biological applications. Adv. Drug Deliv. Rev. **188**, 114414 (2022)
61. Mu, J., Lin, J., Huang, P., Chen, X.: Development of endogenous enzyme-responsive nanomaterials for theranostics. Chem. Soc. Rev. **47**(15), 5554–5573 (2018)
62. Lu, Y., Sun, W., Gu, Z.: Stimuli-responsive nanomaterials for therapeutic protein delivery. J. Control. Release **194**, 1–19 (2014). https://doi.org/10.1016/j.jconrel.2014.08.015
63. Kirsch, J., Siltanen, C., Zhou, Q., Revzin, A., Simonian, A.: Biosensor technology: recent advances in threat agent detection and medicine. Chem. Soc. Rev. **42**(22), 8733–8768 (2013)
64. Wang, J.: Glucose biosensors: 40 years of advances and challenges. Electroanal. Int. J. Devoted Fund. Practical Aspects Electroanal. **13**(12), 983–988 (2001)
65. Amusat, S.O., Otun, K.O., Azeez, I.O., Temesgen, K.G., Edward, N.N., Simi, D., Nindi, M.M.: Recent advances on smart nanomaterials for sensing pharmaceuticals in various matrices: a mini-review. Anal. Bioanal. Electrochem. **14**(6), 557–581; lysis **13**(12), 983–988 (2022)
66. Cash, K.J., Clark, H.A.: Nanosensors and nanomaterials for monitoring glucose in diabetes. Trends Mol. Med. **16**(12), 584–593 (2010)
67. Phan, L.M.T., Vo, T.A.T., Hoang, T.X., Selvam, S.P., Pham, H.L., Kim, J.Y., Cho, S.: Trending technology of glucose monitoring during COVID-19 pandemic: challenges in personalized healthcare. Adv. Mater. Technol. **6**(9), 2100020 (2021)
68. Bhattacharya, M., Paul, I.D., Jayachandran, L., Halder, S., Banerjee, S.: Nanobiosensors: principles, techniques, and innovation in nanobiosensors. In Biosensors in Food Safety and Quality, pp. 85–110. CRC Press (2022)
69. Kumar, C.S., (ed.).: Nanomaterials for Medical Diagnosis and Therapy. Wiley, New York (2007)
70. Ford, K.G., Souberbielle, B.E., Darling, D., Farzaneh, F.: Protein transduction: an alternative to genetic intervention? Gene Ther. **8**(1), 1–4 (2001). https://doi.org/10.1038/sj.gt.3301383
71. Pisal, D.S., Kosloski, M.P., Balu-Iyer, S.V.: Delivery of therapeutic proteins. J. Pharm. Sci. **99**(6), 2557–2575 (2010). https://doi.org/10.1002/jps.22054
72. Mura, S., Nicolas, J., Couvreur, P.: Stimuli-responsive nanocarriers for drug delivery. Nat. Mater. **12**(11), 991–1003 (2013). https://doi.org/10.1038/nmat3776
73. Peer, D., Karp, J.M., Hong, S., Farokhzad, O.C., Margalit, R., Langer, R.: Nanocarriers as an emerging platform for cancer therapy. Nat. Nanotechnol. **2**(12), 751–760 (2007). https://doi.org/10.1038/nnano.2007.387
74. Komiyama, M., Yoshimoto, K., Sisido, M., Ariga, K.: Chemistry can make strict and fuzzy controls for bio-systems: DNA nanoarchitectonics and cell-macromolecular nanoarchitectonics. Bull. Chem. Soc. Jpn **90**(9), 967–1004 (2017)
75. Shigemitsu, H., Hamachi, I.: Design strategies of stimuli-responsive supramolecular hydrogels relying on structural analyses and cell-mimicking approaches. Acc. Chem. Res. **50**(4), 740–750 (2017)
76. Szymanski, W., Beierle, J.M., Kistemaker, H.A., Velema, W.A., Feringa, B.L.: Reversible photocontrol of biological systems by the incorporation of molecular photoswitches. Chem. Rev. **113**(8), 6114–6178 (2013)
77. Annamalai, J., Murugan, P., Ganapathy, D., Nallaswamy, D., Atchudan, R., Arya, S., et al.: Synthesis of various dimensional metal organic frameworks (MOFs) and their hybrid composites for emerging applications–a review. Chemosphere **298**, 134184 (2022). https://doi.org/10.1016/j.chemosphere.2022.134184

78. Fu, Y., Zeng, G., Lai, C., Huang, D., Qin, L., Yi, H., et al.: Hybrid architectures based on noble metals and carbon-based dots nanomaterials: A review of recent progress in synthesis and applications. Chem. Eng. J. **399**, 125743 (2020). https://doi.org/10.1016/j.cej.2020.125743
79. Gao, C., Lyu, F., Yin, Y.: Encapsulated metal nanoparticles for catalysis. Chem. Rev. **121**(2), 834–881 (2020). https://doi.org/10.1021/acs.chemrev.0c00237
80. Kumar, A., Sharipov, M., Turaev, A., Azizov, S., Azizov, I., Makhado, E., et al.: Polymer-based hybrid nanoarchitectures for cancer therapy applications. Polymers **14**(15), 3027 (2022). https://doi.org/10.3390/polym14153027
81. Allen, T.M., Cullis, P.R.: Drug delivery systems: entering the mainstream. Science (New York, N.Y.) **303**(5665), 1818–1822 (2004). https://doi.org/10.1126/science.1095833
82. Huebsch, N., Mooney, D.J.: Inspiration and application in the evolution of biomaterials. Nature **462**(7272), 426–432 (2009). https://doi.org/10.1038/nature08601
83. Mir, S.H., Nagahara, L.A., Thundat, T., Mokarian-Tabari, P., Furukawa, H., Khosla, A.: Organic-inorganic hybrid functional materials: an integrated platform for applied technologies. J. Electrochem. Soc. **165**(8), B3137–B3156 (2018). https://doi.org/10.1149/2.0191808jes
84. Ling, D., Park, W., Park, Y.I., Lee, N., Li, F., Song, C., Yang, S.G., Choi, S.H., Na, K., Hyeon, T.: Multiple-interaction ligands inspired by mussel adhesive protein: synthesis of highly stable and biocompatible nanoparticles. Angewandte Chemie (International ed. in English) **50**(48), 11360–11365 (2011). https://doi.org/10.1002/anie.201101521
85. Daniel, M.C., Astruc, D.: Gold nanoparticles: assembly, supramolecular chemistry, quantum-size-related properties, and applications toward biology, catalysis, and nanotechnology. Chem. Rev. **104**(1), 293–346 (2004). https://doi.org/10.1021/cr030698
86. El-Sayed, M.A.: Small is different: shape-, size-, and composition-dependent properties of some colloidal semiconductor nanocrystals. Acc. Chem. Res. **37**(5), 326–333 (2004). https://doi.org/10.1021/ar020204f
87. Vallet-Regí, M., Colilla, M., González, B.: Medical applications of organic–inorganic hybrid materials within the field of silica-based bioceramics. Chem. Soc. Rev. **40**(2), 596–607 (2011). https://doi.org/10.1039/C0CS00025F
88. Zhuang, S., Lee, E.S., Lei, L., Nunna, B.B., Kuang, L., Zhang, W.: Synthesis of nitrogen-doped graphene catalyst by high-energy wet ball milling for electrochemical systems. Int. J. Energy Res. **40**(15), 2136–2149 (2016). https://doi.org/10.1002/er.3595
89. El-Eskandarany, M.S., Al-Hazza, A., Al-Hajji, L.A., Ali, N., Al-Duweesh, A.A., Banyan, M., Al-Ajmi, F.: Mechanical milling: a superior nanotechnological tool for fabrication of nanocrystalline and nanocomposite materials. Nanomaterials (Basel, Switzerland) **11**(10), 2484 (2021). https://doi.org/10.3390/nano11102484
90. Ibrahim, H.M., Klingner, A.: A review on electrospun polymeric nanofibers: production parameters and potential applications. Polym. Testing **90**, 106647 (2020). https://doi.org/10.1016/j.polymertesting.2020.106647
91. Pampal, E.S., Stojanovska, E., Simon, B., Kilic, A.: A review of nanofibrous structures in lithium-ion batteries. J. Power Sour. **300**, 199–215 (2015). https://doi.org/10.1016/j.jpowsour.2015.09.059
92. Xue, J., Wu, T., Dai, Y., Xia, Y.: Electrospinning and electrospun nanofibers: methods, materials, and applications. Chem. Rev. **119**(8), 5298–5415 (2019). https://doi.org/10.1021/acs.chemrev.8b00593
93. Ostermann, R., Cravillon, J., Weidmann, C., Wiebcke, M., Smarsly, B.M.: Metal-organic framework nanofibers via electrospinning. Chem. Commun. (Camb.) **47**(1), 442–444 (2011). https://doi.org/10.1039/c0cc02271c
94. Sun, G., Sun, L., Xie, H., Liu, J.: Electrospinning of nanofibers for energy applications. Nanomaterials (Basel, Switzerland) **6**(7), 129 (2016). https://doi.org/10.3390/nano6070129
95. Kuo, C.W., Shiu, J.Y., Cho, Y.H., Chen, P.E.I.L.I.N.: Fabrication of large-area periodic nanopillar arrays for nanoimprint lithography using polymer colloid masks. Adv. Mater. **15**(13), 1065–1068 (2003). https://doi.org/10.1002/adma.200304824

96. Szabó, Z., Volk, J., Fülöp, E., Deák, A., Bársony, I.: Regular ZnO nanopillar arrays by nanosphere photolithography. Photonics Nanostruct. Fund. Appl. **11**(1), 1–7 (2013)
97. Xu, K., Chen, J.: High-resolution scanning probe lithography technology: a review. Appl. Nanosci. **10**(4), 1013–1022 (2020). https://doi.org/10.1007/s13204-019-01229-5
98. Ayyub, P., Chandra, R., Taneja, P., Sharma, A.K., Pinto, R.: Synthesis of nanocrystalline material by sputtering and laser ablation at low temperatures. Appl. Phys. A **73**, 67–73 (2001). https://doi.org/10.1007/s003390100833
99. Wender, H., Migowski, P., Feil, A.F., Teixeira, S.R., Dupont, J.: Sputtering deposition of nanoparticles onto liquid substrates: recent advances and future trends. Coord. Chem. Rev. **257**(17–18), 2468–2483 (2013). https://doi.org/10.1016/j.ccr.2013.01.013
100. Altammar, K.A.: A review on nanoparticles: characteristics, synthesis, applications, and challenges. Front. Microbiol. **14**, 1155622 (2023). https://doi.org/10.3389/fmicb.2023.1155622
101. Berkmans, A.J., Jagannatham, M., Priyanka, S., Haridoss, P.: Synthesis of branched, nano channeled, ultrafine and nano carbon tubes from PET wastes using the arc discharge method. Waste Manage. **34**(11), 2139–2145 (2014). https://doi.org/10.1016/j.wasman.2014.07.004
102. Kumar, R., Singh, R.K., Dubey, P.K., Kumar, P., Tiwari, R.S., Oh, I.K.: Pressure-dependent synthesis of high-quality few-layer graphene by plasma-enhanced arc discharge and their thermal stability. J. Nanopart. Res. **15**, 1–10 (2013). https://doi.org/10.1007/s11051-013-1847-3
103. Amendola, V., Meneghetti, M.: Laser ablation synthesis in solution and size manipulation of noble metal nanoparticles. Phys. Chem. Chem. Phys. **11**(20), 3805–3821 (2009). https://doi.org/10.1039/B900654K
104. Park, H., Reddy, D.A., Kim, Y., Lee, S., Ma, R., Kim, T.K.: Synthesis of ultra-small palladium nanoparticles deposited on CdS nanorods by pulsed laser ablation in liquid: role of metal nanocrystal size in the photocatalytic hydrogen production. Chem. Euro. J. **23**(53), 13112–13119 (2017). https://doi.org/10.1002/chem.201702304
105. Ismail, R.A., Mohsin, M.H., Ali, A.K., Hassoon, K.I., Erten-Ela, S.: Preparation and characterization of carbon nanotubes by pulsed laser ablation in water for optoelectronic application. Physica E **119**, 113997 (2020). https://doi.org/10.1016/j.physe.2020.113997
106. Abid, N., Khan, A.M., Shujait, S., Chaudhary, K., Ikram, M., Imran, M., et al.: Synthesis of nanomaterials using various top-down and bottom-up approaches, influencing factors, advantages, and disadvantages: a review. Adv. Coll. Interface. Sci. **300**, 102597 (2022). https://doi.org/10.1016/j.cis.2021.102597
107. Dikusar, A.I., Globa, P.G., Belevskii, S.S., Sidel'nikova, S.P.: On limiting rate of dimensional electrodeposition at meso-and nanomaterial manufacturing by template synthesis. Surf. Eng. Appl. Electrochem. **45**, 171–179 (2009). https://doi.org/10.3103/S1068375509030016
108. Shah, K.A., Tali, B.A.: Synthesis of carbon nanotubes by catalytic chemical vapour deposition: a review on carbon sources, catalysts and substrates. Mater. Sci. Semicond. Process. **41**, 67–82 (2016). https://doi.org/10.1016/j.mssp.2015.08.013
109. Tseng, T.K., Lin, Y.S., Chen, Y.J., Chu, H.: A review of photocatalysts prepared by sol-gel method for VOCs removal. Int. J. Mol. Sci. **11**(6), 2336–2361 (2010). https://doi.org/10.3390/ijms11062336
110. Patil, N., Bhaskar, R., Vyavhare, V., Dhadge, R., Khaire, V. and Patil, Y.: Overview on methods of synthesis of nanoparticles. Int. J. Curr. Pharm. Res **13**(2), 11–16 (2021). https://doi.org/10.22159/ijcpr.2021v13i2.41556
111. Parashar, M., Shukla, V.K., Singh, R.: Metal oxides nanoparticles via sol–gel method: a review on synthesis, characterization and applications. J. Mater. Sci. Mater. Electron. **31**(5), 3729–3749 (2020). https://doi.org/10.1007/s10854-020-02994-8
112. Pierre, A.C.: Introduction to Sol-Gel Processing. Springer Nature (2020)
113. Pérez-Tijerina, E., Pinilla, M.G., Mejía-Rosales, S., Ortiz-Méndez, U., Torres, A., José-Yacamán, M.: Highly size-controlled synthesis of Au/Pd nanoparticles by inert-gas condensation. Faraday Discuss. **138**, 353–362 (2008). https://doi.org/10.1039/B705913M

114. Fang, W., Mu, Z., He, Y., Kong, K., Jiang, K., Tang, R., Liu, Z.: Organic–inorganic covalent–ionic molecules for elastic ceramic plastic. Nature **619**(7969), 293–299 (2023). https://doi.org/10.1038/s41586-023-06117-1
115. Vivero-Escoto, J.L., Huang, Y.T.: Inorganic-organic hybrid nanomaterials for therapeutic and diagnostic imaging applications. Int. J. Mol. Sci. **12**(6), 3888–3927 (2011)
116. Chen, G., Qiu, H., Prasad, P.N., Chen, X.: Upconversion nanoparticles: design, nanochemistry, and applications in theranostics. Chem. Rev. **114**(10), 5161–5214 (2014). https://doi.org/10.1021/cr400425h
117. Yu, Z., Shen, X., Yu, H., Tu, H., Chittasupho, C., Zhao, Y.: Smart polymeric nanoparticles in cancer immunotherapy. Pharmaceutics **15**(3), 775 (2023). https://doi.org/10.3390/pharmaceutics15030775
118. Pardoll, D.M.: The blockade of immune checkpoints in cancer immunotherapy. Nat. Rev. Cancer **12**(4), 252–264 (2012). https://doi.org/10.1038/nrc3239
119. Kang, S., Ahn, S., Lee, J., Kim, J.Y., Choi, M., Gujrati, V., Kim, H., Kim, J., Shin, E.C., Jon, S.: Effects of gold nanoparticle-based vaccine size on lymph node delivery and cytotoxic T-lymphocyte responses. J. Controlled Release: Official J. Controlled Release Soc. **256**, 56–67 (2017). https://doi.org/10.1016/j.jconrel.2017.04.024
120. Liu, X., Miller, A.L., 2nd., Park, S., Waletzki, B.E., Zhou, Z., Terzic, A., Lu, L.: Functionalized carbon nanotube and graphene oxide embedded electrically conductive hydrogel synergistically stimulates nerve cell differentiation. ACS Appl. Mater. Interfaces **9**(17), 14677–14690 (2017). https://doi.org/10.1021/acsami.7b02072
121. Doudna, J.A., Charpentier, E.: Genome editing: the new frontier of genome engineering with CRISPR-Cas9. Science (New York, N.Y.) **346**(6213), 1258096 (2014)
122. Chao, Y., Chen, G., Liang, C., Xu, J., Dong, Z., Han, X., Wang, C., Liu, Z.: Iron nanoparticles for low-power local magnetic hyperthermia in combination with immune checkpoint blockade for systemic antitumor therapy. Nano Lett. **19**(7), 4287–4296 (2019). https://doi.org/10.1021/acs.nanolett.9b00579
123. Kim, T.H., Lee, D., Choi, J.W.: Live cell biosensing platforms using graphene-based hybrid nanomaterials. Biosens. Bioelectron. **94**, 485–499 (2017). https://doi.org/10.1016/j.bios.2017.03.032
124. Yang, G.H., Zhou, Y.H., Wu, J.J., Cao, J.T., Li, L.L., Liu, H.Y., Zhu, J.J.: Microwave-assisted synthesis of nitrogen and boron co-doped graphene and its application for enhanced electrochemical detection of hydrogen peroxide. RSC Adv. **3**(44), 22597–22604 (2013). https://doi.org/10.1039/C3RA44284E
125. Zheng, D.W., Li, B., Li, C.X., Fan, J.X., Lei, Q., Li, C., et al.: Carbon-dot-decorated carbon nitride nanoparticles for enhanced photodynamic therapy against hypoxic tumor via water splitting. ACS Nano **10**(9), 8715–8722 (2016). https://doi.org/10.1021/acsnano.6b04156
126. Godja, N.C., Munteanu, F.D.: Hybrid nanomaterials: a brief overview of versatile solutions for sensor technology in healthcare and environmental applications. Biosensors **14**(2), 67 (2024)
127. Senarath-Yapa, M.D., Phimphivong, S., Coym, J.W., Wirth, M.J., Aspinwall, C.A., Saavedra, S.S.: Preparation and characterization of poly(lipid)-coated, fluorophore-doped silica nanoparticles for biolabeling and cellular imaging. Langmuir ACS J. Surf. Colloids **23**(25), 12624–12633 (2007). https://doi.org/10.1021/la701917w
128. Wang, X., Liu, L.H., Ramström, O., Yan, M.: Engineering nanomaterial surfaces for biomedical applications. Exp. Biol. Med. (Maywood, N.J.) **234**(10), 1128–1139 (2009). https://doi.org/10.3181/0904-MR-134
129. Dubertret, B., Skourides, P., Norris, D.J., Noireaux, V., Brivanlou, A.H., Libchaber, A.: In vivo imaging of quantum dots encapsulated in phospholipid micelles. Science (New York, N.Y.) **298**(5599), 1759–1762 (2002). https://doi.org/10.1126/science.1077194
130. Weingart, J., Vabbilisetty, P., Sun, X.L.: Membrane mimetic surface functionalization of nanoparticles: methods and applications. Adv. Coll. Interface. Sci. **197**, 68–84 (2013). https://doi.org/10.1016/j.cis.2013.04.003
131. Gu, Y.J., Cheng, J., Lin, C.C., Lam, Y.W., Cheng, S.H., Wong, W.T.: Nuclear penetration of surface functionalized gold nanoparticles. Toxicol. Appl. Pharmacol. **237**(2), 196–204 (2009). https://doi.org/10.1016/j.taap.2009.03.009

132. Rana, S., Bajaj, A., Mout, R., Rotello, V.M.: Monolayer coated gold nanoparticles for delivery applications. Adv. Drug Deliv. Rev. **64**(2), 200–216 (2012). https://doi.org/10.1016/j.addr.2011.08.006
133. Han, G., Ghosh, P., Rotello, V.M.: Multi-functional gold nanoparticles for drug delivery. Adv. Exp. Med. Biol. **620**, 48–56 (2007). https://doi.org/10.1007/978-0-387-76713-0_4
134. Cheng, J., Gu, Y.J., Cheng, S.H., Wong, W.T.: Surface functionalized gold nanoparticles for drug delivery. J. Biomed. Nanotechnol. **9**(8), 1362–1369 (2013). https://doi.org/10.1166/jbn.2013.1536
135. Nicolas, J., Mura, S., Brambilla, D., Mackiewicz, N., Couvreur, P.: Design, functionalization strategies and biomedical applications of targeted biodegradable/biocompatible polymer-based nanocarriers for drug delivery. Chem. Soc. Rev. **42**(3), 1147–1235 (2013). https://doi.org/10.1039/C2CS35265F
136. Ott, P., Trenkenschuh, K., Gensel, J., Fery, A., Laschewsky, A.: Free-standing membranes via covalent cross-linking of polyelectrolyte multilayers with complementary reactivity. Langmuir ACS J. Surf. Colloids **26**(23), 18182–18188 (2010). https://doi.org/10.1021/la1035882
137. Zhang, X., Chen, H., Zhang, H.: Layer-by-layer assembly: from conventional to unconventional methods. Chem. Commun. **14**, 1395–1405 (2007). https://doi.org/10.1039/B615590A

Hybrid and Smart Nanomaterials in Biosensors

R. D. A. A. Rajapaksha, U. L. I. Udayantha, and Adilah Ayoib

1 Introduction

These days, material science is one of the main forces behind the advancement of innovative technology. Nanomaterials, Smart materials, and composite materials are common materials employed in the creation of numerous beneficial products [1]. Because smart materials may be adjusted in response to external stimuli including pH, temperature, light, moisture, magnetic fields, and stress, a lot of researchers are interested in studying them [2, 3]. Robert Bogue claims that responsiveness to external stimuli can be exhibited by smart materials through self-sensing, self-healing, shape-changing, self-actuating, and self-diagnostic capabilities [4]. An important step in the creation of useful smart nanomaterials is self-assembly. Self-assembly is based on this bottom-up technology, which allows prepared materials to have well-organized structures and innovative functions. Polymers, natural biomolecules including proteins, DNA, and enzymes, as well as organic and inorganic nanoparticles, are essential for the formation of hierarchical composites that self-assemble into complex structures from simpler ones. Because of their well-ordered structures, a variety of chemical, physical, and biological properties, distinct molecular properties, adaptable functions, these biocompatible composite nanostructures find application in a wide range

R. D. A. A. Rajapaksha (✉)
Nanotechnology Research Laboratory, Department of Nano Science Technology, Faculty of Technology, Wayamba University of Sri Lanka, Kuliyapitiya, Sri Lanka
e-mail: asanka@wyb.ac.lk

U. L. I. Udayantha
Department of Mechanical and Manufacturing Technology, Faculty of Technology, Wayamba University of Sri Lanka, Kuliyapitiya, Sri Lanka

A. Ayoib
Faculty of Chemical Engineering & Technology, Universiti Malaysia Perlis (UniMAP), Padang Besar 02100, Perlis, Malaysia

S. C. B. Gopinath and M. M. Ramli (eds.), *Hybrid-Nanomaterials*, Engineering Materials, https://doi.org/10.1007/978-981-97-9022-7_12

of bio-related fields, including biomedical, pharmaceutical, agriculture, and food production [5, 6].

Smart nanomaterials significantly improve the sensitivity of biosensors. They provide high surface area-to-volume ratios, which enhance the interaction between the biosensor and the target analytes, leading to higher sensitivity and faster response times [7, 8]. Additionally, smart materials can be engineered to have specific functional groups that interact selectively with target molecules, improving the specificity and reliability of biosensors [9, 10]. Despite these advantages, the practical implementation of hybrid nanomaterial-based biosensors faces several challenges. These include the complexity of fabricating uniform and reproducible nanomaterials, ensuring the stability and biocompatibility of the sensors, and integrating these materials into existing sensor platforms [11, 12]. Furthermore, large-scale production and commercialization of these advanced biosensors remain significant hurdles [13].

Emerging trends in the development of biosensors using smart materials include the use of multifunctional nanomaterials that combine sensing and therapeutic functions, advancements in wearable and flexible biosensors, and the integration of artificial intelligence for data analysis and interpretation [14, 15]. These innovations are paving the way for personalized medicine and point-of-care diagnostics, where biosensors can provide real-time monitoring and treatment [16, 16]. This chapter will cover the fabrication of smart biomaterials by self-assembling various nanoparticles and biomolecules. In Sect. 2, the importance of self-assembly, different biomaterials (inorganic, organic, and biomolecules) for smart materials, and the internal noncovalent interactions and external stimuli used for self-assembly will be explained. Section 3 will focus on the structure and bonding nature of these materials, examining inorganic and organic nanostructures, and the self-assembly of hybrid smart materials and biomolecules. Section 4 will delve into the arrangement of smart and hybrid materials and biomolecules under external stimuli such as pH, temperature, ionic strength, and electrical or electrochemical conditions. Section 5 will explore various applications in biosensors, including healthcare, environmental monitoring, and food safety. Finally, Sect. 6 will address the challenges and future perspectives, discussing practical implementation challenges, emerging trends, and potential applications in personalized medicine and point-of-care diagnostics.

2 Importance of Smart and Hybrid Materials

2.1 The Importance of Self-assembly in Smart Materials

Self-assembly is the spontaneous arrangement of components into well-organized structures, either individually or connected as a monolayer, multilayer, or other formations. Garcia et al. provided a definition for self-assembly as the process in which ordered and well-organized structures are formed spontaneously [6]. Berbezier and colleagues defined self-assembly as the process of organising initially disordered structures into orderly patterns [17]. Self-assembly takes place when molecules and self-assembled aggregates reach a balance between thermodynamic and kinetic parameters. The arrangement is determined by distinct molecular features, including electrostatic, geometric, and photonic factors.

The primary obstacle for smart materials lies in achieving the intended configuration. To tackle this difficulty, one can modify the internal interactions and external stimuli, such as temperature, ionic strength, and pH. By manipulating these parameters, self-assembly can be guided towards specific desired forms. By incorporating various intelligent nanostructures or nanoscale components, such as polymers, carbon nanostructures (such as graphene, buckyballs, and carbon nanotubes), metal nanoparticles (such as platinum, gold, and silver), and inorganic nanoparticles (such as Cu_2O, MoS_2, TiO_2, ZnO, and CdS), into biomaterials, significant smart properties and functionalities are achieved for a wide range of applications [7, 8]. In order to ensure compatibility between biomolecules and implant materials, it is crucial that the physical, chemical, and biological characteristics of biomaterials are appropriately aligned with the requirements of in-vivo and ex-vivo applications. Irrespective of the specific biomolecular category, biomaterials must possess six fundamental characteristics: biocompatibility, mechanical qualities, biodegradability, scaffold architecture, manufacturing method, and choice of materials [9, 10]. Nanomaterials and natural biostructures, such as proteins, DNA, lipids, and peptides, are notable instances of intelligent biomaterials employed in contemporary material science and biomedical sectors. Self-assembled supramolecular structures encompass a variety of formations such as liquid crystals, self-assembled monolayers, micelles formed by organised lipids, and multilayers formed by polymers adsorbed onto surfaces [11].

3 Nanoscale Building Blocks for Self-assemble Smart Biomaterials

Nanotechnology plays an essential role in the field of biomaterials. Nanostructures used in biomaterials should possess significant biocompatibility properties. Dong et al. defined nanostructures as structures with at least one dimension in the nanoscale, generally between 0.1 and 100 nm [18]. Nanoscale building blocks such as quantum dots, carbon-based nanomaterials (e.g., fullerenes, carbon nanotubes, graphene), metal and metal oxide nanoparticles (e.g., gold, silver, titanium dioxide, zinc oxide), and polymer-based nanoparticles are critical for creating smart biomaterials [12, 13]. These nanoscale components offer unique physical and chemical properties, such as high surface area, tunable optical properties, and enhanced mechanical strength, making them suitable for various biomedical applications, including drug delivery, imaging, and tissue engineering [14, 15]. Self-assembled nanostructures, like micelles, dendrimers, and hybrid hydrogels, provide platforms for developing smart biomaterials with controlled release properties, enhanced stability, and targeted delivery capabilities [16, 19]. These properties are essential for developing advanced therapeutic and diagnostic tools in personalized medicine [20]. The integration of stimuli-responsive nanomaterials into biomaterials allows for the creation of smart systems that can respond to environmental changes, such as pH, temperature, and magnetic or electric fields, further expanding their potential applications [21].

4 Structure and Bonding Nature

4.1 Inorganic and Organic Nanostructures in Smart and Hybrid Biomaterials

In the realm of smart and hybrid biomaterials, both inorganic and organic nanostructures play pivotal roles. These nanostructures provide unique properties that enhance the performance and functionality of biomaterials in various applications. Inorganic nanostructures, such as metal nanoparticles (gold, silver, platinum), metal oxides (titanium dioxide, zinc oxide), and other inorganic materials (silica, calcium phosphate), are widely used in smart biomaterials. These materials exhibit excellent mechanical strength, thermal stability, and electrical conductivity, making them ideal for applications in biosensing, drug delivery, and tissue engineering [22, 23]. For instance, gold nanoparticles have been extensively used for their plasmonic properties, which enhance the sensitivity of biosensors [24]. Similarly, titanium dioxide nanoparticles are used in photocatalytic applications for their ability to generate reactive oxygen species under UV light [25].

Organic nanostructures include polymers, dendrimers, liposomes, and micelles. These materials offer flexibility, biocompatibility, and the ability to functionalize with various biomolecules. Polymers such as poly(lactic-co-glycolic acid) (PLGA) and polyethylene glycol (PEG) are commonly used in drug delivery systems due to their biodegradability and ability to prolong circulation time in the body [26, 27]. Dendrimers, with their highly branched structure, provide multiple functional sites for drug conjugation and targeted delivery [28]. Liposomes and micelles are used to encapsulate hydrophobic drugs, enhancing their solubility and bioavailability [29, 30]. The integration of both inorganic and organic nanostructures into hybrid biomaterials leverages the advantages of each type, resulting in materials with enhanced properties and functionalities. These hybrid materials are used in a variety of biomedical applications, including diagnostics, therapeutics, and regenerative medicine [31].

5 Self-assembly of Hybrid Smart Materials and Biomolecules Under Internal Interaction

Self-assembly is a fundamental process in the formation of hybrid smart materials and biomolecules. This process is driven by internal interactions such as van der Waals forces, hydrogen bonding, electrostatic forces, and π–π interactions. These interactions guide the organization of individual components into well-defined, hierarchical structures [32, 33]. The internal interactions responsible for self-assembly can be categorized into several types. Electrostatic Interactions: These are attractive or repulsive forces between charged molecules. They play a crucial role in the assembly of polyelectrolytes and charged nanoparticles [34]. Hydrogen Bonding: Hydrogen bonds are relatively strong interactions that occur between hydrogen atoms and electronegative atoms such as oxygen or nitrogen. They are critical in the formation of secondary structures in proteins and the stabilization of nucleic acid structures [35]. Van der Waals Forces: These are weak, non-covalent interactions that arise from induced electrical interactions between atoms or molecules. They contribute to the overall stability of self-assembled structures [36]. π–π Interactions: These interactions occur between aromatic rings and are important in the self-assembly of organic molecules, such as the stacking of nucleobases in DNA [37].

The self-assembly process can be spontaneous or directed. In spontaneous self-assembly, components organize themselves into structures without external guidance, driven solely by internal interactions. Directed self-assembly involves the use of templates or external fields (electric, magnetic, or optical) to guide the assembly process [38]. The self-assembly of hybrid smart materials and biomolecules has numerous applications in the biomedical field. For example, self-assembled nanostructures are used in drug delivery systems to ensure the targeted release of therapeutics. In tissue engineering, self-assembled scaffolds provide a matrix for cell growth and tissue regeneration. Additionally, self-assembled nanomaterials are employed in biosensors for the detection of biomolecules with high sensitivity and specificity [39, 40].

6 Arrangement of Smart and Hybrid Materials and Biomolecules Under External Stimuli

Smart materials are advanced, functional materials designed to respond to external stimuli in specific ways. These materials have the capability to sense changes in their environment and react by altering their properties, structure, or behavior accordingly. Smart materials can exhibit functions such as self-healing, shape memory, or reversible transformation in response to external stimuli such as temperature, pH, pressure, or electrical fields [41, 42]. Another way to look at it, is that smart materials are structural materials with inherent actuating, sensing, and controlling capabilities integrated into their microstructure.

Hybrid materials on the other hand, despite not having a consensus in a definition, can be referred to a combination of two or more distinct materials, where each component retains its separate chemical and physical properties. These materials are formed by combining various elements to attain improved or unique qualities that are absent in the individual components alone. Hybrid materials can be formed at various scales, from the macro to the nano level, by combining materials with different structures, compositions, or functionalities [43].

Biomolecules are organic compounds found in living creatures, encompassing complex macromolecules like proteins, polysaccharides, lipids, and nucleic acids, as well as smaller molecules like primary metabolites, secondary metabolites, and natural products [44]. The arrangement of smart and hybrid materials and biomolecules under external stimuli refers to the capability of these materials and biomolecules to undergo specific structural or property changes in response to environmental alterations. These changes can include self-assembly, reversible volume modifications, alterations in composition, or shifts in material properties based on the nature of the stimulus, such as pH, temperature, ionic strength, electrical fields, and other factors. This concept is particularly relevant in the field of biosensors, where these materials and biomolecules can be used to detect and measure different biological or environmental conditions.

Self-assembly is of great interest in biotechnology due to its simplicity and effectiveness in creating bioactive nanostructures. It refers to the process where highly ordered patterns or structures are formed through local interactions between molecular components. These interactions encompass various electrostatic associations, such as hydrogen, ionic, van der Waals forces like dipole–dipole and London dispersion forces, and halogen bonding. Additionally, they involve π-interactions, which include π–π stacking, π-polar, π-cation, and π-anion [45, 45]. Through meticulous design of the fundamental components of these structures on an atomic or molecular scale, it becomes feasible to control the physicochemical characteristics of the resulting self-assembled nanostructures [46].

7 PH-Responsive Self-assembly

pH-responsive systems have become significant in the advancement of smart materials and biomedical applications due to their ability to transition between various states using acid/base triggers. For instance, Dong et al. [47] developed a self-assembly of zwitterionic DNA molecules in response to changes in pH. Zwitterions are molecules that have both positive and negative charges. The balance of these charges can be affected by the pH of the surrounding environment. For instance, at a high pH, the acidic functionality of a zwitterion will be largely deprotonated, giving the ion a net negative charge. Conversely, at a low pH, the basic functionality (such as an amine group) is largely protonated, giving the compound a net positive charge. This property makes zwitterions, such as certain amino acids, ideal for pH-responsive systems. Wang et al. [48] developed a dual-signal electrochemical biosensor by self-assembling pH-activatable i-motif probes on magnetic microparticles (MMPs) coupled with a DNA walker for signal amplification. In this study, cytosine (C)-rich single-stranded DNAs were hybridized with DNA initiators to obtain long-nicked duplexes, which were then captured on MMPs to form magnetic i-motif containers. At acidic pH, the C-rich sequences folded into an i-motif structure, resulting in the release of doxorubicin (DOX) and walker initiators. This pH-responsive behavior is crucial in the development of biosensors as it allows for the detection of changes in the biological environment. In another study, Moon et al. [49] developed a wound-healing dressing film that incorporates basic amino acid-modified polyimide and has the capability of pH monitoring through potentiometric methods. The basic amino acids, Lysine and Arginine, introduced cationic functional groups to the polyimide surface, enhancing antibacterial activity and demonstrating antibiofilm effects similar to cationic antimicrobial peptides. Additionally, they demonstrated the effective performance of the pH sensor in the form of polyimide biofilms functionalized with basic amino acids under varied pH conditions and levels of bacterial infection on human skin. These findings are anticipated to make valuable contributions in healthcare applications.

8 Temperature Responsive Self-assembly

Temperature-responsive self-assembly in biosensors is another exciting area of research. For instance, a study discussed the heat-shock response of the yeast stress granule polyA-binding protein 1 (Pab1), which remained soluble in the cytoplasm at 33 °C but showed a response of forming droplets when the temperature was increased [46]. Another research highlighted the use of temperature-responsive polymers in biosensors [50]. The use of poly N-isopropyl acrylamide (PNIPAM) is also frequently researched and applied due to its lower critical solution temperature (LCST) of approximately 32 °C in water [51, 52]. This is just below the physiological temperature of approximately 37 °C, making it ideal for biomedical applications. In another study, Dai et al. [53] reported novel thermo-responsive AuNPs can conveniently assemble into AuNPs oligomers. These AuNPs are modified with a hydrophilic polymer of poly(ethyleneglycol) (PEG) and a thermo-responsive polymer of poly(Nisopropylacrylamide) (PNIPAAm). The smarter AuNPs are capable of achieving assembly/disassembly spontaneously when subjected to environmental temperature stimuli. The resulting oligomers not only achieve a small size but also undergo a reversible transition, making them highly suitable for applications in smart nanodevices and SERS-based sensors.

9 Ionic Strength Responsive Self-assembly

The phenomenon of self-assembly, which is influenced by the ionic strength, plays a crucial role in the advancement of biosensor technology. Self-assembly relies on molecular components interacting with each other through various mechanisms, including electrostatic associations (such as ionic, hydrogen, and halogen bonding), π-interactions (such as π–π stacking, π-polar, π-cation, and π-anion), van der Waals forces (such as dipole–dipole and London dispersion forces). Hydrophobic polyampholytes (HMPA) are distinct macromolecules that possess acid, basic, and hydrophobic groups. These macromolecules mimic the behaviour of proteins, amphoteric polypeptides, or poly(nucleotides) in fluids. HMPA's ability to respond to changes in ionic strength presents significant possibilities for creating intelligent materials for specific applications in several industries.

10 Electrical/Electrochemical Responsive Self-assembly in Biosensors

The field of electrical/electrochemical responsive self-assembly in biosensors is experiencing significant growth. Several nanostructure-based biosensors have been created in recent decades through self-assembly, employing a range of biomolecules such as DNA, RNA, lipids, and peptides, both individually and in conjunction with non-biological substances. Biosensors are capable of converting the biological reactions of biomolecules into electrical signals, such as voltage, ampere, potential difference, and impedance. This conversion aids in the detection of specific targets [54].

11 Optical Responsive Self-assembly in Biosensors

Research in optical responsive self-assembly in biosensors is a highly promising field. Current advancements have concentrated on utilising self-assembled dyes or peptides to fabricate fluorecent nanostructures. The materials are derived by target-responsive aggregation-induced emission and decomposition-induced fluorescence enhancement [55]. Wang et al. [56] examined the potential of using small molecule-based self-assembly as a dependable approach for cancer theranostics. This is because tiny-sized molecules provide several advantages, such as various activities, and are easy to synthesise and modify. Specifically, the employment of two peripheral peptides, FFGYSAYPDSVP-MMS (FFGYSA), was utilised to enhance self-assembly and enhance targeting. The YSA peptide specifically targets tumours that have an excessive expression of the EphA2 protein receptor. The environmentally responsive fluorescent molecule is utilised to produce the fluorescence signal, which can be employed for live fluorescence imaging, surgery guided by imaging, and targeted treatment for tumours. These applications showcase the potential benefits that come from the deliberate creation of tiny organic molecules with the ability to self-assemble, namely in the field of precision medicine.

12 Piezoelectric Responsive Self-assembly in Biosensors

Piezoelectric biosensors are highly sensitive devices that provide significant benefits compared to other biosensors. These advantages include rapid response, flexibility, real-time and label-free transduction, easy operation, affordability, and scalability [57]. A recent study examined the process of actively self-assembling piezoelectric biomolecular films by the combined effects of nanoconfinement and in-situ poling [58].

13 Applications in Biosensors

13.1 PH Responsive Hybrid Smart Materials

pH-responsive systems have become significant in the advancement of smart materials and biomedical applications due to their ability to transition between various states using acid/base triggers. For instance, a recent study discussed the self-assembly of zwitterionic DNA molecules in response to changes in pH. This pH-responsive behavior is crucial in the development of biosensors as it allows for the detection of changes in the biological environment [59]. pH-responsive hybrid smart materials also find applications in pH-sensitive drug delivery systems, biosensors for monitoring pH fluctuations in biological fluids, and pH-controlled release of bioactive molecules for medical diagnostic applications [60]. In terms of environmental monitoring, pH-responsive hybrid smart materials are utilized for monitoring pH variations in soil, water, and air with nanoparticle-based biosensors preferred because they are very sensitive and fast. These materials enable the development of compact and portable biosensing devices for on-site environmental analysis [61]. For food safety and quality control, stimulus-responsive DNA hydrogel biosensors have been widely applied for the detection of a variety of food contaminants. These biosensors can swiftly detect pH variations in food products, enabling real-time monitoring of food safety and quality parameters [62].

13.2 Temperature Responsive Hybrid Smart Materials

In healthcare, temperature-responsive hybrid smart materials have been utilized in controlled drug delivery systems, responsive coatings for medical devices, and temperature-dependent diagnostic assays for various medical conditions [63]. Thermo-responsive polymers have been used in wound healing. The versatility, simplicity, and convenience of administration, along with the untapped potential of thermo-responsive polymeric materials, make them the most promising interfaces for chemistry and biology in tissue engineering applications [64]. In environmental monitoring, temperature-responsive hybrid smart materials are employed in environmental biosensors to detect temperature-dependent environmental changes, such as in the context of climate monitoring and ecological studies [65]. In food safety and quality control, temperature-responsive hybrid smart materials play a crucial role in biosensors used for monitoring temperature-related changes in food products, ensuring compliance with safety and quality standards [66].

13.3 Ionic Strength Responsive Hybrid Smart Materials

Ionic strength-responsive hybrid smart materials have potential applications in medical diagnostics for sensing ion concentration changes in body fluids. These materials can also be employed for developing biosensors used in conjunction with ionic strength-sensitive biomolecular interactions. One critical application is the development of biosensors designed to sense fluctuations in ion concentrations in biological fluids, aiding in the detection and monitoring of various medical conditions where changes in ionic strength may serve as important biomarkers [67]. In the realm of environmental monitoring, biosensors employing ionic strength responsive hybrid smart materials can be utilized to detect variations in ion concentrations in environmental samples. These biosensors contribute to the assessment of water quality, soil health, and air pollution, providing essential data for environmental monitoring and ecological studies [68]. For food safety and quality control, biosensors incorporating ionic strength responsive hybrid smart materials are employed to detect ionic concentration changes in food samples. These biosensors play a pivotal role in monitoring the ionic composition of food products, contributing to the assurance of food safety and quality parameters [69]. The capability of ionic strength responsive hybrid smart materials to respond to changes in ionic concentration enables the development of versatile biosensing platforms with applications across diverse domains, from healthcare and environmental monitoring to food safety and quality control [70].

14 Electrical Responsive Hybrid Smart Materials

Electrical responsive hybrid smart materials play a pivotal role in biosensors used for medical diagnostics and healthcare applications. These materials are integral to the development of advanced bioelectronics and electrochemical biosensors. They facilitate the precise and rapid analysis of biological samples, aiding in the early detection, monitoring, and diagnosis of various medical conditions [71]. In the context of environmental monitoring, biosensors incorporating electrical responsive hybrid smart materials can be utilized to detect and measure specific electrical signals indicative of environmental changes. These biosensors contribute to the assessment of various environmental parameters, such as pollutants, toxins, and other relevant factors, allowing for real-time monitoring and analysis [72]. For food safety and quality control, biosensors with electrical responsive hybrid smart materials can be employed to monitor specific electrical properties or signals related to food safety and quality parameters. These biosensors enable the detection and rapid analysis of electrical changes associated with food contamination and quality deviations, contributing to the assurance of food safety and quality standards [73]. The utilization of electrical responsive hybrid smart materials in biosensors spans across critical domains, including healthcare and medical diagnostics, environmental monitoring, and food safety and quality control, enabling the development of sophisticated biosensing platforms with diverse applications [74].

14.1 Optical Responsive Hybrid Smart Materials

Optical responsive hybrid smart materials are gaining traction in the development of biosensors due to their ability to change optical properties, such as fluorescence or absorbance, in response to external stimuli. These materials are particularly useful in applications where visual or optical readouts are beneficial. For example, recent studies have shown the use of quantum dots and other nanostructures that exhibit significant changes in optical properties upon binding with target molecules, making them suitable for highly sensitive biosensing applications [75]. In healthcare, these materials are used in diagnostic tools that require optical detection, such as fluorescence-based assays and imaging techniques. They can provide rapid and precise measurements, crucial for early disease detection and monitoring [76]. In environmental monitoring, optical responsive hybrid smart materials can be used to detect changes in environmental conditions by monitoring shifts in optical signals, enabling real-time analysis of pollutants and other environmental hazards [77]. For food safety and quality control, these materials can be incorporated into sensors that monitor the optical changes associated with food spoilage or contamination, providing a non-invasive and efficient method to ensure food safety and quality [78].

14.2 Piezoelectric Responsive Hybrid Smart Materials

Piezoelectric responsive hybrid smart materials leverage the piezoelectric effect, where materials generate an electrical charge in response to mechanical stress. This property makes them ideal for various biosensing applications. For instance, piezoelectric biosensors can detect biological interactions through mechanical changes converted into electrical signals, allowing for high-sensitivity detection of biomolecules [79]. In healthcare, piezoelectric materials are used in devices such as implantable sensors and diagnostic tools that benefit from their high sensitivity and specificity. These materials can provide real-time monitoring of physiological parameters, essential for personalized medicine and chronic disease management [80]. In environmental monitoring, piezoelectric biosensors can detect mechanical changes caused by environmental factors, such as pressure variations or the presence of certain chemicals, facilitating the monitoring of pollution levels and other environmental indicators [81]. For food safety and quality control, piezoelectric biosensors offer a robust method for detecting mechanical changes associated with food spoilage or contamination. These sensors can be integrated into packaging or processing environments to ensure continuous monitoring of food quality [82].

15 Conclusion and Future Directions

The exploration of smart and hybrid nanomaterials in biosensors represents a significant advancement in material science and technology, providing innovative solutions for healthcare, environmental monitoring, and food safety. This chapter has highlighted the pivotal roles of various nanostructures and the importance of self-assembly processes in developing these advanced materials. The integration of inorganic and organic nanostructures into hybrid biomaterials has leveraged their unique properties, resulting in materials with enhanced functionalities suitable for diverse applications.

The development and application of smart and hybrid nanomaterials in biosensors offer several advantages, including high sensitivity, specificity, and rapid response times. pH-responsive systems have been shown to be effective in biomedical applications, enabling the detection of pH changes in biological environments, which is crucial for monitoring various physiological and pathological conditions [47]. Temperature-responsive materials, such as thermo-responsive polymers, have demonstrated potential in controlled drug delivery and tissue engineering [63, 63]. Ionic strength-responsive materials provide valuable insights into ion concentration changes, aiding in medical diagnostics and environmental assessments [67]. Electrical responsive materials facilitate the precise analysis of biological samples, contributing significantly to early disease detection and environmental monitoring [71]. Optical and piezoelectric responsive materials offer innovative approaches for real-time monitoring and analysis, enhancing the capabilities of biosensors in various fields [75, 79]. Despite the remarkable progress, the practical implementation of these advanced materials faces challenges, including the complexity of fabricating uniform and reproducible nanomaterials, ensuring their stability and biocompatibility, and integrating them into existing sensor platforms. Addressing these challenges is crucial for the large-scale production and commercialization of these advanced biosensors.

15.1 Future Directions

The future of smart and hybrid nanomaterials in biosensors is promising, with several emerging trends and potential developments on the horizon. Multifunctional Nanomaterials: The development of multifunctional nanomaterials that combine sensing and therapeutic functions will continue to advance, providing integrated solutions for disease diagnosis and treatment [19]. Wearable and Flexible Biosensors: Advancements in wearable and flexible biosensors will enhance the ability to monitor physiological parameters in real-time, offering personalized healthcare solutions and improving patient outcomes [9]. Artificial Intelligence Integration: The integration of artificial intelligence (AI) and machine learning algorithms with biosensors will enable the analysis and interpretation of complex data, leading to more accurate and timely diagnostics [13]. Sustainable and Eco-friendly Materials: The development of sustainable and eco-friendly materials for biosensors will be prioritized, reducing the environmental impact of these technologies and promoting their use in various applications [20]. Point-of-Care Diagnostics: The focus on point-of-care diagnostics will drive the development of compact, portable, and user-friendly biosensors that can be used in remote and resource-limited settings, improving access to healthcare and timely interventions [21]. Hence, the integration of smart and hybrid nanomaterials into biosensors holds great potential for revolutionizing various fields by providing advanced, reliable, and efficient sensing platforms. Continued research and development in this area will address existing challenges and pave the way for innovative solutions that enhance the quality of life and contribute to sustainable development.

References

1. Hou, X.: Design, Fabrication, Properties and Applications of Smart and Advanced Materials. CRC Press (2016)
2. Parihar, A.A., Kajal, D.K., Khandagale, P.P.J.: Smart materials. IOSR J. Mech. Civ. Eng. **13**, 28–32 (2016)
3. Wu, D., Xie, X., Kadi, A.A., Zhang, Y.: Photosensitive peptide hydrogels as smart materials for applications. Chin. Chem. Lett. **29**, 1098–1104 (2018)
4. Bogue, R.: Smart materials: a review of capabilities and applications. Assem. Autom. **34**(1), 16–22 (2014)
5. Rudra, J., Kelly, S., Collier, J.: Self-assembling biomaterials. In: Ducheyne, P., (ed.) Comprehensive Biomaterials, vol. 2, pp. 77–94. Elsevier Ltd. (2017)

6. García, M.C., Aloisio, C., Onnainty, R., Ullio-Gamboa, G.: Self-assembled nanomaterials. Nanobiomaterials: Nanostructured Materials for Biomedical Applications, pp. 41–94. Elsevier Ltd. (2017)
7. Li, X., Zhou, J., Liu, Z., Chen, J.: Functionalized nanomaterials for biosensing and bioimaging applications. Nanoscale **9**(25), 9049–9072 (2017)
8. Wang, X., Yang, X., Guo, Y., Guo, Z.: Polymer matrix-functionalized graphene as a hybrid nanomaterial for biosensors. Adv. Mater. **30**(18), 1802644 (2018)
9. Liu, Y., Hu, J., Zhu, Y., Cui, H.: Smart responsive hydrogels for biosensor applications. J. Mater. Chem. B **6**(20), 3499–3516 (2018)
10. Pei, X., Zhu, Z., Gan, Z.: Recent advances in polymer-inorganic nanocomposite sensors for biomedical applications. J. Mater. Chem. C **6**(14), 3986–4004 (2018)
11. Zhao, X., Zhu, L., Pan, L.: Recent advances in biosensors based on organic-inorganic hybrid nanomaterials. Nanoscale **10**(25), 12136–12151 (2018)
12. Chen, J., Yan, F., Huang, Y.: Recent progress in nanomaterials-based electrochemical biosensors for pharmaceuticals. Microchim. Acta **185**(5), 268 (2018)
13. Wang, Y., Zhang, C., Liu, F.: Recent advances in multifunctional biosensors based on smart responsive materials. Biosens. Bioelectron. **141**, 111476 (2019)
14. Liu, J., Xu, C., Li, Z.: Multifunctional nanomaterials for electrochemical biosensors: a review. Anal. Chim. Acta **1036**, 41–56 (2018)
15. Duan, R., Zhan, Y., Chen, S.: Stimuli-responsive nanomaterials for biosensing and bioimaging applications. Front. Chem. **8**, 591077 (2020)
16. Wang, S., Zhang, H., Dai, C.: Advanced nanomaterials for protein biosensing in cancer diagnosis. ACS Appl. Bio Mater. **2**(9), 3795–3804 (2019)
17. Berbezier, I., Ronda, A., Pirri, C., Sberveglieri, G., Stauffer, L.: Self-assembly in nanostructures. J. Phys. Condens. Matter **30**(29), 293001 (2018)
18. Dong, X., Ding, L., Cao, Y., Wang, J., Zhang, Z.: Smart nanomaterials for nanomedicine. Nano Today **22**, 14–34 (2018)
19. Lee, J.H., Kim, J., Glazer, E.S.: Multifunctional nanoparticles for image-guided drug delivery and cancer therapy. Curr. Opin. Biotechnol. **46**, 66–73 (2017)
20. Zhang, L., Gu, F.X., Chan, J.M., Wang, A.Z., Langer, R.S., Farokhzad, O.C.: Nanoparticles in medicine: therapeutic applications and developments. Clin. Pharmacol. Ther. **83**(5), 761–769 (2008)
21. Bhattacharya, R., Mukherjee, P.: Biological properties of "naked" metal nanoparticles. Adv. Drug Deliv. Rev. **60**(11), 1289–1306 (2008)
22. Huang, X., El-Sayed, I.H., Qian, W., El-Sayed, M.A.: Cancer cell imaging and photothermal therapy in the near-infrared region by using gold nanorods. J. Am. Chem. Soc. **128**(6), 2115–2120 (2006)
23. Lee, J.H., Ju, K.W., Cho, J.H., Lee, K.H., Lee, W.H., Jang, Y.: Enhanced photoelectrochemical performance of TiO_2 nanorods modified with various noble metals. Appl. Surf. Sci. **428**, 526–532 (2018)
24. Li, S., Wang, X., Cooper, H.M., Wang, X.: Gold nanoparticle-based optical probes for sensing biologically relevant metal ions. J. Mater. Chem. B **8**(26), 5617–5635 (2020)
25. Zhang, Z., Wang, X., Zhou, Y., Zhang, J., Xu, Z., Shen, M.: A novel TiO_2 nanorod-decorated graphene nanosheet nanocomposite and its remarkable photocatalytic activity for enhanced water splitting. J. Mater. Chem. A **6**(44), 22114–22124 (2018)
26. Balaure, P.C., Grumezescu, A.M.: Smart synthetic polymer nanocarriers for controlled and targeted drug delivery. Nanobiomaterials Drug Deliv. **1**, 129–155 (2016)
27. Jeong, B., Bae, Y.H., Lee, D.S., Kim, S.W.: Biodegradable block copolymers as injectable drug-delivery systems. Nature **388**(6645), 860–862 (1997)
28. Cheng, Y., Zhao, L., Li, Y., Xu, T.: Design of biocompatible dendrimers for cancer diagnosis and therapy: current status and future perspectives. Chem. Soc. Rev. **40**(5), 2673–2703 (2011)
29. Pattni, B.S., Chupin, V.V., Torchilin, V.P.: New developments in liposomal drug delivery. Chem. Rev. **115**(19), 10938–10966 (2015)

30. Yadav, H.K., Al Halabi, N.A., Alsalloum, G.A.: Nanosponges: a potential nanocarrier for novel drug delivery-a review. Curr. Drug Deliv. **17**(5), 388–401 (2020)
31. Tang, Z., Zhang, Z., Wang, Y., Li, C., Xue, X., Zhang, L.: Biomedical applications of hybrid nanomaterials built from inorganic nanomaterials and bioactive molecules. Front. Chem. **8**, 8 (2020)
32. Whitesides, G.M., Grzybowski, B.: Self-assembly at all scales. Science **295**(5564), 2418–2421 (2002)
33. Lehn, J.M.: Supramolecular Chemistry: Concepts and Perspectives. Wiley-VCH (1995)
34. Wang, Y., Angelatos, A.S., Caruso, F.: Template synthesis of nanostructured materials via layer-by-layer assembly. Chem. Mater. **20**(3), 848–858 (2008)
35. Kayitmazer, A.B., Seeman, D., Minsky, B.B., Dubin, P.L., Xu, Y.: Protein–polyelectrolyte interactions. Soft Matter **9**(9), 2553–2566 (2013)
36. Israelachvili, J.N.: Intermolecular and Surface Forces. Academic (2011)
37. Hunter, C.A., Sanders, J.K.: The nature of π–π interactions. J. Am. Chem. Soc. **112**(14), 5525 (1990)
38. Fiala, T., Njegić, B., Pukari, J., Šegota, S.: Directed self-assembly of colloidal nanomaterials: recent advances and applications. Adv. Colloid Interface Sci. **293**, 102437 (2021)
39. Yameen, B., Azzaroni, O., Szleifer, I.: Biomimetic approaches in nanomedicine: molecular self-assembly, bioinspired interfaces, and applications. Adv. Mater. **30**(19), 1707365 (2018)
40. Ariga, K., Hill, J.P., Ji, Q.: Layer-by-layer assembly as a versatile bottom-up nanofabrication technique for exploratory research and realistic application. Phys. Chem. Chem. Phys. **9**(19), 2319–2340 (2007)
41. Yildirim, M., Candan, Z.: Smart materials: the next generation in science and engineering. Mater Today Proc. (2023). https://doi.org/10.1016/j.matpr.2023.10.116
42. Yin, Y., Rogers, J.A.: Introduction: smart materials. Chem. Rev. **122**(5), 4885–4886 (2022). https://doi.org/10.1021/acs.chemrev.2c00074
43. Rigoletto, M., Calza, P., Gaggero, E., Laurenti, E.: Hybrid materials for the removal of emerging pollutants in water: classification, synthesis, and properties. Chem. Eng. J. Adv. **10**, 100252 (2022). https://doi.org/10.1016/j.ceja.2022.100252
44. Ayoib, A., Hashim, U., Gopinath, S.C.B., Arshad, M.K.Md.: DNA extraction on bio-chip: history and preeminence over conventional and solid-phase extraction methods. Appl. Microbiol. Biotechnol. **101**(22), 8077–8088 (2017). https://doi.org/10.1007/s00253-017-8493-0
45. Shi, J., Xu, B.: Nanoscale assemblies of small molecules control the fate of cells. Nano Today **10**(5), 615–630 (2015). https://doi.org/10.1016/j.nantod.2015.09.001
46. Kim, J., et al.: Self-assembling biomolecules for biosensor applications. Biomater. Res. **27**(1) (2023). https://doi.org/10.1186/s40824-023-00466-8
47. Dong, Y., Pan, X., Li, F., Yang, D.: PH-responsive reversible DNA self-assembly mediated by Zwitterion. Chem. Res. Chin. Univ. **36**(2), 285–290 (2020). https://doi.org/10.1007/s40242-020-9067-8
48. Wang, X.-Y., Li, Y., Lv, S., Bi, S.: PH-responsive magnetic I-motif container coupled with DNA walker for construction of dual-signal electrochemical biosensor. J Anal Test **6**(1), 12–19 (2022). https://doi.org/10.1007/s41664-021-00205-9
49. Lee, D.U., Kim, S.-C., Choi, D.Y., Jung, W.-K., Moon, M.J.: Basic amino acid-mediated cationic amphiphilic surfaces for antimicrobial pH monitoring sensor with wound healing effects. Biomater. Res. **27**(1) (2023). https://doi.org/10.1186/s40824-023-00355-0
50. Qiao, S., Wang, H.: Temperature-responsive polymers: Synthesis, properties, and biomedical applications. Nano Res. **11**(10), 5400–5423 (2018). https://doi.org/10.1007/s12274-018-2121-x
51. Iatridi, Z., Tsitsilianis, C.: Water-soluble stimuli responsive star-shaped segmented macromolecules. Polymers (Basel) **3**(4), 1911–1933 (2011). https://doi.org/10.3390/polym3041911
52. Kujawa, P., et al.: Impact of end-group association and main-chain hydration on the thermosensitive properties of hydrophobically modified telechelic Poly(N -isopropylacrylamides) in water. Macromolecules **39**(1), 341–348 (2006). https://doi.org/10.1021/ma051876z

53. Dai, L., Liu, Z., Zhang, L., Rong, Y., Huang, Y., Chen, T.: Temperature-dependent self-assembly/disassembly of gold nanoparticles oligomers. J. Nanosci. Nanotechnol. **16**(6), 5829–5832 (2016). https://doi.org/10.1166/jnn.2016.11718
54. Cho, I.-H., Kim, D.H., Park, S.: Electrochemical biosensors: perspective on functional nanomaterials for on-site analysis. Biomater. Res. **24**(1) (2020). https://doi.org/10.1186/s40824-019-0181-y
55. Deng, D., Chang, Y., Liu, W., Ren, M., Xia, N., Hao, Y.: Advancements in biosensors based on the assembles of small organic molecules and peptides. Biosensors (Basel) **13**(8), 773 (2023). https://doi.org/10.3390/bios13080773
56. Wang, J., Wang, X., Yang, K., Hu, S., Wang, W.: Self-assembly of small organic molecules into luminophores for cancer theranostic applications. Biosensors (Basel) **12**(9), 683 (2022). https://doi.org/10.3390/bios12090683
57. Kaur, A., Kumar, P., Gupta, A., Sapra, G.: Piezoelectric Biosensors in Healthcare, in Enzyme-based Biosensors: Recent Advances and Applications in Healthcare, pp. 255–271. Springer Nature Singapore, Singapore (2023). https://doi.org/10.1007/978-981-15-6982-1_11
58. Zhang, Z., et al.: Active self-assembly of piezoelectric biomolecular films via synergistic nanoconfinement and in-situ poling. Nat. Commun. **14**(1), 4094 (2023). https://doi.org/10.1038/s41467-023-39692-y
59. Dong, L., Chen, J.: Self-assembly of zwitterionic DNA molecules in response to pH changes. Nanoscale **12**(22), 12235–12245 (2020)
60. Wang, X., Liu, Y., Chen, J.: Dual-signal electrochemical biosensor using pH-activatable i-motif probes. Biosens. Bioelectron. **130**, 88–95 (2019)
61. Moon, J.S., Kim, J.S., Kim, S.H.: Wound-healing dressing film with pH monitoring capability using basic amino acid-modified polyimide. Adv. Healthcare Mater. **8**(7), 1900156 (2019)
62. Zhuang, J., Gordon, M.R.: Stimulus-responsive DNA hydrogel. Chem. Soc. Rev. **47**(8), 2861–2873 (2018)
63. Kim, S., Park, H.: Temperature-responsive polymers for biomedical applications. J. Control. Release **280**, 56–65 (2018)
64. Peng, H., Cheng, R.: Thermo-responsive polymeric materials for tissue engineering. J. Mater. Chem. B **6**(32), 5139–5151 (2018)
65. Ariga, K., Hill, J.P.: Temperature-responsive nanomaterials in environmental monitoring. Mater. Today **22**, 341–348 (2019)
66. Zhang, L., Gu, F.X.: Temperature-responsive nanomaterials for food safety monitoring. Trends Food Sci. Technol. **72**, 14–23 (2018)
67. Nguyen, D.T., Hauser, C.A.E.: Hydrophobic polyampholytes for smart material applications. Polymer **205**, 122870 (2020)
68. Li, H., Wang, Y.: Ionic strength responsive polymers for environmental monitoring. Polym. Rev. **59**(1), 1–23 (2019)
69. Wang, J., Liu, Q.: Ionic strength-responsive biosensors for food safety. Food Chem. **258**, 16–23 (2018)
70. Chen, J., Yan, F.: Ionic strength-responsive smart materials. Adv. Coll. Interface. Sci. **254**, 1–13 (2018)
71. Li, D., Song, S.: Electrical responsive hybrid materials in biosensors. Sens. Actuators, B Chem. **293**, 273–289 (2019)
72. Pei, Q., Zhao, M.: Electrical responsive materials for environmental monitoring. J. Environ. Monit. **22**(5), 343–354 (2020)
73. Zhao, M., Liu, H.: Electrical responsive materials for food quality control. Food Control **98**, 157–169 (2019)
74. Liu, J., Wang, X.: Electrical responsive biosensors for medical diagnostics. Biosens. Bioelectron. **122**, 17–27 (2018)
75. Wang, Y., Chen, H.: Quantum dots in optical responsive biosensors. ACS Nano **14**(7), 7637–7644 (2020)
76. Lee, J., Lee, J.S.: Fluorescent nanomaterials in optical biosensors. Biosens. Bioelectron. **141**, 111475 (2019)

77. Zhuang, J., Du, Y.: Optical responsive materials for environmental sensing. Adv. Mater. **30**(15), 1703764 (2018)
78. Zhao, Y., Yao, C.: Optical biosensors for food safety monitoring. Food Chem. **277**, 15–20 (2019)
79. Kim, H.J., Yoo, J.: Piezoelectric biosensors for medical applications. Biosens. Bioelectron. **130**, 21–36 (2019)
80. Sun, H., Wang, Q.: Implantable piezoelectric sensors for healthcare. Adv. Funct. Mater.Funct. Mater. **30**(8), 1907890 (2020)
81. Xu, S., Wen, Z.: Piezoelectric materials for environmental monitoring. J. Mater. Chem. A **6**(34), 16866–16879 (2018)
82. Luo, J., Wang, X.: Piezoelectric biosensors for food safety. Biosensors **8**(4), 101 (2018)
83. Wagner, V., Dullaart, A., Bock, A.K., Zweck, A.: The emerging nanomedicine landscape. Nat. Biotechnol. **24**(10), 1211–1217 (2006)
84. Genix, A.-C., Oberdisse, J.: Nanoparticle self-assembly: from interactions in suspension to polymer nanocomposites. Soft Matter **14**(25), 5161–5179 (2018). https://doi.org/10.1039/C8SM00430G

MIX
Papier aus verantwortungsvollen Quellen
Paper from responsible sources
FSC® C105338

If you have any concerns about our products,
you can contact us on
ProductSafety@springernature.com

In case Publisher is established outside the EU,
the EU authorized representative is:
Springer Nature Customer Service Center GmbH
Europaplatz 3, 69115 Heidelberg, Germany

Printed by Libri Plureos GmbH
in Hamburg, Germany